Signals and Communication Technology

For further volumes:
http://www.springer.com/series/4748

K.R. Rao · D.N. Kim ·
J.J. Hwang

Fast Fourier Transform:
Algorithms and Applications

 Springer

Dr. K.R. Rao
Univ. of Texas at Arlington
Electr. Engineering
Nedderman Hall
Yates St. 416
76013 Arlington Texas
USA
rao@uta.edu

Dr. D.N. Kim
Univ. of Texas at Arlington
Electr. Engineering
Nedderman Hall
Yates St. 416
76013 Arlington Texas
USA
cooldnk@yahoo.com

Dr. J.J. Hwang
Kunsan National Univ.
School of Electron. & Inform.
Engineering
68 Miryong-dong
573-701 Kunsan
Korea, Republic of (South Korea)
hwang@kunsan.ac.kr

ISSN 1860-4862
ISBN 978-1-4020-6628-3   e-ISBN 978-1-4020-6629-0
DOI 10.1007/978-1-4020-6629-0
Springer Dordrecht Heidelberg London New York

Library of Congress Control Number: 2010934857

Cover design: SPi Publisher Services

Printed on acid-free paper

Springer is part of Springer Science+Business Media (www.springer.com)

Preface

This book presents an introduction to the principles of the fast Fourier transform (FFT). It covers FFTs, frequency domain filtering, and applications to video and audio signal processing.

As fields like communications, speech and image processing, and related areas are rapidly developing, the FFT as one of the essential parts in digital signal processing has been widely used. Thus there is a pressing need from instructors and students for a book dealing with the latest FFT topics.

This book provides a thorough and detailed explanation of important or up-to-date FFTs. It also has adopted modern approaches like MATLAB examples and projects for better understanding of diverse FFTs.

Fast Fourier transform (FFT) is an efficient implementation of the discrete Fourier transform (DFT). Of all the discrete transforms, DFT is most widely used in digital signal processing. The DFT maps a sequence either in the time domain or in the spatial domain into the frequency domain. The development of the DFT originally by Cooley and Tukey [A1] followed by various enhancements/modifications by other researchers has provided the incentive and the impetus for its rapid and widespread utilization in a number of diverse disciplines. Independent of the Cooley-Tukey approach, several algorithms such as prime factor, split radix, vector radix, split vector radix, Winograd Fourier transform, and integer FFT have been developed. The emphasis of this book is on various FFTs such as the decimation-in-time FFT, decimation-in-frequency FFT algorithms, integer FFT, prime factor DFT, etc.

In some applications such as dual-tone multi-frequency detection and certain pattern recognition, their spectra are skewed to some regions that are not uniformly distributed. With this basic concept we briefly introduce the nonuniform DFT (NDFT), dealing with arbitrarily spaced samples in the Z-plane, while the DFT deals with equally spaced samples on the unit circle with the center at the origin in the Z-plane.

A number of companies provide software for implementing FFT and related basic applications such as convolution/correlation, filtering, spectral analysis, etc. on various platforms. Also general-purpose DSP chips can be programmed to implement the FFT and other discrete transforms.

This book is designed for senior undergraduate and graduate students, faculty, engineers, and scientists in the field, and self-learners to understand FFTs and directly apply them to their fields, efficiently. It is designed to be both a text and a reference. Thus examples, projects and problems all tied with MATLAB, are provided for grasping the concepts concretely. It also includes references to books and review papers and lists of applications, hardware/software, and useful websites. By including many figures, tables, bock diagrams and graphs, this book helps the reader understand the concepts of fast algorithms readily and intuitively. It provides new MATLAB functions and MATLAB source codes.

The material in this book is presented without assuming any prior knowledge of FFT. This book is for any professional who wants to have a basic understanding of the latest developments in and applications of FFT. It provides a good reference for any engineer planning to work in this field, either in basic implementation or in research and development.

D.N. Kim acknowledges the support by the National Information Technology (IT) Industry Promotion Agency (NIPA) and the Ministry of Knowledge Economy, Republic of Korea, under the IT Scholarship Program.

Organization of the Book

Chapter 1 introduces various applications of the discrete Fourier transform. Chapter 2 is devoted to introductory material on the properties of the DFT for the equally spaced samples. Chapter 3 presents fast algorithms to be mainly categorized as decimation-in-time (DIT) or decimation-in-frequency (DIF) approaches. Based on these, it introduces fast algorithms like split-radix, Winograd algorithm and others. Chapter 4 is devoted to integer FFT which approximates the discrete Fourier transform. One-dimensional DFT is extended to the two-dimensional signal and then to the multi-dimensional signal in Chapter 5. Applications to filtering are presented in this chapter. Variance distribution in the DFT domain is covered. It also introduces how we can diagonalize a circulant matrix using the DFT matrix. Fast algorithms for the 2-D DFT are covered in Chapter 6. Chapter 7 is devoted to introductory material on the properties of nonuniform DFT (NDFT) for the nonequally spaced samples. Numerous applications of the FFT are presented in Chapter 8. Appendix A covers performance comparison of discrete transforms. Appendix B covers spectral distance measures of image quality. Appendix C covers Integer DCTs. DCTs and DSTs are derived in Appendix D (DCT – discrete cosine transform, DST – discrete sine transform). Kronecker products and separability are briefly covered in Appendix E. Appendix F describes mathematical relations. Appendices G and H include MATLAB basics and M files. The bibliography contains lists of references to books and review papers, software/hardware, and websites. Numerous problems and projects are listed at the end of each chapter.

Arlington, TX   K.R. Rao
August 2010

Contents

Abbreviations

AAC	Advanced audio coder, AAC-LD (low delay) AAC-LC (low complexity)
AC, AC-2, AC-3	Audio coder
ACATS	FCC advisory committee on advanced television service
ACM	Association for computing machinery
ADSL	Asymmetric digital subscriber line
AES	Audio engineering society
ANN	Artificial neural network
ANSI	American National Standards Institute
APCCAS	IEEE Asia-Pacific Conference on Circuits and Systems
ASIC	Application specific integrated circuit
ASPEC	Adaptive spectral perceptual entropy coding of high quality music signals
ASSP	Acoustics, speech, and signal processing
ATC	Adaptive transform coding
ATRAC	Adaptive transform acoustic coding
ATSC	Advanced television systems committee
AVC	Advanced video coding, MPEG-4 AVC (MPEG-4 Part 10)
AVS	Audio video standard
BER	Bit error rate
BF	Butterfly
BIFORE	Binary Fourier representation
BPF	Band pass filter
BRO	Bit reversed order
BV	Basis vector
CAS	Circuits and systems
CBT	Complex BIFORE transform
CCETT	*Centre Commun d'Etudes de Télédiffusion et Télécommuications* in French (Common Study Center of Telediffusion and Telecommunications)

CD	Compact disc
CDMA	Code division multiple access
CE	Consumer electronics
CELP	Code-excited linear prediction
CF	Continuous flow
CFA	Common factor algorithms
CGFFT	Conjugate gradient FFT
CICC	Custom integrated circuits conference
CIPR	Center for image processing research
CMFB	Cosine modulated filter bank
CODEC	Coder and decoder
COFDM	Coded orthogonal frequency division multiplex
CM	Circulant matrix
CR	Compression ratio
CRT	Chinese reminder theorem
CSVT	Circuits and systems for video technology
DA	Distributed arithmetic
DAB	Digital audio broadcasting
DCC	Digital compact cassette
DCT	Discrete cosine transform
DEMUX	Demultiplexer
DF	Decision feedback
DFT	Discrete Fourier transform
DHT	Discrete Hartley transform
DIS	Draft International Standard
DMT	Discrete multitone modulation
DOS	Disc operating systems
DOT	Discrete orthogonal transform
DPCM	Differential pulse code modulation
D-PTS	Decomposition partial transmit sequence
DSP	Digital signal processing/processor
DST	Discrete sine transform
DTMF	Dual-tone multifrequency. DTMF signals/tones are used for telephone touch keypads [B40]
DTT	Discrete trigonometric transform
DVB-T	Digital video broadcasting standard for terrestrial transmission, using COFDM modulation
DVD	Digital video/versatile disc
DWT	Discrete wavelet transform
ECCTD	Biennial European Conference on Circuit Theory and Design
EDN	Electrical design news
EECON	Electrical Engineering Conference of Thailand
EEG	Electroencephalograph
EKG	Electrocardiograph, ECG

EMC	Electromagnetic compatibility, IEEE transactions on
EURASIP	European Association for Signal Processing
EUSIPCO	European Signal Processing Conference
FAQ	Frequently asked questions
FCC	The Federal Communications Commission
FFT	Fast Fourier transform
FIR	Finite impulse response
FMM	Fast multipole method
FPGA	Field programmable gate array
FRAT	Finite radon transform
FRExt	Fidelity range extensions
FRIT	Finite ridgelet transform
FSK	Frequency shift keying
FTP	File transfer protocol
FUDCuT	Fast uniform discrete curvelet transform
FxpFFT	Fixed point FFT
GDFHT	Generalized discrete Fourier Hartley transform
GDFT	Generalized DFT
GLOBECOM	IEEE Global Telecommunications Conference
GZS	Geometrical zonal sampling
H.263	Standard for visual communication via telephone lines
HDTV	High-definition television
HPF	High pass filter
HT	Hadamard transform
HTTP	Hyper text transfer protocol
HVS	Human visual sensitivity
Hz	Hertz cycles/sec, or cycles/meter
IASTED	The International Association of Science and Technology for Development
IBM	International business machines
IC	Integrated circuit(s)
ICA	Independent component analysis
ICASSP	IEEE International Conference on Acoustics, Speech, and Signal Processing
ICC	IEEE International Conference on Communications
ICCE	IEEE International Conference on Consumer Electronics
ICCS	IEEE International Conference on Circuits and Systems
ICECS	International Conference on Electronics, Circuits and Systems
ICIP	IEEE International Conference on Image Processing
ICME	IEEE International Conference on Multimedia and Expo
ICSPAT	International Conference on Signal Processing Applications and Technology
IDFT	Inverse DFT
IEC	International Electrotechnical Commission

IEEE	The Institute of Electrical and Electronics Engineers
IEICE	Institute of Electronics, Information and Communication Engineers
IFFT	Inverse FFT
IGF	Inverse Gaussian filter
IJG	Independent JPEG group
ILPM	Inverse LPM (log polar mapping)
IMTC	IEEE Instrumentation and Measurement Technology Conference
IntFFT	Integer FFT
IP	Image processing / intellectual property
IRE	Institute of radio engineers
IS	International standard
IS&T	The Society for Imaging Science and Technology
ISCAS	IEEE International Symposium on Circuits and Systems
ISCIT	IEEE International Symposium on Communications and Information Technologies
ISDN	Integrated services digital network
ISO	International Organization for Standardization
ISPACS	IEEE International Symposium on Intelligent Signal Processing and Communication Systems
IT	Information theory
JPEG	Joint photographic experts group
JSAC	Journal on selected areas in communications, IEEE
JTC	Joint technical committee
KLT	Karhunen–Loève transform
LAN	Local area networks
LC	Low complexity
LMS	Least mean square
LO	Lexicographic ordering
LPF	Low pass filter
LPM	Log polar mapping
LPOS	Left point of symmetry
LS	Least square, lifting scheme
LSI	Linear shift invariant
LUT	Look up tables
LW	Long window
MCM	Multichannel carrier modulation
MD	MiniDisc
MDCT	Modified discrete cosine transform
MDST	Modified discrete sine transform
ML	Maximum likelihood/multiplierless
MLS	Maximum length sequence
MLT	Modulated lapped transform
MMSE	Minimum mean square error

MoM	The method of moments
MOPS	Million operations per second
MOS	Mean opinion score
MOV	Model output variable
MPEG	Moving picture experts group
MR	Mixed radix
MRI	Magnetic resonance imaging
ms	Millisecond
M/S	Mid/side, middle and side, or sum and difference
MSB	Most significant bit
MUSICAM	Masking pattern universal subband integrated coding and multiplexing (MPEG-1 Level 2, MP2)
MUX	Multiplexer
MVP	Multimedia video processor
MWSCAS	Midwest Symposium on Circuits and Systems
NAB	National Association of Broadcasters
NBC	Nonbackward compatible (with MPEG-1 audio)
NMR	Noise-to-mask ratio
NNMF	Nonnegative matrix factorization
NTC	National telecommunications conference
OCF	Optimum coding in the frequency domain
OEM	Original equipment manufacturer
OFDM	Orthogonal frequency-division multiplexing
ONB	Orthonormal basis
PAC	Perceptual audio coder
PAMI	Pattern analysis and machine intelligence
PAPR	Peak-to-average power ratio
PC	Personal computer
PCM	Pulse code modulation
PDPTA	International Conference on Parallel and Distributed Processing Techniques and Applications
PE	Perceptual entropy
PFA	Prime factor algorithm
PFM	Prime factor map
PoS	Point of symmetry
PQF	Polyphase quadrature filter
PR	Perfect reconstruction
PRNG	Pseudorandom number generator
P/S	Parallel to serial converter
PSF	Point spread function
PSK	Phase shift keying
PSNR	Peak-to-peak signal-to-noise ratio
QAM	Quadrature amplitude modulation
QMF	Quadrature mirror filter

QPSK	Quadrature phase-shift keying
RA	Radon transform
RAM	Random access memory
RELP	Residual excited linear prediction
RF	Radio frequency
RFFT	Real valued FFT
RI	Ridgelet transform
RMA	Royal Military Academy of Belgium
RPI	Rensselear Polytechnic Institute
RPOS	Right point of symmetry
R−S	Reed Solomon
RST	Rotation, scaling and translation
SDDS	Sony dynamic digital sound
SEPXFM	Stereo-entropy-coded perceptual transform coder
SIAM	Society for Industrial and Applied Mathematics
SiPS	Signal processing systems
SMPTE	Society of Motion Picture and Television Engineers
SMR	Signal-to-mask ratio
SNR	Signal to noise ratio
SOPOT	Sum-of-powers-of-two
SP	Signal processing
S/P	Serial to parallel converter
SPIE	Society of Photooptical and Instrumentation Engineers
SPS	Symmetric periodic sequence
SR	Split radix
SS	Spread spectrum
SSST	Southeastern Symposium on System Theory
STBC	Space-time block code
SW	Short window
TDAC	Time domain aliasing cancellation
T/F	Time-to-frequency
UDFHT	Unified discrete Fourier-Hartley transform
USC	University of Southern California
VCIP	SPIE and IS&T visual communications and image processing
VCIR	Visual communication and image representation
VLSI	Very large scale integration
VSP	Vector signal processor
WD	Working draft
WHT	Walsh-Hadamard transform
WLAN	Wireless LAN
WMV	Window media video
WPMC	International Symposium on Wireless Personal Multimedia Communications

Chapter 1
Introduction

Fast Fourier transform (FFT) [A1, LA23] is an efficient implementation of the discrete Fourier transform (DFT) [A42]. Of all the discrete transforms, DFT is most widely used in digital signal processing. The DFT maps a sequence $x(n)$ into the frequency domain. Many of its properties are similar to those of the Fourier transform of an analog signal. The development of the DFT originally by Cooley and Tukey [A1] followed by various enhancements/modifications by other researchers (some of them tailored to specific software/hardware) has provided the incentive and the impetus for its rapid and widespread utilization in a number of diverse disciplines. Independent of the Cooley–Tukey approach, several algorithms such as prime factor [B29, A42], split radix [SR1, O9, A12, A42], vector radix [A16, B41], split vector radix [DS1, SR2, SR3] and Winograd Fourier transform algorithm (WFTA) [A35, A36, A37] have been developed. A number of companies provide software for implementing FFT and related basic applications such as convolution/correlation, filtering, spectral analysis etc. on various platforms. Also general purpose DSP chips can be programmed to implement the FFT and other discrete transforms.

Chapter 2 defines the DFT and its inverse (IDFT) and describes their properties. This is followed by development of the fast algorithms which are basically identical for both DFT/IDFT. DFT is complex, orthogonal and separable. Because of the separable property, extension from 1-D (one-dimensional) DFT/IDFT to multi-D (multi-dimensional) DFT/IDFT is simple and straight forward. By applying series of 1-D DFTs/IDFTs, multi-D DFTs/IDFTs can be executed. Their actual implementation of course is via the fast algorithms as these result in reduced memory, reduced computational complexity and reduced roundoff/truncation errors due to finite word length (bit size) arithmetic. The computational complexity can be further reduced when data sequences are real. Additional advantages of the fast algorithms are: they are recursive (multiple size DFTs/IDFTs can be implemented by the same algorithm), and modular. Also different algorithms can be combined in a flexible manner to yield an optimal approach. Specific algorithms for implementing 2-D DFTs/IDFTs directly (bypassing the 1-D approach) have also been

K.R. Rao et al., *Fast Fourier Transform: Algorithms and Applications*,
Signals and Communication Technology,
DOI 10.1007/978-1-4020-6629-0_1, © Springer Science+Business Media B.V. 2010

developed [A42]. References reflecting the plethora of algorithms and applications are provided at the end of the book.

1.1 Applications of Discrete Fourier Transform

The applications of DFT are extensive. Algorithmic evolution of FFT/IFFT (software/hardware) has accelerated these applications. Some of these are listed below:

- Array antenna analysis
- Autocorrelation and cross correlation
- Bandwidth compression
- Channel separation and combination
- Chirp z transform
- Convolution
- Decomposition of convolved signals
- EKG and EEG signal processing
- Filter banks
- Filter simulation
- Forensic science
- Fourier spectroscopy
- Fractal image coding
- FSK demodulation
- Generalized spectrum and homomorphic filtering
- Ghost cancellation
- Image quality measures
- Image registration
- Interpolation and decimation
- Linear estimation
- LMS adaptive filters
- MDCT/MDST via FFT (Dolby AC-3 [audio coder, 5.1 channel surround sound], DVD, MPEG-2 AAC [Advanced Audio Coder], MPEG-4 Audio [> 64 Kbps])
- Magnetic resonance imaging
- Motion estimation
- Multichannel carrier modulation
- Multiple frequency detection
- Multiple time series analysis and filtering
- Noise filtering
- Numerical solution of differential equations
- OFDM modulation
- Optical signal processing
- Pattern recognition
- Phase correlation based motion estimation

- POC (phase only correlation) in medical imaging
- Power spectrum analysis
- PSK classification
- Psychoacoustic model for audio coding
- Radar signal processing
- Signal representation and discrimination
- Sonar signal processing
- Spectral estimation
- Speech encryption
- Speech signal processing
- Speech spectrograms
- Spread spectrum
- Surface texture analysis
- Video/image compression
- Watermarking
- Wiener filtering (image denoising)
- 2-D and 3-D image rotation

Chapter 2
Discrete Fourier Transform

2.1 Definitions

The DFT and IDFT can be defined as follows.

2.1.1 DFT

$$X^F(k) = \sum_{n=0}^{N-1} x(n) W_N^{kn}, \quad k = 0, 1, \ldots, N-1, \quad N \text{ DFT coefficients} \qquad (2.1a)$$

$$W_N = \exp\left(\frac{-j2\pi}{N}\right)$$

$$W_N^{kn} = \exp\left[\left(\frac{-j2\pi}{N}\right)kn\right]$$

where $x(n)$, $n = 0, 1, \ldots, N-1$ is a uniformly sampled sequence, T is sampling interval. $W_N = \exp(-j2\pi/N)$ is the N-th root of unity, and $X^F(k)$, $k = 0, 1, \ldots, N-1$ is the k-th DFT coefficient. $j = \sqrt{-1}$.

2.1.2 IDFT

$$x(n) = \frac{1}{N} \sum_{k=0}^{N-1} X^F(k) W_N^{-kn}, \quad n = 0, 1, \ldots, N-1, \quad N \text{ data samples} \qquad (2.1b)$$

$$\left(W_N^{kn}\right)^* = W_N^{-kn} = \exp[(j2\pi/N)kn] \qquad e^{\pm j0} = \cos\theta \pm j\sin\theta$$

K.R. Rao et al., *Fast Fourier Transform: Algorithms and Applications*,
Signals and Communication Technology,
DOI 10.1007/978-1-4020-6629-0_2, © Springer Science+Business Media B.V. 2010

Superscript $*$ indicates complex conjugate operation. The DFT pair can be symbolically represented as:

$$x(n) \Leftrightarrow X^F(k) \qquad (2.2)$$

The normalization factor $1/N$ in (2.1b) can be equally distributed between DFT and IDFT (this is called unitary DFT) or it can be moved to the forward DFT i.e.,

2.1.3 Unitary DFT (Normalized)

$$\text{Forward} \quad X^F(k) = \frac{1}{\sqrt{N}} \sum_{n=0}^{N-1} x(n) W_N^{kn}, \quad k = 0, 1, \ldots, N-1 \qquad (2.3a)$$

$$\text{Inverse} \quad x(n) = \frac{1}{\sqrt{N}} \sum_{k=0}^{N-1} X^F(k) W_N^{-kn}, \quad n = 0, 1, \ldots, N-1 \qquad (2.3b)$$

Alternatively,

$$\text{Forward} \quad X^F(k) = \frac{1}{N} \sum_{n=0}^{N-1} x(n) \exp\left(\frac{-j2\pi kn}{N}\right), \quad k = 0, 1, \ldots, N-1 \qquad (2.4a)$$

$$\text{Inverse} \quad x(n) = \sum_{k=0}^{N-1} X^F(k) \exp\left(\frac{j2\pi kn}{N}\right), \quad n = 0, 1, \ldots, N-1 \qquad (2.4b)$$

While the DFT/IDFT as defined in (2.1), (2.3) and (2.4) is equally valid, for the sake of consistency, we will henceforth adopt (2.1). Hence

$$\text{DFT} \quad X^F(k) = \sum_{n=0}^{N-1} x(n) \left[\cos\frac{2\pi kn}{N} - j\sin\frac{2\pi kn}{N} \right], \quad k = 0, 1, \ldots, N-1 \quad (2.5a)$$

$$\text{IDFT} \quad x(n) = \frac{1}{N} \sum_{k=0}^{N-1} X^F(k) \left[\cos\frac{2\pi kn}{N} + j\sin\frac{2\pi kn}{N} \right], \quad n = 0, 1, \ldots, N-1 \quad (2.5b)$$

Both $x(n)$ and $X^F(k)$ are N-point sequences or length N sequences.

$$\underline{x}(n) = [x(0), x(1), \ldots, x(N-1)]^T \qquad N\text{-point data vector, and}$$
$$(N \times 1)$$

$$\underline{X}^F(k) = [X^F(0), X^F(1), \ldots, X^F(N-1)]^T \qquad N\text{-point DFT vector.}$$
$$(N \times 1)$$

$$W_N = \exp\left(\frac{-j2\pi}{N}\right) \qquad N\text{-th root of unity}$$

Fig. 2.1 The eight roots of unity distributed uniformly along the unit circle with the center at the origin in the z-plane

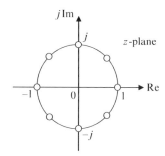

$$W_N^k = W_N^{k \bmod N}$$

$k \bmod N = k$ modulo $N = $ Remainder of $\left(\frac{k}{N}\right)$. For example 23 mod 5 $= 3$, $\frac{23}{5} = 4 + \frac{3}{5}$. Superscript T implies transpose.

$\sum_{k=0}^{N-1} W_N^k = 0$. All the N roots are distributed uniformly on the unit circle with the center at the origin. The sum of all the N roots of unity is zero. For example

$\sum_{k=0}^{7} W_8^k = 0$ (Fig. 2.1). In general, $\sum_{k=0}^{N-1} W_N^{pk} = N\delta(p)$ where p is an integer.

The relationship between the Z-transform and DFT can be established as follows.

2.2 The Z-Transform

$X(z)$, the Z-transform of $x(n)$ is defined as

$$X(z) = \sum_{n=0}^{N-1} x(n) z^{-n} \tag{2.6a}$$

$f_s = \dfrac{1}{T} = $ sampling rate, $\left(\dfrac{\# \text{ of samples}}{\sec}\right)$, or $T = \dfrac{1}{f_s}$, sampling interval in seconds.

$$X(e^{j\omega T}) = \sum_{n=0}^{N-1} x(n) \exp\left[\frac{-j2\pi fn}{f_s}\right] \tag{2.6b}$$

is $X(z)$ evaluated on the unit circle with center at the origin in the z-plane.

By choosing equally distributed points on this unit circle (2.6b) can be expressed as

$$X^F(k) = \sum_{n=0}^{N-1} x(n) \exp\left[\frac{-j2\pi kn}{N}\right], \quad k = 0, 1, \ldots, N-1 \tag{2.7}$$

where $k = Nf/f_s$.

The Z-transform of $x(n)$ evaluated on the unit circle with center at the origin in the z-plane at equally distributed points, is the DFT of $x(n)$ (Fig. 2.3).

$X^F(k)$, $k = 0, 1, \ldots, N-1$ represents the DFT of $\{x(n)\}$ at frequency $f = kf_s/N$. It should be noted that because of the frequency folding, the highest frequency representation of $x(n)$ is at $X^F\left(\frac{N}{2}\right)$ i.e., at $f = f_s/2$. For example, let $N = 100, T = 1$ μs. Then $f_s = 1$ MHz and the resolution in the frequency domain f_0 is 10^4 Hz. The highest frequency content is 0.5 MHz. These are illustrated below: (Figs. 2.2 through 2.4)

The resolution in the frequency domain is

$$f_0 = \frac{1}{NT} = \frac{1}{T_R} = \frac{f_s}{N} \tag{2.8}$$

$T_R = NT$ is the record length. For a given N, one can observe the inverse relationship between the resolution in time T and resolution in frequency f_0. Note that $x(n)$ can be a uniformly sampled sequence in space also in which case T is in meters and $f_s = \frac{1}{T} = \#$ of samples/meter.

The periodicity of the DFT can be observed from Fig. 2.3. As we trace $X^F(k)$ at equally distributed points around the unit circle with center at the origin in the z-plane, *the DFT repeats itself* every time we go over the unit circle. Hence

$$X^F(k) = X^F(k + l\,N) \tag{2.9a}$$

where l is an integer. *The DFT assumes that $x(n)$ is also periodic* with period N i.e.,

$$x(n) = x(n + l\,N) \tag{2.9b}$$

For a practical example of Fig. 2.4, if a signal $x_1(n)$ is given as Fig. 2.5a, then its magnitude spectrum is Fig. 2.5b.

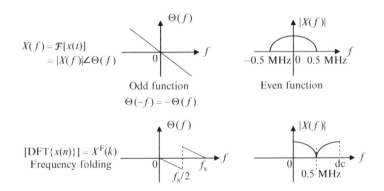

Fig. 2.2 Effect of frequency folding at $f_s/2$

a

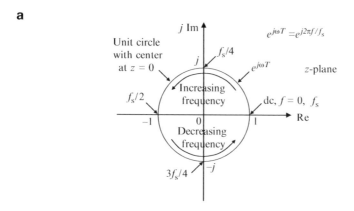

b

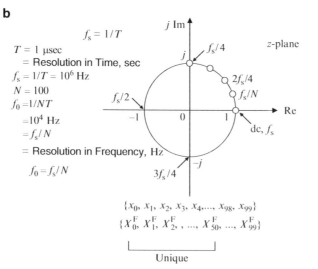

$$\{x_0, x_1, x_2, x_3, x_4, ..., x_{98}, x_{99}\}$$
$$\{X_0^F, X_1^F, X_2^F, , ..., X_{50}^F, ..., X_{99}^F\}$$

|_____|

Unique

Frequency points along the unit circle
with the center at the origin in the z-plane

Fig. 2.3 a DFT of $x(n)$ is the Z-transform at N equally distributed points on the unit circle with the center at the origin in the z-plane. **b** specific example for $T = 1\ \mu s$ and $N = 100$

$$x_1(n) = \sin\left(\frac{2\pi}{T_1}n\right) = \sin\left(\frac{2\pi}{20}n\right), \quad n = 0, 1, \ldots, N-1, \quad N = 100$$

$$T_1 = 20 \times T = 20\ \mu s \qquad \text{Record length,} \quad T_R = N \times T = 100\ \mu s$$

$$f_s = 1/T = 1\ \text{MHz}, \ f_0 = \frac{f_s}{N} = \frac{1}{100}\ \text{MHz}, \ f_1 = 5f_0 = \frac{5}{100}\ \text{MHz}$$

$$\frac{f_s}{2} = 50f_0 = \frac{1}{2}\ \text{MHz}$$

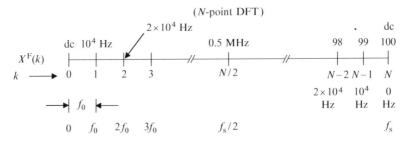

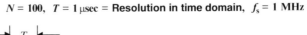

Fig. 2.4 Highest frequency representation of $x(n)$ is at $X^F(N/2)$ i.e., at $f = f_s/2$

The DFT/IDFT expressed in summation form in (2.1) can be expressed in vector-matrix form as

$$[\underline{X}^F(k)] = [F][\underline{x}(n)] \tag{2.10a}$$

where $[F]$ is the $(N \times N)$ DFT matrix described in (2.11).

$$
\begin{array}{ccc}
& \begin{array}{c} \text{Columns} \\ n \to 0\ 1\ 2\ \cdots\ n\ \cdots\ N-1 \end{array} & \begin{array}{c} \text{Rows} \\ k \\ \downarrow \end{array}
\end{array}
$$

$$
\begin{bmatrix} X^F(0) \\ X^F(1) \\ \vdots \\ X^F(k) \\ \vdots \\ X^F(N-1) \end{bmatrix}
=
\begin{bmatrix} & & \\ & W_N^{nk} & \\ & (n,k=0,1,\ldots,N-1) & \\ & & \end{bmatrix}
\begin{bmatrix} x(0) \\ x(1) \\ \vdots \\ x(n) \\ \vdots \\ x(N-1) \end{bmatrix}
\begin{array}{c} 0 \\ 1 \\ \vdots \\ k \\ \vdots \\ (N-1) \end{array}
$$

$$
\begin{array}{ccc}
(N \times 1) & (N \times N) & (N \times 1)
\end{array}
$$

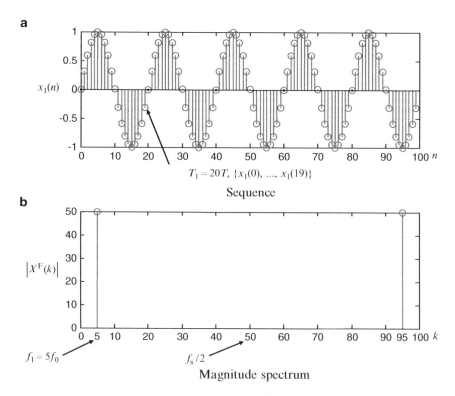

Fig. 2.5 Highest frequency representation of $x_1(n)$ is at $X^F(N/2)$ i.e., at $f = f_s/2$ and not at $f = f_s$ because of frequency folding (Fig. 2.2)

$$\text{IDFT} \quad [\underline{x}(n)] = \frac{1}{N}[F]^*[\underline{X}^F(k)] \quad (2.10\text{b})$$

$$
\underbrace{\begin{bmatrix} x(0) \\ x(1) \\ \vdots \\ x(n) \\ \vdots \\ x(N-1) \end{bmatrix}}_{(N \times 1)}
= \frac{1}{N}
\underbrace{\begin{bmatrix} & & \\ & W_N^{-nk} & \\ & (n,k=0,1,\ldots,N-1) & \\ & & \end{bmatrix}}_{(N \times N)}
\underbrace{\begin{bmatrix} X^F(0) \\ X^F(1) \\ \vdots \\ X^F(k) \\ \vdots \\ X^F(N-1) \end{bmatrix}}_{(N \times 1)}
$$

Columns $k \to 0 \ 1 \ 2 \ \cdots \ k \ \cdots \ N-1$ Rows n $\downarrow$ 0 1 $\vdots$ n $\vdots$ $(N-1)$

The $(N \times N)$ DFT matrix $[F]$ is:

$$
\begin{array}{c}
[F] \\
(N \times N)
\end{array}
=
\begin{array}{cccccc}
\text{Columns} & 0 & 1 & 2 & \cdots\ n\ \cdots & (N-1) & \text{Rows} \\
\rightarrow & & & & & & \downarrow
\end{array}
$$

$$
\begin{bmatrix}
W_N^0 & W_N^0 & W_N^0 & \cdots & W_N^0 \\
W_N^0 & W_N^1 & W_N^2 & \cdots & W_N^{(N-1)} \\
W_N^0 & W_N^2 & W_N^4 & \cdots & W_N^{2(N-1)} \\
\vdots & \vdots & \vdots & \cdots & \vdots \\
W_N^0 & W_N^k & W_N^{2k} & \cdots & W_N^{k(N-1)} \\
\vdots & \vdots & \vdots & \cdots & \vdots \\
W_N^0 & W_N^{(N-1)} & W_N^{2(N-1)} & \cdots & W_N^{(N-1)(N-1)}
\end{bmatrix}
\begin{matrix}
0 \\ 1 \\ 2 \\ \vdots \\ k \\ \vdots \\ (N-1)
\end{matrix}
$$

$$\text{DFT Matrix}$$

$$(2.11)$$

Note that

$$
\left(\frac{1}{\sqrt{N}} [F]^* \right) \left(\frac{1}{\sqrt{N}} [F] \right) = [I_N] = (N \times N) \text{ unit matrix.}
$$

Each row of $[F]$ is a basis vector (BV). The elements of $[F]$ in row l and column k are $W_N^{lk}, l, k = 0, 1, \ldots, N-1$. As $W_N = \exp(-j2\pi/N)$ is the Nth root of unity, of the N^2 elements in $[F]$, only N elements are unique, i.e., $W_N^l = W_N^{l \bmod N}$, where $l \bmod N$ implies remainder of l divided by N. For example, 5 mod 8 is 5, 10 mod 5 is 0, and 11 mod 8 is 3. Mod is the abbreviation for modulo. The following observations can be made about the DFT matrix:

1. $[F]$ is symmetric, $[F] = [F]^T$.
2. $[F]$ is unitary, $[F][F]^* = N[I_N]$, where $[I_N]$ is an unit matrix of size $(N \times N)$.

$$
[F]^{-1} = \tfrac{1}{N}[F]^*, \quad [F][F]^{-1} = [I_N] = \text{unit matrix} =
\begin{bmatrix}
1 & & & \\
& 1 & & O \\
& & \ddots & \\
O & & & 1
\end{bmatrix}
\quad (2.12)
$$

For example the (8×8) DFT matrix can be simplified as (here $W = W_8 = \exp(-j2\pi/8)$)

$$
\begin{array}{c}
\text{Columns} \quad 0 \quad 1 \quad 2 \quad 3 \quad 4 \quad 5 \quad 6 \quad 7 \quad \text{Rows} \\
\rightarrow \qquad\qquad\qquad\qquad\qquad\qquad\qquad\qquad \downarrow \\
\begin{bmatrix}
1 & 1 & 1 & 1 & 1 & 1 & 1 & 1 \\
1 & W & W^2 & W^3 & -1 & -W & -W^2 & -W^3 \\
1 & W^2 & -1 & -W^2 & 1 & W^2 & -1 & -W^2 \\
1 & W^3 & -W^2 & W & -1 & -W^3 & W^2 & -W \\
1 & -1 & 1 & -1 & 1 & -1 & 1 & -1 \\
1 & -W & W^2 & -W^3 & -1 & W & -W^2 & W^3 \\
1 & -W^2 & -1 & W^2 & 1 & -W^2 & -1 & W^2 \\
1 & -W^3 & -W^2 & -W & -1 & W^3 & W^2 & W
\end{bmatrix}
\begin{array}{c}
0 \\ 1 \\ 2 \\ 3 \\ 4 \\ 5 \\ 6 \\ 7
\end{array}
\end{array}
$$

$$(2.13)$$

Observe that $W_N^{N/2} = -1$ and $W_N^{N/4} = -j$. Also $\sum_{k=0}^{N-1} W_N^k = 0 = $ sum of all the N distinct roots of unity. These roots are uniformly distributed on the unit circle with center at the origin in the z-plane (Fig. 2.3).

2.3 Properties of the DFT

From the definition of the DFT, several properties can be developed.

1. *Linearity*: Given $x_1(n) \Leftrightarrow X_1^F(k)$ and $x_2(n) \Leftrightarrow X_2^F(k)$ then

$$
[a_1 x_1(n) + a_2 x_2(n)] \Leftrightarrow [a_1 X_1^F(k) + a_2 X_2^F(k)] \tag{2.14}
$$

where a_1 and a_2 are constants.

2. Complex conjugate theorem: For an N-point DFT, when $x(n)$ *is a real sequence*,

$$
X^F\left(\frac{N}{2} + k\right) = X^{F*}\left(\frac{N}{2} - k\right), \qquad k = 0, 1, \ldots, \frac{N}{2} \tag{2.15}
$$

This implies that both $X^F(0)$ and $X^F(N/2)$ are real. By expressing $X^F(k)$ in polar form as $X^F(k) = |X^F(k)| \exp[j\Theta(k)]$, it is evident that $|X^F(k)|$ versus k is an even function and $\Theta(k)$ versus k is an odd function around $N/2$ in the frequency domain (Fig. 2.4). $|X^F(k)|$ and $\Theta(k)$ are called the magnitude spectrum and phase spectrum respectively. Of the N DFT coefficients only $(N/2) + 1$ coefficients are independent. $|X^F(k)|^2, k = 0, 1, \ldots, N - 1$ is the power spectrum. This is an even function around $N/2$ in the frequency domain (Fig. 2.6).

(When $x(n)$ is real)

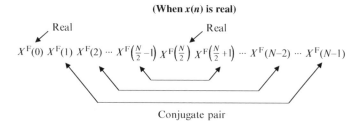

Conjugate pair

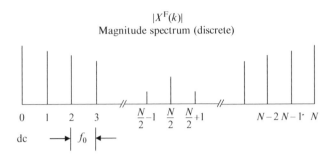

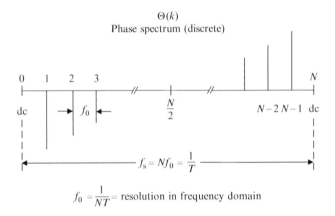

Fig. 2.6 Magnitude and phase spectra when $x(n)$ is a real sequence

$$X^F\left(\frac{N}{2} + k\right) = \sum_{n=0}^{N-1} x(n) W_N^{\left(\frac{N}{2}+k\right)n}$$

$$X^{F^*}\left(\frac{N}{2} - k\right) = \left[\sum_{n=0}^{N-1} x(n) W_N^{\left(\frac{N}{2}-k\right)n}\right]^* = \sum_{n=0}^{N-1} x(n) W_N^{\left(\frac{N}{2}+k\right)n} = X^F\left(\frac{N}{2} + k\right)$$

since $W_N^{N/2} = \exp\left(\dfrac{-j2\pi}{N}\dfrac{N}{2}\right) = e^{-j\pi} = -1, \quad \left(W_N^{-kn}\right)^* = W_N^{kn}.$

$$X^F(0) = \sum_{n=0}^{N-1} x(n), \text{ dc coefficient} \qquad \boxed{\dfrac{1}{N}\sum_{n=0}^{N-1} x(n) : \text{ Mean of } x(n)}$$

$$X^F\left(\frac{N}{2}\right) = \sum_{n=0}^{N-1} x(n) W_N^{n\frac{N}{2}} = \sum_{n=0}^{N-1} x(n)(-1)^n$$

Figure 2.7 illustrates the even and odd function properties of $X^F(k)$ at $k = N/2$, when $x(n)$ is real.

Real data sequence ($N = 32$)

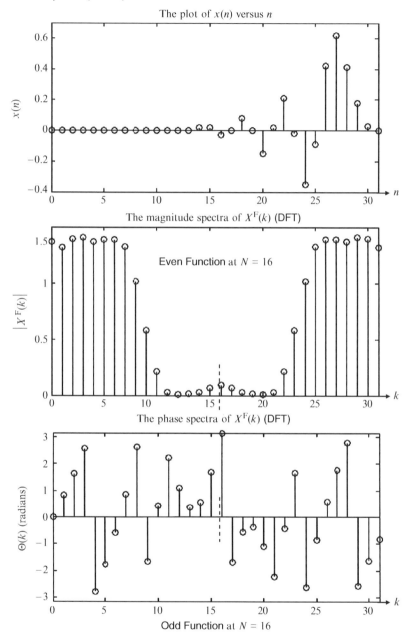

Fig. 2.7 Even and odd function properties of $X^F(k)$ at $k = N/2$, when $x(n)$ is real

3. Parseval's theorem

This property is valid for all unitary transforms.

$$\sum_{n=0}^{N-1} x(n)x^*(n) = \frac{1}{N} \sum_{k=0}^{N-1} X^F(k)X^{F^*}(k) \qquad (2.16a)$$

Energy of the sequence $\{x(n)\}$ is preserved in the DFT domain.

Proof:

$$\sum_{n=0}^{N-1} |x(n)|^2 = \sum_{n=0}^{N-1} x(n)x^*(n) = \frac{1}{N} \sum_{n=0}^{N-1} x^*(n) \sum_{k=0}^{N-1} X^F(k) \; W_N^{-nk}$$

$$= \frac{1}{N} \sum_{k=0}^{N-1} X^F(k) \sum_{n=0}^{N-1} x^*(n) W_N^{-nk} = \frac{1}{N} \sum_{k=0}^{N-1} X^F(k) \left(\sum_{n=0}^{N-1} x(n)W_N^{nk} \right)^*$$

$$= \frac{1}{N} \sum_{k=0}^{N-1} X^F(k)X^{F^*}(k) = \frac{1}{N} \sum_{k=0}^{N-1} |X^F(k)|^2$$

$$(2.16b)$$

4. Circular shift

Given: $x(n) \Leftrightarrow X^F(k)$, then

$$x(n + h) \Leftrightarrow X^F(k) \; W_N^{-hk} \qquad (2.17)$$

$x(n + h)$ is shifted circularly to the left by h sampling intervals in the time domain i.e., $\{x_{n+h}\}$ is $(x_h, x_{h+1}, \; x_{h+2}, \ldots, x_{N-1}, \; x_0, \; x_1, \ldots, x_{h-1})$.

Proof:

$$\mathrm{DFT}\,[x(n + h)] = \sum_{n=0}^{N-1} x(n + h) \; W_N^{nk}, \qquad let \; m = n + h$$

$$= \sum_{m=h}^{N+h-1} x(m) \; W_N^{(m-h)k} \qquad (2.18)$$

$$= \left(\sum_{m=h}^{N+h-1} x(m)W_N^{mk} \right) W_N^{-hk}$$

$$= X^F(k)W_N^{-hk}$$

since $\sum_{m=h}^{N+h-1} x(m) \; W_N^{mk} = X^F(k), \; k = 0, 1, \ldots, N - 1.$

As $\left| W_N^{-hk} \right| = 1$, the DFT of a circularly shifted sequence and the DFT of the original sequence are related only by phase. Magnitude spectrum and hence power spectrum are invariant to the circular shift of a sequence.

5. DFT of $\left[x(n) \exp\left(\frac{j2\pi hn}{N} \right) \right]$

$$\text{DFT}\left[x(n) \exp\left(\frac{j2\pi hn}{N} \right) \right] = \sum_{n=0}^{N-1} (x(n)W_N^{-hn})W_N^{kn} = \sum_{n=0}^{N-1} x(n)\, W_N^{(k-h)n} = X^F(k-h)$$

(2.19)

Since $X^F(k) = \sum_{n=0}^{N-1} x(n)\, W_N^{kn}$, $k = 0, 1, \ldots, N-1$, $X^F(k-h)$ is $X^F(k)$ circularly shifted by h sampling intervals in the frequency domain. For the special case when $h = N/2$

$$\text{DFT}\left[x(n) \exp\left(\frac{j2\pi}{N}\frac{N}{2}n \right) \right] = \text{DFT}\left[(-1)^n x(n) \right], \qquad e^{\pm j\pi n} = (-1)^n$$

$$= \text{DFT}\{x(0), -x(1), x(2), -x(3), x(4), \ldots, (-1)^{N-1}x(N-1)\} = X^F\left(k - \frac{N}{2} \right)$$

(2.20)

Here $\{x(0), x(1), x(2), x(3), \ldots, x(N-1)\}$ is the original N-point sequence.

In the discrete frequency spectrum, the dc component is now shifted to its midpoint (Fig. 2.8). To the left and right sides of the midpoint, frequency increases (consider these as positive and negative frequencies).

6. DFT of a permuted sequence [B1]

$$\{x(pn)\} \Leftrightarrow \{X^F(qk)\} \qquad 0 \leq p, q \leq N-1$$

(2.21)

Let the sequence $x(n)$ be permuted, with n replaced by pn modulo N, where $0 \leq p \leq N-1$ and p is an integer relatively prime to N. Then the DFT of $x(pn)$ is given by

$$A^F(k) = \sum_{n=0}^{N-1} x(pn)W^{nk}$$

(2.22)

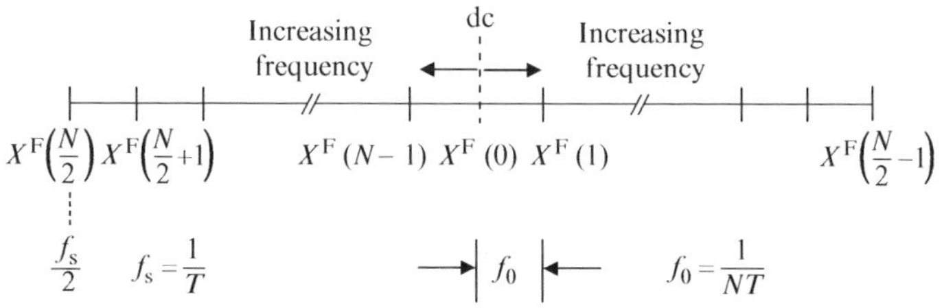

Fig. 2.8 DFT of $[(-1)^n x(n)]$. The dc coefficient is at the center with increasing frequencies to its left and right

When a and b have no common factors other than 1, they are said to be relatively prime and denoted as $(a, b) = 1$. Since $(p, N) = 1$, we can find an integer q such that $0 \leq q \leq N - 1$ and $qp \equiv (1 \text{ modulo } N)$.[1] Equation (2.22) is not changed if n is replaced by qn modulo N. We then have

$$
\begin{aligned}
A^{\mathrm{F}}(k) &= \sum_{n=0}^{N-1} x(pqn)W^{nqk} \\
&= \sum_{n=0}^{N-1} x(aNn + n)W^{nqk} \quad \text{since } qp \equiv (1 \text{ modulo } N), \text{ or } qp = aN + 1 \\
&= \sum_{n=0}^{N-1} x(n)W^{n(qk)} = X^{\mathrm{F}}(qk) \qquad \text{as } x(n) \text{ is periodic from (2.9b)}
\end{aligned}
$$

$$(2.23)$$

where a is an integer.

Example 2.1   For $N = 8$, p is $\{3, 5, 7\}$. $q = 3$ for $p = 3$. $q = 5$ for $p = 5$. $q = 7$ for $p = 7$. For $p = 3$, $pn = \{0, 3, 6, 1, 4, 7, 2, 5\}$. Since $p = q = 3$, $pn = qk$ when $n = k$. For $p = 5$, $pn = \{0, 5, 2, 7, 4, 1, 6, 3\}$. For $p = 7$, $pn = \{0, 7, 6, 5, 4, 3, 2, 1\}$.

Example 2.2   Let $N = 8$ and $p = 3$. Then $q = 3$. Let $x(n) = \{0, 1, 2, 3, 4, 5, 6, 7\}$ and let

$$
A^{\mathrm{F}}(K) = X^{\mathrm{F}}(qk) = \{X^{\mathrm{F}}(0), X^{\mathrm{F}}(3), X^{\mathrm{F}}(6), X^{\mathrm{F}}(1), X^{\mathrm{F}}(4), X^{\mathrm{F}}(7), X^{\mathrm{F}}(2), X^{\mathrm{F}}(5)\}
$$

Then

$$
\begin{aligned}
a(n) &= N\text{-point IDFT of } [A^{\mathrm{F}}(k)] \\
&= \{0, 3, 6, 1, 4, 7, 2, 5\} = x(pn)
\end{aligned}
$$

2.4   Convolution Theorem

Circular convolution of two periodic sequences in time/spatial domain is equivalent to multiplication in the DFT domain. Let $x(n)$ and $y(n)$ be two real periodic sequences with period N. Their circular convolution is given by

$$
\begin{aligned}
z_{\mathrm{con}}(m) &= \frac{1}{N} \sum_{n=0}^{N-1} x(n)\, y(m - n) \qquad m = 0, 1, \ldots, N - 1 \\
&= x(n) * y(n)
\end{aligned}
$$

$$(2.24a)$$

---

[1]See problem 2.21(a).

In the DFT domain this is equivalent to

$$Z_{\mathrm{con}}^{\mathrm{F}}(k) = \frac{1}{N} X^{\mathrm{F}}(k)\, Y^{\mathrm{F}}(k) \qquad (2.24\mathrm{b})$$

where $x(n) \Leftrightarrow X^{\mathrm{F}}(k)$, $y(n) \Leftrightarrow Y^{\mathrm{F}}(k)$, and $z_{\mathrm{con}}(m) \Leftrightarrow Z_{\mathrm{con}}^{\mathrm{F}}(k)$.

Proof: DFT of $z_{\mathrm{con}}(m)$ is

$$\sum_{m=0}^{N-1} \left[\frac{1}{N} \sum_{n=0}^{N-1} x(n) y(m-n) \right] W_N^{mk}$$

$$= \frac{1}{N} \sum_{n=0}^{N-1} x(n)\, W_N^{nk} \sum_{m=0}^{N-1} y(m-n)\, W_N^{(m-n)k}$$

$$= \frac{1}{N}\, X^{\mathrm{F}}(k)\, Y^{\mathrm{F}}(k)$$

$$\mathrm{IDFT}\left[\frac{1}{N}\, X^{\mathrm{F}}(k)\, X^{\mathrm{F}}(k) \right] = z_{\mathrm{con}}(m)$$

To obtain a noncircular or aperiodic convolution using the DFT, the two sequences $x(n)$ and $y(n)$ have to be extended by adding zeros. Even though this results in a circular convolution, for one period it is the same as the noncircular convolution (Fig. 2.9). This technique can be illustrated as follows:

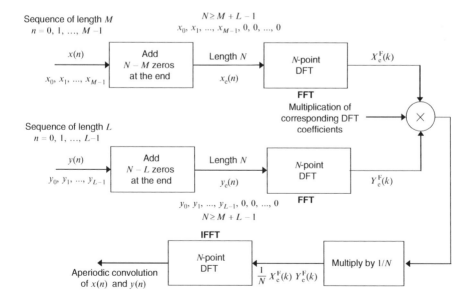

Fig. 2.9 Aperiodic convolution using the DFT/IDFT. $x_{\mathrm{e}}(n)$ and $y_{\mathrm{e}}(n)$ are extended sequences

Given: $\{x(n)\} = \{x_0, x_1, \ldots, x_{M-1}\}$, an M-point sequence and $\{y(n)\} = \{y_0, y_1, \ldots, y_{L-1}\}$, an L-point sequence, obtain their circular convolution.

1. Extend $x(n)$ by adding $N - M$ zeros at the end i.e., $\{x_e(n)\} = \{x_0, x_1, \ldots, x_{M-1}, 0, 0, \ldots, 0\}$, where $N \geq M + L - 1$.
2. Extend $y(n)$ by adding $N - L$ zeros at the end i.e., $\{y_e(n)\} = \{y_0, y_1, \ldots, y_{L-1}, 0, 0, \ldots, 0\}$.
3. Apply N-point DFT to $\{x_e(n)\}$ to get $\{X_e^F(k)\}$, $k = 0, 1, \ldots, N - 1$.
4. Carry out step 3 on $\{y_e(n)\}$ to get $\{Y_e^F(k)\}$, $k = 0, 1, \ldots, N - 1$.
5. Multiply $\frac{1}{N} X_e^F(k)$ and $Y_e^F(k)$ to get $\frac{1}{N} X_e^F(k) Y_e^F(k)$, $k = 0, 1, \ldots, N - 1$.
6. Apply N-point IDFT to $\{\frac{1}{N} X_e^F(k) Y_e^F(k)\}$ to get $z_{con}(m)$, $m = 0, 1, \ldots, N - 1$, which is the aperiodic convolution of $\{x(n)\}$ and $\{y(n)\}$.

Needless to say all the DFTs/IDFTs are implemented via the fast algorithms (see Chapter 3). Note that $\{X_e^F(k)\}$ and $\{Y_e^F(k)\}$ are N-point DFTs of the extended sequences $\{x_e(n)\}$ and $\{y_e(n)\}$ respectively.

Example 2.3   Periodic and Nonperiodic (Fig. 2.10)
Discrete convolution of two sequences $\{x(n)\}$ and $\{y(n)\}$.

$$z_{con}(m) = \frac{1}{N} \sum_{n=0}^{N-1} x(n) y(m-n)$$

$x(n)$ $n = 0, 1, \ldots, M - 1$
$y(n)$ $n = 0, 1, \ldots, L - 1$ (where $N = L + M - 1$)
$z_{con}(m)$ $m = 0, 1, \ldots, N - 1$

This is illustrated with a sample example as follows:
Example 2.4
Let $x(n)$ be $\{1, 1, 1, 1\}$ $n = 0, 1, 2, 3$ i.e., $L = 4$. Convolve $x(n)$ with itself.

$$z_{con}(m) = \frac{1}{N} \sum_{n=0}^{N-1} x(n) x(m-n)$$

$$z_{con}(0) = \frac{1}{8} \sum_{n=0}^{7} x(n) x(-n)$$
$$= 1/8$$

1. Keep $\{x(n)\}$ as it is.
2. Reflect $\{y(n)\}$ around 0 to get $\{y(-n)\}$.

a

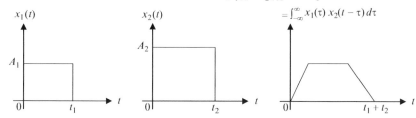

b

Periodic Extension:

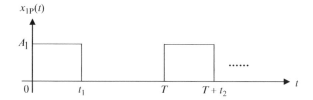

$x_{1P}(t)$ is $x_1(t)$ repeated periodically with a period T where $T \geq t_1 + t_2$.

Similarly $x_{2P}(t)$.

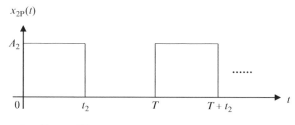

$[x_{1P}(t) * x_{2P}(t)]$

This is same as $x_1(t) * x_2(t)$ for $T > (t_1 + t_2)$.

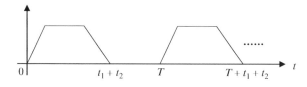

Fig. 2.10 Nonperiodic convolution **a** of $x_1(t)$ and $x_2(t)$ is same as periodic convolution. **b** of $x_{1P}(t)$ and $x_{2P}(t)$ over one period

3. Multiply $x(n)$ and $y(-n)$ and add.
4. Shift $\{y(-n)\}$ to the right by one sampling interval. Multiply $x(n)$ and $y(1-n)$ and add.
5. Shift $\{y(-n)\}$ by two sampling intervals. Multiply and add.
6. Shift $\{y(-n)\}$ by three sampling intervals. Multiply and add, and so on.

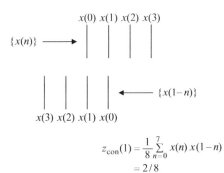

$$z_{\text{con}}(1) = \frac{1}{8} \sum_{n=0}^{7} x(n)\, x(1-n)$$
$$= 2/8$$

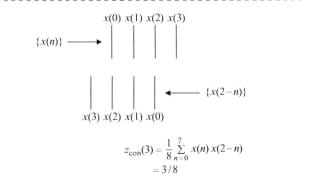

$$z_{\text{con}}(3) = \frac{1}{8} \sum_{n=0}^{7} x(n)\, x(2-n)$$
$$= 3/8$$

$$z_{\text{con}}(3) = \frac{1}{8} \sum_{n=0}^{7} x(n)\, x(3-n)$$
$$= 4/8$$

$x(0)\ x(1)\ x(2)\ x(3)$

$\{x(n)\} \longrightarrow$

$\longleftarrow \{x(4-n)\}$

$x(3)\ x(2)\ x(1)\ x(0)$

$$z_{\text{con}}(4) = \frac{1}{8} \sum_{n=0}^{7} x(n)\, x(4-n)$$
$$= 3/8$$

- -

$x(0)\ x(1)\ x(2)\ x(3)$

$\{x(n)\} \longrightarrow$

$\longleftarrow \{x(5-n)\}$

$x(3)\ x(2)\ x(1)\ x(0)$

$$z_{\text{con}}(5) = \frac{1}{8} \sum_{n=0}^{7} x(n)\, x(5-n)$$
$$= 2/8$$

- -

$x(0)\ x(1)\ x(2)\ x(3)$

$\{x(n)\} \longrightarrow$

$\longleftarrow \{x(6-n)\}$

$x(3)\ x(2)\ x(1)\ x(0)$

$$z_{\text{con}}(6) = \frac{1}{8} \sum_{n=0}^{7} x(n)\, x(6-n)$$
$$= 1/8$$

- -

$x(0)\ x(1)\ x(2)\ x(3)$

$\{x(n)\} \longrightarrow$

$\longleftarrow \{x(7-n)\}$

$x(3)\ x(2)\ x(1)\ x(0)$

$$z_{\text{con}}(7) = \frac{1}{8} \sum_{n=0}^{7} x(n)\, x(7-n) = 0$$

$$z_{\text{con}}(m) = 0, \qquad m \geq 7 \text{ and } m \leq (-1)$$

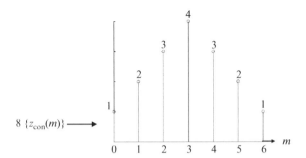

Convolution of a uniform sequence with itself yields a triangular sequence. This is an aperiodic or noncircular convolution. To obtain this through the DFT/IDFT, extend both $\{x(n)\}, n = 0, 1, \ldots, L - 1$ and $\{y(n)\}, n = 0, 1, \ldots, M - 1$ by adding zeros at the end such that $N \geq (L + M - 1)$ (Fig. 2.9).

2.4.1 Multiplication Theorem

Multiplication of two periodic sequences in time/spatial domain is equivalent to circular convolution in the DFT domain (see Problem 2.16).

2.5 Correlation Theorem

Similar to the convolution theorem, an analogous theorem exists for the correlation (Fig. 2.11). Circular correlation of two real periodic sequences $x(n)$ and $y(n)$ is given by

$$z_{\text{cor}}(m) = \frac{1}{N} \sum_{n=0}^{N-1} x^*(n) y(m+n) \qquad \text{when } x(n) \text{ is complex}$$

$$= \frac{1}{N} \sum_{n=0}^{N-1} x(n) y(m+n) \qquad m = 0, 1, \ldots, N - 1$$

(2.25a)

as $x(n)$ is usually real. This is the same process as in discrete convolution except $\{y(n)\}$ is not reflected. All other properties hold. In the DFT domain this is equivalent to

$$Z_{\text{cor}}^{\text{F}}(k) = \frac{1}{N} X^{\text{F}^*}(k) Y^{\text{F}}(k). \quad \text{Note } Z_{\text{cor}}^{\text{F}}(k) \neq \frac{1}{N} X^{\text{F}}(k) Y^{\text{F}^*}(k)$$

(2.25b)

$x(n)$	1	2	3	0	0
$y(m-n)$					
$m=0$	1	0	0	3	2
$m=1$	2	1			3
$m=2$	3	2	1		
$m=3$		3	2	1	
$m=4$			3	2	1

$z_{con}(m)$	1	4	10	12	9

	1	1	3	0	0
					$y(m+n)$
1	2	3	0	0	$m=0$
2	3			1	$m=1$
3			1	2	$m=2$
	1	2	3		$-m=3=3\pm N=-2$
	1	2	3		$m=4=4\pm N=-1$

12	5	3	3	7	$z_{cor}(m)$

$x = [1\ 1\ 3]$

3 7 12 5 3 $= \mathrm{xcorr}\,(y, x)$
wrapped around

$$z_{con}(m) = \frac{1}{N}\sum_{n=0}^{N-1}x(n)y(m-n)$$

$$z_{cor}(m) = \frac{1}{N}\sum_{n=0}^{N-1}x(n)y(m+n)$$

$$m = 0, 1, ..., N-1$$

$$Z_{con}^{F}(k) = \frac{1}{N}X^{F}(k)Y^{F}(k)$$

$$Z_{cor}^{F}(k) = \frac{1}{N}X^{F^{*}}(k)Y^{F}(k)$$

Aperiodic convolution
by using circular convolution
from multiplication in DFT

Aperiodic correlation
by using circular correlation
from multiplication in DFT

Fig. 2.11 Relationship between convolution and correlation theorems

Proof.

$$\text{DFT of } z_{cor}(m) \text{ is } \sum_{m=0}^{N-1}\left[\frac{1}{N}\sum_{n=0}^{N-1}x(n)y(m+n)\right]W_{N}^{mk}$$

$$= \frac{1}{N}\sum_{n=0}^{N-1}x(n)W_{N}^{-nk}\sum_{m=0}^{N-1}y(m+n)\,W_{N}^{(m+n)k}, \quad \text{let } m+n=l$$

$$= \frac{1}{N}\sum_{n=0}^{N-1}x(n)\,W_{N}^{-nk}\sum_{l=n}^{N-1+n}y(l)\,W_{N}^{lk}$$

$$= \frac{1}{N}X^{F^{*}}(k)Y^{F}(k). \qquad \text{Here } z_{cor}(m) \Leftrightarrow Z_{cor}^{F}(k)$$

As in the case of the convolution, to obtain a noncircular (aperiodic) correlation through the DFT/IDFT, both $\{x(n)\}$ and $\{y(n)\}$ must be extended by adding zeros at the end such that $N \geq M + L - 1$, where M and L are lengths of the sequences $\{x(n)\}$ and $\{y(n)\}$ respectively. Even though this results in a circular correlation, for one period it is the same as the noncircular correlation.

By taking complex conjugation of $X^{F}(k)$ in Fig. 2.9, the same block diagram can be used to obtain an aperiodic correlation.

$$\text{IDFT}\left[\frac{1}{N}X^{F^{*}}(k)Y^{F}(k)\right] = z_{cor}(m) \qquad (2.26)$$

This technique can be illustrated as follows.

Given: $\{x(n)\} = \{x_0, x_1, \ldots, x_{M-1}\}$, an M-point sequence and $\{y(n)\} = \{y_0, y_1, \ldots, y_{L-1}\}$, an L-point sequence, obtain their circular correlation.

1. Extend $x(n)$ by adding $N - M$ zeros at the end i.e., $\{x_e(n)\} = \{x_0, x_1, \ldots, x_{M-1}, 0, 0, \ldots, 0\}$, where $N \geq M + L - 1$.
2. Extend $y(n)$ by adding $N - L$ zeros at the end i.e., $\{y_e(n)\} = \{y_0, y_1, \ldots, y_{L-1}, 0, 0, \ldots, 0\}$.
3. Apply N-point DFT to $\{x_e(n)\}$ to get $\{X_e^F(k)\}$, $k = 0, 1, \ldots, N - 1$.
4. Carry out step 3 on $\{y_e(n)\}$ to get $\{Y_e^F(k)\}$, $k = 0, 1, \ldots, N - 1$.
5. Multiply $\frac{1}{N}X_e^{F^*}(k)$ and $Y_e^F(k)$ to get $\frac{1}{N}X_e^{F^*}(k)\,Y_e^F(k)$, $k = 0, 1, \ldots, N - 1$.
6. Apply N-point IDFT to $\{\frac{1}{N}X_e^{F^*}(k)\,Y_e^F(k)\}$ to get $z_{cor}(m)$, $m = 0, 1, \ldots, N - 1$ which is the aperiodic correlation of $\{x(n)\}$ and $\{y(n)\}$.
7. Since $z_{cor}(m)$ for $0 \leq m \leq N - 1$ are the values of correlation at different lags, with positive and negative lags stored in a wrap-around order, we can get $z(m)$ as follows. If $M \geq L$ and $N = M + L - 1$

$$
\begin{aligned}
z(m) &= z_{cor}(m), & 0 \leq m \leq L - 1 \\
z(m - N) &= z_{cor}(m), & M \leq m \leq N - 1
\end{aligned}
\tag{2.27}
$$

Example 2.5

Given: $\{x(n)\} = \{1, 2, 3, 1\}$, an M-point sequence and $\{y(n)\} = \{1, 1, 3\}$, an L-point sequence, obtain their aperiodic correlation.

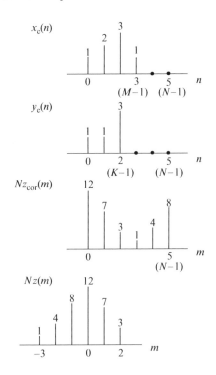

1. Extend $x(n)$ by adding $N - M$ zeros at the end i.e., $\{x_e(n)\} = \{1, 2, 3, 1, 0, 0\}$, where $N = L + M - 1$.
2. Extend $y(n)$ by adding $N - L$ zeros at the end i.e., $\{y_e(n)\} = \{1, 1, 3, 0, 0, 0\}$.
3. Apply N-point DFT to $\{x_e(n)\}$ to get $\{X_e^F(k)\}$, $k = 0, 1, \ldots, N - 1$.
4. Carry out step (3) on $\{y_e(n)\}$ to get $\{Y_e^F(k)\}$, $k = 0, 1, \ldots, N - 1$.
5. Multiply $X_e^{F^*}(k)$ and $Y_e^F(k)$ to get $X_e^{F^*}(k)Y_e^F(k)$, $k = 0, 1, \ldots, N - 1$.
6. Apply N-point IDFT to $\{X_e^{F^*}(k)Y_e^F(k)\}$ to get $N\{z_{cor}(m)\} = \{12, 7, 3, 1, 4, 8\}$ which is the aperiodic correlation of $\{x(n)\}$ and $\{y(n)\}$.
7. Since $z_{cor}(m)$ for $0 \le m \le 6$ are the values of correlation in a wrap-around order, we can get $N\{z(m)\} = \{1, 4, 8, 12, 7, 3\}$ for $-2 \le m \le 3$ according to (2.27).

Example 2.6: Implement Example 2.5 by MATLAB.

$x = [1\ 2\ 3\ 1]$;	
$y = [1\ 1\ 3]$;	
$x\_e = [1\ 2\ 3\ 1\ 0\ 0]$;	% (1)
$y\_e = [1\ 1\ 3\ 0\ 0\ 0]$;	% (2)
$Z = \text{conj (fft } (x\_e))\ .*\ \text{fft } (y\_e)$;	% (5)
$z = \text{ifft } (Z)$	% (6) [12 7 3 1 4 8]

If we wrap around the above data, we get the same with the result of the following aperiodic correlation.

xcorr(y, x)	% [1 4 8 12 7 3]

2.6 Overlap-Add and Overlap-Save Methods

Circular convolution of two sequences in time/spatial domain is equivalent to multiplication in the DFT domain. To obtain an aperiodic convolution using the DFT, the two sequences have to be extended by adding zeros as described in Section 2.3.

2.6.1 The Overlap-Add Method

When an input sequence of infinite duration $x(n)$ is convolved with the finite-length impulse response of a filter $y(n)$, the sequence to be filtered is segmented into sections $x_r(n)$ and the filtered sections $z_r(m)$ are fitted together in an appropriate way. Let the rth section of an input sequence be $x_r(n)$, an impulse response be $y(n) = \{\frac{1}{4}, \frac{1}{2}, \frac{1}{4}\}$, and the rth section of the filtered sequence be $z_r(m)$. The procedure to get the filtered sequence $z(m)$ is illustrated with an example in Fig. 2.12.

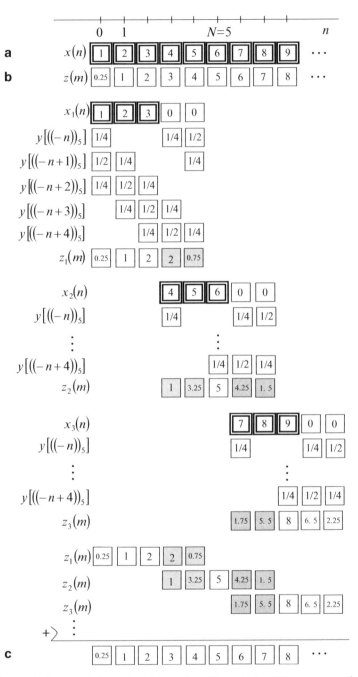

Fig. 2.12 Aperiodic convolution using the overlap-add method. **a** The sequence $x(n)$ to be convolved with the sequence $y(n)$. **b** The aperiodic convolution of $x(n)$ and $y(n)$. **c** Aperiodic convolution using the DFT/IDFT. Figure 2.13 shows how to use the DFT/IDFT for this example. $N = L + M - 1 = 5, L = M = 3$

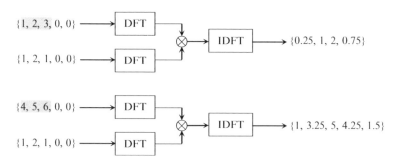

Fig. 2.13 Obtaining aperiodic convolution from circular convolution using the overlap-add method. $N = L + M - 1$, $L = M = 3$. Note that the DFTs and IDFTs are implemented via fast algorithms (see Chapter 3)

The filtering of each segment can then be implemented using the DFT/IDFT as shown in Fig. 2.13.

Discrete convolution of two sequences $\{x_r(n)\}$ and $\{y(n)\}$ can be computed as follows:

$$z_r(m) = \sum_{n=0}^{N-1} x_r(n)y[((m-n))_N]$$
$$x_r(n) \quad n = 0, 1, \ldots, L - 1 \qquad (2.28)$$
$$y[((n))_N] \quad n = 0, 1, \ldots, M - 1 \quad (\text{where } N = L + M - 1)$$
$$z_r(m) \quad m = 0, 1, \ldots, N - 1$$

In (2.28) the second sequence $y[((m-n))_N]$ is circularly time reversed and circularly shifted with respect to the first sequence $x_r(n)$. The sequence $y[((n))_N]$ is shifted modulo N. The sequences $x_r(n)$ have L nonzero points and $(M-1)$ zeros to have a length of $(L + M - 1)$ points. Since the beginning of each input section is separated from the next by L points and each filtered section has length $(L + M - 1)$, the filtered sections will overlap by $(M - 1)$ points, and the overlap samples must be added (Fig. 2.12). So this procedure is referred to as the *overlap-add method*. This method can be done in MATLAB by using the command $z = $ FFTFILT (y, x).

An alternative fast convolution procedure, called the *overlap-save method*, corresponds to carrying out an L-point circular convolution of an M-point impulse response $y(n)$ with an L-point segment $x_r(n)$ and identifying the part of circular convolution that corresponds to an aperiodic convolution. After the first $(M - 1)$ points of each output segment are discarded, the consecutive output segments are combined to form the output (Fig. 2.14). Each consecutive input section consists of $(L - M + 1)$ new points and $(M - 1)$ points so that the input sections overlap (see [G2]).

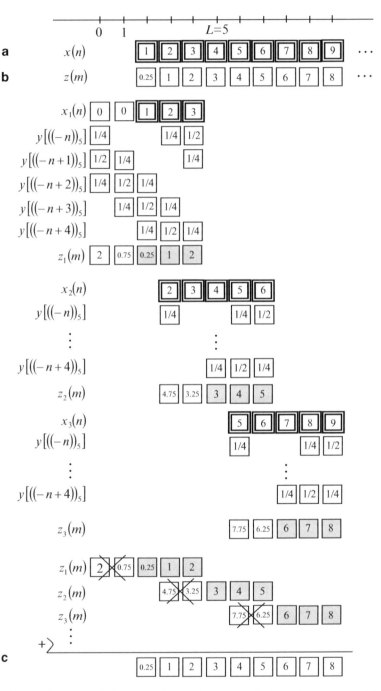

Fig. 2.14 Aperiodic convolution using the overlap-save method. **a** The sequence $x(n)$ to be convolved with the sequence $y(n)$. **b** The aperiodic convolution of $x(n)$ and $y(n)$. **c** Aperiodic convolution using the DFT/IDFT. $L = 5$, $M = 3$

2.7 Zero Padding in the Data Domain

To obtain the nonperiodic convolution/correlation of two (one of length L and another of length M) sequences using the DFT/IDFT approach, we have seen, the two sequences must be extended by adding zeros at the end such that their lengths are $N \geq L + M - 1$ (Fig. 2.9). It is appropriate to query the effects of zero padding in the frequency (DFT) domain. Let $\{x(n)\} = \{x_0, x_1, \ldots, x_{M-1}\}$, an M-point sequence be extended by adding $M - N$ zeros at the end of $\{x(n)\}$. The extended sequence is $\{x_e(n)\} = \{x_0, x_1, \ldots, x_{M-1}, 0, \ldots, 0\}$, where

$$x_e(n) = 0, \quad M \leq n \leq N - 1$$

The DFT of $\{x_e(n)\}$ is

$$X_e^F(k) = \sum_{n=0}^{N-1} x_e(n) W_N^{nk}, \quad k = 0, 1, \ldots, N - 1 \tag{2.29a}$$

$$= \sum_{n=0}^{M-1} x(n) W_N^{nk} \tag{2.29b}$$

Note that the subscript e in $X_e^F(k)$ stands for the DFT of the extended sequence $\{x_e(n)\}$. The DFT of $\{x(n)\}$ is

$$X^F(k) = \sum_{m=0}^{M-1} x(m) W_M^{mk}, \quad k = 0, 1, \ldots, M - 1 \tag{2.30}$$

Inspection of (2.29) and (2.30) shows that adding zeros to $\{x(n)\}$ at the end results in interpolation in the frequency domain.

Indeed when $N = PM$ where P is an integer $X_e^F(k)$ is an interpolated version $X^F(k)$ by the factor P. Also from (2.29b) and (2.30), $X_e^F(kP) = X^F(k)$ i.e.,

$$\sum_{n=0}^{M-1} x(n) \exp\left(\frac{-j2\pi nkP}{PM}\right) = \sum_{n=0}^{M-1} x(n) \exp\left(\frac{-j2\pi nk}{M}\right)$$

The process of adding zeros at the end of a data sequence $\{x(n)\}$ is called zero padding and is useful in a detailed representation in the frequency domain. No additional insight in the data domain is gained as IDFT of (2.29b) and (2.30) yields $\{x(n)\}$ and $\{x_e(n)\}$ respectively. Zero padding in the frequency domain (care must be taken in adding zeros to $X^F(k)$ because of the conjugate symmetry property Eq. [2.15]) is however not useful.

Example 2.7 If $N = 8$ and $M = 4$, then $P = 2$. Let

$$\{x(n)\} = \{1, 2, 3, 4\}$$

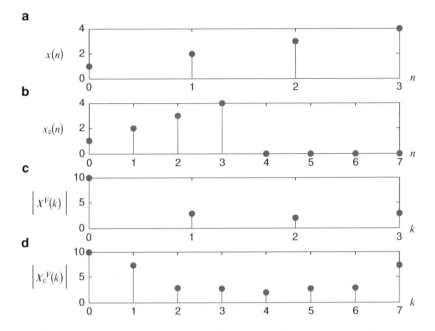

Fig. 2.15 Magnitude spectra of **a** a sequence and **b** the extended sequence are **c** and **d** respectively

then

$$\{X^{\mathrm{F}}(k)\} = \{10, -2 + j2, -2, -2 - j2\}$$

$$\{x_{\mathrm{e}}(n)\} = \{1, 2, 3, 4, 0, 0, 0, 0\}$$

$$\{X_{\mathrm{e}}^{\mathrm{F}}(k)\} = \{10, -0.414 - j7.243, -2 + j2, 2.414 - j1.243, -2,$$
$$2.414 + j1.243, -2 - j2, -0.414 + j7.243\}$$

See Fig. 2.15. For a detailed representation of frequency response, zeros are added to input data (Fig. 2.16).

2.8 Computation of DFTs of Two Real Sequences Using One Complex FFT

Given two real sequences $x(n)$ and $y(n)$, $0 \le n \le N - 1$, their DFTs can be computed using one complex FFT. Details are as follows:

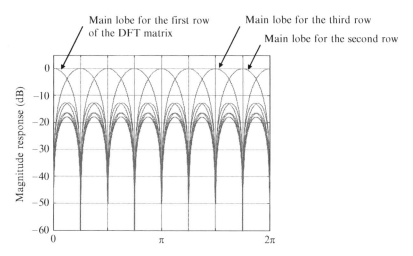

Fig. 2.16 Zeros are added at the end of each eight-point basis vector (BV) of the DFT for frequency response of size 1,024

Form a complex sequence

$$p(n) = x(n) + jy(n), \quad n = 0, 1, \dots, N - 1 \tag{2.31}$$

N-point DFT of $p(n)$ is

$$
\begin{aligned}
P^{\mathrm{F}}(k) &= \sum_{n=0}^{N-1} p(n) W_N^{nk}, \quad k = 0, 1, \dots, N-1 \\
&= \sum_{n=0}^{N-1} x(n) W_N^{nk} + j \sum_{n=0}^{N-1} y(n) W_N^{nk} \\
&= X^{\mathrm{F}}(k) + jY^{\mathrm{F}}(k) \tag{2.32}
\end{aligned}
$$

$$
\begin{aligned}
\text{Then} \quad P^{\mathrm{F}^*}(N - k) &= X^{\mathrm{F}^*}(N - k) - jY^{\mathrm{F}^*}(N - k) \\
&= X^{\mathrm{F}}(k) - jY^{\mathrm{F}}(k) \tag{2.33}
\end{aligned}
$$

as $x(n)$ and $y(n)$ are real sequences. Hence

$$X^{\mathrm{F}}(k) = \frac{1}{2} \left[P^{\mathrm{F}}(k) + P^{\mathrm{F}^*}(N - k) \right] \tag{2.34a}$$

$$Y^{\mathrm{F}}(k) = \frac{1}{2} \left[P^{\mathrm{F}^*}(N - k) - P^{\mathrm{F}}(k) \right] \tag{2.34b}$$

Here we used the property that when $x(n)$ is real

$$X^{\mathrm{F}^*}(N - k) = X^{\mathrm{F}}(k) \tag{2.35}$$

Proof:

$$X^{F^*}(N-k) = \sum_{n=0}^{N-1}\left[x(n)W_N^{(N-k)n}\right]^* = \sum_{n=0}^{N-1}\left[x(n)W_N^{-kn}\right]^*$$

$$= \sum_{n=0}^{N-1} x(n)W_N^{kn} = X^F(k), \quad \text{since } W_N^{Nn} = 1$$

2.9 A Circulant Matrix Is Diagonalized by the DFT Matrix

2.9.1 *Toeplitz Matrix*

Toeplitz matrix is a square matrix. On any NW to SE diagonal column the elements are the same.
Example:

NW   NE

$$\begin{bmatrix} a_1 & a_2 & a_3 & a_4 \\ a_5 & a_1 & a_2 & a_3 \\ a_6 & a_5 & a_1 & a_2 \\ a_7 & a_6 & a_5 & a_1 \end{bmatrix}$$

SW   SE

(4×4) Toeplitz matrix

e.g., c = [1 5 6 7]; first column, $N = 4$
r = [1 2 3 4]; first row
Toeplitz (c, r)

2.9.2 *Circulant Matrix*

A circulant matrix (CM) is such that each row is a circular shift of the previous row. A CM can be classified as left CM or right CM depending on the circular shift is to the right or left respectively. In our development we will consider only the right circular shift.

NW   NE

$$[H] = \begin{bmatrix} h_0 & h_1 & h_2 & \cdots & h_{N-1} \\ h_{N-1} & h_0 & h_1 & \cdots & h_{N-2} \\ h_{N-2} & h_{N-1} & h_0 & \cdots & h_{N-3} \\ \vdots & \vdots & \vdots & \ddots & \vdots \\ h_1 & h_2 & h_3 & \cdots & h_0 \end{bmatrix} \qquad (2.36)$$

SW   SE

$[H]_{m,n} = [h_{(m-n) \bmod N}]$ element of $[H]$ in row m and column $n, 0 \leq m, n \leq N - 1$.

e.g., $c = [\,0\ 2\ 1\,]$; first column, $N = 3$
$\quad\ r = [\,0\ 1\ 2\,]$; first row
$\quad$ Toeplitz (c, r)

2.9.3   A Circulant Matrix Is Diagonalized by the DFT Matrix

Let $[H]$ be an $(N \times N)$ circulant matrix defined in (2.36) [B6, J13]. Define $\boldsymbol{\phi}_k$ as the basis vectors of the DFT matrix $[W_N^{nk}]$, n, $k = 0, 1, \ldots, N - 1$

$$\underset{(N \times 1)}{\boldsymbol{\phi}_k} = \left(1, W_N^{-k}, W_N^{-2k}, \cdots, W_N^{-(N-1)k}\right)^T \qquad W_N = \exp\left(\tfrac{-j2\pi}{N}\right) \qquad (2.37)$$

$$k\text{th basis vector}, \quad k = 0, 1, \ldots, N - 1$$

Note that the basis vectors $\boldsymbol{\phi}_k$ are columns of $\left[W_N^{nk}\right]^*$ (see Problem 2.24(b)). Define

$$[[H]\boldsymbol{\phi}_k]_m = \sum_{n=0}^{N-1} h_{m-n} W_N^{-kn}, \qquad \boxed{\begin{array}{c} [W_N^{kn}], \quad (N \times N) \text{ DFT matrix} \\ (N \times N) \\ n, k = 0, 1, \ldots, N - 1 \end{array}} \qquad (2.38)$$

$$m, k = 0, 1, \ldots, N - 1$$

This is the mth row of $[H]$ post multiplied by $\boldsymbol{\phi}_k$. Thus it is a scalar. Let $m - n = l$, then

$$[[H]\boldsymbol{\phi}_k]_m = \sum_{l=m}^{m-N+1} h_l W_N^{-k(m-l)}$$

$$= W_N^{-km}\left(\sum_{l=m}^{m-N+1} h_l W_N^{kl}\right) \qquad (2.39)$$

$$[[H]\boldsymbol{\phi}_k]_m = W_N^{-km}\left(\sum_{l=-N+m+1}^{-1} h_l W_N^{kl} + \sum_{l=0}^{m} h_l W_N^{kl}\right)$$

$$= W_N^{-km}\left(\sum_{l=-N+m+1}^{-1} h_l W_N^{kl} + \sum_{l=0}^{N-1} h_l W_N^{kl} - \sum_{l=m+1}^{N-1} h_l W_N^{kl}\right)$$

$$= W_N^{-km}(I + II - III)$$

$$= W_N^{-km}\sum_{l=0}^{N-1} h_l W_N^{kl}, \qquad \text{since} \quad I = III \qquad (2.40)$$

Proof.   In I let $p = l + N$, $l = p - N$

$$I = \sum_{p=m+1}^{N-1} h_{p-N} W_N^{k(p-N)} = \sum_{p=m+1}^{N-1} h_p W_N^{kp} \tag{2.41}$$

since $h_{p-N} = h_p$ and $W_N^{-kN} = 1$.

$$[[H]\boldsymbol{\phi}_k]_m = \sum_{n=0}^{N-1} h_{m-n} W_N^{-kn}, \qquad m, k = 0, 1, \ldots, N-1.$$

$$= W_N^{-km} \sum_{l=0}^{N-1} h_l W_N^{kn},$$

$$= \boldsymbol{\phi}_k(m) \, \lambda_k \tag{2.42}$$

where $\lambda_k = \sum_{l=0}^{N-1} h_l W_N^{kl}, k = 0, 1, \ldots, N-1$, is the kth eigenvalue of $[H]$.
This is the N-point DFT of the first row of $[H]$.

$$\therefore \quad \underset{(N \times N)(N \times 1)}{[H] \quad \boldsymbol{\phi}_k} = \underset{(1 \times 1)(N \times 1)}{\lambda_k \quad \boldsymbol{\phi}_k}, \qquad k = 0, 1, \ldots, N-1 \tag{2.43}$$

The N column vectors are combined to from

$$[H](\boldsymbol{\phi}_0, \boldsymbol{\phi}_1, \ldots) = (\lambda_0 \boldsymbol{\phi}_0, \lambda_1 \boldsymbol{\phi}_1, \ldots) = (\boldsymbol{\phi}_0, \boldsymbol{\phi}_1, \ldots) \, \text{diag} \, (\lambda_0, \lambda_1, \ldots)$$
$$[H][\Phi] = [\Phi] \, \text{diag} \, (\lambda_0, \lambda_1, \ldots, \lambda_{N-1}) \tag{2.44}$$

Pre multiply both sides of (2.44) by $[\Phi]^{-1}$.

$$[\Phi]^{-1}[H][\Phi] = \frac{1}{N} [\Phi]^* [H][\Phi]$$

$$= \frac{1}{N} [W_N^{nk}][H][W_N^{nk}]^*$$

$$= \text{diag} \, (\lambda_0, \lambda_1, \ldots, \lambda_{N-1}) \tag{2.45}$$

where

$$\underset{(N \times N)}{[\Phi]} = (\underset{(N \times 1)}{\boldsymbol{\phi}_0}, \underset{(N \times 1)}{\boldsymbol{\phi}_1}, \ldots, \underset{(N \times 1)}{\boldsymbol{\phi}_k}, \ldots, \underset{(N \times 1)}{\boldsymbol{\phi}_{N-1}}) = [W_N^{nk}]^*$$

$$\boldsymbol{\phi}_k = \left(1, W_N^{-k}, W_N^{-2k}, \ldots, W_N^{-(N-1)k} \right)^T, \qquad k = 0, 1, \ldots, N-1$$

$$[\Phi]^{-1} = \frac{1}{N} [\Phi]^*, \quad [\Phi] \text{ is a unitary matrix}$$

$$[H] = [\Phi] \, \text{diag}(\lambda_0, \lambda_1, \ldots, \lambda_{N-1})[\Phi]^{-1} \tag{2.46}$$

$$\underset{(N \times N)}{[\Phi]^*} = (\underset{(N \times 1)}{\boldsymbol{\phi}_0^*}, \underset{(N \times 1)}{\boldsymbol{\phi}_1^*}, \ldots, \underset{(N \times 1)}{\boldsymbol{\phi}_{N-1}^*}) = [W_N^{nk}] \qquad (N \times N) \text{ DFT matrix}$$

Equation (2.45) shows that the basis vectors of the DFT are eigenvectors of a circular matrix. Equation (2.45) is similar to the 2-D DFT of $[H]$ (see Eg. [5.6a]) except a complex conjugate operation is applied to the second DFT matrix in (2.45).

2.10 Summary

This chapter has defined the discrete Fourier transform (DFT) and several of its properties. Fast algorithms for efficient computation of the DFT, called fast Fourier transform (FFT), are addressed in the next chapter. These algorithms have been instrumental in ever increasing applications in diverse disciplines. They cover the gamut from radix-2/3/4 DIT, DIF, DIT/DIF to mixed radix, split-radix and prime factor algorithms.

2.11 Problems

Given $x(n) \Leftrightarrow X^{\mathrm{F}}(k)$ and $y(n) \Leftrightarrow Y^{\mathrm{F}}(k)$. Here both are N point DFTs.

2.1 Show that the DFT is a unitary transform i.e.,

$$[W]^{-1} = \frac{1}{N}[W]^*$$

where $[W]$ is the $(N \times N)$ DFT matrix.

2.2 Let $x(n)$ be real. $N = 2^n$. Then show that $X^{\mathrm{F}}\left(\frac{N}{2} - k\right) = X^{\mathrm{F}^*}\left(\frac{N}{2} + k\right)$, $k = 0, 1, \ldots, N/2$. What is the implication of this?

2.3 Show that $\sum_{n=0}^{N-1} x(n)y^*(n) = \frac{1}{N}\sum_{k=0}^{N-1} X^{\mathrm{F}}(k)Y^{\mathrm{F}^*}(k)$.

2.4 Show that $\sum_{n=0}^{N-1} x^2(n) = \sum_{k=0}^{N-1} |X^{\mathrm{F}}(k)|^2$. There may be a constant. Energy is preserved under a unitary transformation.

2.5 Show that $(-1)^n x(n) \Leftrightarrow X^{\mathrm{F}}\left(k - \frac{N}{2}\right)$.

2.6 Let $N = 4$. Explain fftshift(fft(fftshift(x))) in terms of the DFT coefficients $X^{\mathrm{F}}(k)$ or fft(x).

2.7 Show that $\sum_{n=0}^{N-1} W_N^{(r-k)n} = \begin{cases} N, & r = k \\ 0, & r \neq k \end{cases}$.

2.8 *Modulation/Frequency Shifting* Show that $\left[x(n)\exp\left(\frac{j2\pi k_0 n}{N}\right)\right] \Leftrightarrow X^{\mathrm{F}}(k - k_0)$.

$X^{\mathrm{F}}(k - k_0)$ is $X^{\mathrm{F}}(k)$ shifted circularly to the right by k_0 along k (frequency domain).

2.9 *Circular Shift* Show that $x(n - n_0) \Leftrightarrow X^{\mathrm{F}}(k)\exp\left(\frac{-j2\pi k n_0}{N}\right)$. What is the implication of this?

2.10 *Circular Shift* Show that $\delta(n + n_0) \Leftrightarrow \exp\left(\frac{j2\pi k n_0}{N}\right)$ with $N = 4$ and $n_0 = 2$.

Here Kronecker delta function $\delta(n) = 1$ for $n = 0$ and $\delta(n) = 0$ for $n \neq 0$.

2.11 *Time Scaling* Show that $x(an) \Leftrightarrow \frac{1}{a}X^F\left(\frac{k}{a}\right)$, '$a$' is a constant. What is the implication of this?

2.12 Show that $X^F(k) = X^{F^*}(-k)$, when $x(n)$ is real.

2.13 Show that $x^*(n) \Leftrightarrow X^{F^*}(-k)$.

2.14 *Time Reversal* In two different ways show that $x(-n) \Leftrightarrow X^F(-k)$.
   (a) Use the DFT permutation property of (2.21) and $(N, N-1) = 1$ for any integer N.
   (b) Use the definition of the DFT pair in (2.1).

Use the definition of the DFT pair in (2.1) for Problems 2.15–2.17.

2.15 The discrete convolution of two sequences $x_1(n)$ and $x_2(n)$ is defined as

$$y(n) = \frac{1}{N}\sum_{m=0}^{N-1} x_1(m)\, x_2(n-m), \quad n = 0, 1, \ldots, N-1$$

Both $x_1(n)$ and $x_2(n)$ are N-point sequences. Show that $Y^F(k) = \frac{1}{N}X_1^F(k)X_2^F(k)$, $(k = 0, 1, \ldots, N-1)$. $Y^F(k), X_1^F(k)$ and $X_2^F(k)$ are the N-point DFTs of $y(n), x_1(n)$ and $x_2(n)$, respectively.

2.16 The circular convolution of $X_1^F(k)$ and $X_2^F(k)$ is defined as

$$Y^F(k) = \frac{1}{N}\sum_{m=0}^{N-1} X_1^F(m)X_2^F(k-m), \quad k = 0, 1, \ldots, N-1$$

Show that $y(n) = \frac{1}{N}x_1(n)x_2(n), n = 0, 1, \ldots, N-1$. $Y^F(k), X_1^F(k)$ and $X_2^F(k)$ are the N-point DFTs of $y(n), x_1(n)$ and $x_2(n)$, respectively.

2.17 The discrete correlation of $x_1(n)$ and $x_2(n)$ is defined as

$$z(n) = \frac{1}{N}\sum_{m=0}^{N-1} x_1(m)\, x_2(n+m), \quad n = 0, 1, \ldots, N-1$$

Show that $Z^F(k) = \frac{1}{N}X_1^{F^*}(k)X_2^F(k)$, $k = 0, 1, \ldots, N-1$, where $Z^F(k)$ is the DFT of $z(n)$.

2.18 Given

$$\underline{b} = [A]\underline{x} = \begin{pmatrix} 1 & 2 & -1 & -2 \\ 2 & 1 & 2 & -1 \\ -1 & 2 & 1 & 2 \\ -2 & -1 & 2 & 1 \end{pmatrix}\begin{pmatrix} x_0 \\ x_1 \\ x_2 \\ x_3 \end{pmatrix} \tag{P2.1}$$

Let $\underline{a} = (1, -2, -1, 2)^T$. To compute (P2.1), we use four-point FFTs as

$$\text{IFFT}[\ \text{FFT}[\underline{a}]\ \times\ \text{FFT}\ [\underline{x}]\ ] \tag{P2.2}$$

where $\times$ denotes the element-by-element multiplication of the two vectors. Is this statement true? Explain your answer.

2.19 Using the unitary 1D-DFT, prove the following theorems:
   (a) Convolution theorem
   (b) Multiplication theorem
   (c) Correlation theorem

2.20 Derive (2.25b) from the DFT of (2.25a).

2.21 Regarding the DFT of a permuted sequence
   (a) When $(p,N) = 1$, we can find an integer q such that $0 \leq q \leq N - 1$ and $qp \equiv (1 \text{ modulo } N)$. If $(p,N) \neq 1$, we cannot find such an integer q. Explain the latter with examples.
   (b) Repeat Example 2.1 for $N = 9$.
   (c) Check the DFTs of permuted sequences for $N = 8$ and $N = 9$ using MATLAB.
   (d) Repeat Example 2.1 for $N = 16$.

2.22 Let $[\Lambda] = \text{diag}\,(\lambda_0, \lambda_1, \ldots, \lambda_{N-1})$. Derive (2.45) from (2.43).

2.23 Given the circulant matrix

$$[H] = \begin{bmatrix} 1 & 2 & 3 & 4 \\ 4 & 1 & 2 & 3 \\ 3 & 4 & 1 & 2 \\ 2 & 3 & 4 & 1 \end{bmatrix}$$

show that the diagonal elements of the DFT of $[H]$ equal the eigenvalues of the matrix by MATLAB.

2.24 *An orthonormal basis (ONB) for the DFT*
   (a) By using the relation:

$$\begin{bmatrix} 1 & 2 \\ 3 & 4 \end{bmatrix}\begin{bmatrix} x_0 \\ x_1 \end{bmatrix} = \begin{bmatrix} 1 \\ 3 \end{bmatrix}x_0 + \begin{bmatrix} 2 \\ 4 \end{bmatrix}x_1, \quad \text{or} \quad [A]\underline{x} = \underline{a}_0\,x_0 + \underline{a}_1 x_1 \tag{P2.3}$$

where $[A] = (\underline{a}_0, \underline{a}_1)$, show that

$$\underline{a}_0^T \underline{x}\, \underline{a}_0 + \underline{a}_1^T \underline{x}\, \underline{a}_1 = [A][A]^T \underline{x} \tag{P2.4}$$

   (b) If vector $\underline{a}_k$ represents a column of the normalized inverse DFT matrix, then it is called a *basis vector* for the DFT.

$$\left(\frac{1}{\sqrt{N}}([F]^T)^*\right) = (\underline{a}_0, \underline{a}_1, \ldots, \underline{a}_{N-1}) \tag{P2.5}$$

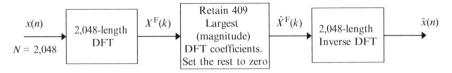

Fig. P2.1 Reconstruct $\hat{x}(n)$ from truncated DFT

where $[F]$ is a DFT matrix. DFT coefficient can be expressed as

$$X_k^{\mathrm{F}} = \langle \underline{a}_k, \underline{x} \rangle = (\underline{a}_k^T)^* \underline{x} \qquad k = 0, 1, \ldots, N-1 \qquad (P2.6)$$

Show that

$$\underline{x} = \sum_{k=0}^{N-1} X_k^{\mathrm{F}} \underline{a}_k = \sum_{k=0}^{N-1} \langle \underline{a}_k, \underline{x} \rangle \underline{a}_k = \left(\frac{1}{\sqrt{N}} \left([F]^T \right)^* \right) \left(\frac{1}{\sqrt{N}} [F] \right) \underline{x} \qquad (P2.7)$$

where X_k^{F} is a scalar and DFT coefficient.

2.12   Projects

2.1 Access dog ECG data from the Signal Processing Information Base (SPIB) at
     URL   http://spib.rice.edu/spib/data/signals/medical/dog_heart.html   ($N =$
     2,048). Sketch this $x(n)$ versus n. Take DFT of this data and sketch $X^{\mathrm{F}}(k)$
     versus k (both magnitude and phase spectra). Retain 409 DFT coefficients
     (Largest in magnitude) and set the remaining 1,639 DFT coefficients to zeros
     (truncated DFT). Reconstruct $\hat{x}(n)$ from this truncated DFT (Fig. P2.1).
     (1) Sketch $\hat{x}(n)$ versus n.
     (2) Compute MSE $= \frac{1}{2048} \sum_{n=0}^{2047} |x(n) - \hat{x}(n)|^2$.
     (3) Compute DFT of $[(-1)^n x(n)]$ and sketch the magnitude and phase spectra.
     (4) Summarize your conclusions (DFT properties etc.). See Chapter 2 in
         [B23].

2.2 Let

$$x(n) = \{1, 2, 3, 4, 3, 2, 1\}$$
$$y(n) = \{-0.0001, 0.0007, -0.0004, -0.0049, 0.0087, 0.0140,$$
$$-0.0441, -0.0174, 0.1287, 0.0005, -0.2840,$$
$$-0.0158, 0.5854, 0.6756, 0.3129, 0.0544\}$$

     (1) Compute directly the discrete convolution of the two sequences. Sketch
         the results.

     (2) Use DFT/FFT approach (Fig. 2.9). Show that both give the same results.

Chapter 3
Fast Algorithms

Direct computation of an N-point DFT requires nearly $O(N^2)$ complex arithmetic operations. An arithmetic operation implies a multiplication and an addition. However, this complexity can be significantly reduced by developing efficient algorithms. The key to this reduction in computational complexity is that in an $(N \times N)$ DFT matrix (see (2.11) and (2.13)) of the N^2 elements, only N elements are distinct. These algorithms are denoted as FFT (fast Fourier transform) algorithms [A1]. Several techniques are developed for the FFT. We will initially develop the decimation-in-time (DIT) and decimation-in-frequency (DIF) FFT algorithms. The detailed development will be based on radix-2. This will then be extended to other radices such as radix-3, radix-4, etc. The reader can then foresee that innumerable combinations of fast algorithms exist for the FFT, i.e., mixed-radix, split-radix, DIT, DIF, DIT/DIF, vector radix, vector-split-radix, etc.

Also FFT can be implemented via other discrete transforms such as discrete Hartley (DHT) [I-28, I-31, I-32], Walsh-Hadamard (WHT) [B6, T5, T8], etc. Other variations such as FFT of two real sequences via FFT of a single complex sequence are described. In the literature, terms such as complex FFT and real FFT are mentioned. DFT/IDFT as well as FFT/IFFT are inherently complex. Complex FFT means the input sequence $x(n)$ is complex. Real FFT signifies that $x(n)$ is real. The fast algorithms provide a conduit for implementing transforms such as modified DCT (MDCT), modified DST (MDST) and affine transformations [D1, D2].

Advantages

In general the fast algorithms reduce the computational complexity of an N-point DFT to about $N\log_2 N$ complex arithmetic operations. Additional advantages are reduced storage requirements and reduced computational error due to finite bit length arithmetic (multiplication/division, and addition/subtraction, to be practical, are implemented with finite word lengths). Needless to say, the fast algorithms have contributed to the DFT implementation by DSP chips [DS1, DS2, DS3, DS4, DS5, DS6, DS7, DS8, DS9, DS10, DS11, DS12]. Also ASIC VLSI chips [V1, V2, V3, V4, V5, V6, V7, V8, V9, V10, V11, V12, V13, V14, V15, V16, V17, V18, V19,

V20, V21, V22, V23, V24, V25, V26, V27, V28, V29, V30, V31, V32, V33, V34, V35, V36, V37, V38, V39, V40, L1, L2, L3, L4, L5, L6, L7, L8, L9, L10, O9] have been designed and fabricated that can perform 1-D or 2-D DFTs (also IDFTs) at very high speeds. These chips are versatile in the sense that several length DFTs can be implemented by the same chips. We will describe in detail radix-2 DIT and DIF FFT algorithms [A42].

3.1 Radix-2 DIT-FFT Algorithm

This algorithm is based on decomposing an N-point sequence (assume $N = 2^l$, $l = $ integer) into two $N/2$-point sequences (one of even samples and another of odd samples) and obtaining the N-point DFT in terms of the DFTs of these two sequences. This operation by itself results in some savings of the arithmetic operations. Further savings can be achieved by decomposing each of the two $N/2$-point sequences into two $N/4$-point sequences (one of even samples and another of odd samples) and obtaining the $N/2$-point DFTs in terms of the corresponding two $N/4$-point DFTs. This process is repeated till two-point sequences are obtained.

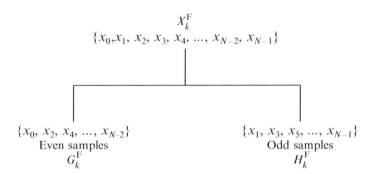

Length N DFT

$$X^{\mathrm{F}}(k) = \sum_{n=0}^{N-1} x(n)\, W_N^{nk}, \quad k = 0, 1, \ldots, N-1$$

$$= \sum_{\substack{n\,\text{even}\\ \text{integer}}} x(n)\, W_N^{nk} + \sum_{\substack{n\,\text{odd}\\ \text{integer}}} x(n)\, W_N^{nk}$$

$$= \sum_{r=0}^{(N/2)-1} x(2r)\, W_N^{2rk} + \sum_{r=0}^{(N/2)-1} x(2r+1)\, W_N^{(2r+1)k}$$

$$= \sum_{r=0}^{(N/2)-1} x(2r)\, \left(W_N^2\right)^{rk} + W_N^k \sum_{r=0}^{(N/2)-1} x(2r+1)\, \left(W_N^2\right)^{rk} \tag{3.1a}$$

Note that

$$W_N^2 = \exp\left[\frac{-j2(2\pi)}{N}\right] = \exp\left(\frac{-j2\pi}{N/2}\right) = W_{N/2}$$

$$X^F(k) = \sum_{r=0}^{(N/2)-1} x(2r)\, W_{N/2}^{rk} + W_N^k \sum_{r=0}^{(N/2)-1} x(2r+1)\, W_{N/2}^{rk} \tag{3.1b}$$

$$= G^F(k) + W_N^k H^F(k), \quad k = 0, 1, \ldots, \tfrac{N}{2} - 1$$

Here $X^F(k)$ the N-point DFT of $x(n)$ is expressed in terms of $N/2$-point DFTs, $G^F(k)$ and $H^F(k)$, which are DFTs of even samples and odd samples of $x(n)$ respectively.

$X^F(k)$: periodic with period N, $X^F(k) = X^F(k + N)$.

$G^F(k)$, $H^F(k)$: periodic with period $\frac{N}{2}$, $G^F(k) = G^F\left(k + \frac{N}{2}\right)$ and $H^F(k) = H^F\left(k + \frac{N}{2}\right)$.

$$X^F(k) = G^F(k) + W_N^k H^F(k), \quad k = 0, 1, \ldots, \tfrac{N}{2} - 1 \tag{3.2a}$$

$$X^F(k + N/2) = G^F(k) + W_N^{k+N/2} H^F(k)$$

$$W_N^{N/2} = \exp\left(\frac{-j2\pi}{N}\frac{N}{2}\right) = \exp(-j\pi) = -1$$

Since $W_N^{k+N/2} = \exp\left[\frac{-j2\pi}{N}\left(k + \frac{N}{2}\right)\right] = W_N^k W_N^{N/2} = -W_N^k$, it follows that

$$X^F\left(k + \tfrac{N}{2}\right) = G^F(k) - W_N^k H^F(k), \quad k = 0, 1, \ldots, \tfrac{N}{2} - 1 \tag{3.2b}$$

Equation (3.2) is shown below as a butterfly (Fig. 3.1).

For each k, Fig. 3.1a requires two multiplies and two adds whereas Fig. 3.1b requires one multiply and two adds. Repeat this process iteratively till two-point DFTs are obtained, i.e., $G^F(k)$ the $N/2$-point DFT of $\{x_0, x_2, x_4, \ldots, x_{N-2}\}$ can be implemented using two $N/4$-point DFTs. Similarly for $H^F(k)$.

$G^F(k)$ and $H^F(k)$, each requires $(N/2)^2$ complex adds and $(N/2)^2$ complex multiplies. $X^F(k)$ N-point DFT requires N^2 complex adds and multiplies.

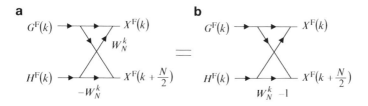

Fig. 3.1 Butterfly to compute (3.2)

Let $N = 16$. Direct computation of a 16-point DFT requires $N^2 = 256$ adds and multiplies. Through $G^F(k)$ and $H^F(k)$, it requires only $128 + 16 = 144$ adds and multiplies, resulting in saving of 112 adds and multiplies. Additional savings can be achieved by decomposing the eight-point DFTs into two four-point DFTs and finally into four two-point DFTs. $G^F(k)$ and $H^F(k)$ are eight-point DFTs. Each requires 64 adds and multiplies. The algorithm based on this technique is called radix-2 DIT-FFT.

N-point DFT, $f_0 = (1/NT)$ resolution in frequency domain

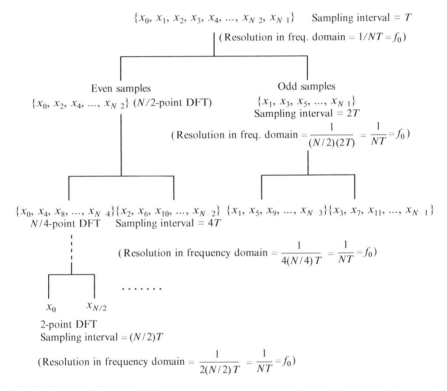

To obtain the N-point DFT we have to go back starting with two-point DFTs. At each stage of going back the sampling interval is decimated by a factor of 2. Hence the name decimation in time (DIT). The frequency resolution, f_0 however, stays the same.

The radix-2 DIT-FFT algorithm is illustrated below for $N = 8$.

$$\text{Forward } X^F(k) = \sum_{n=0}^{7} x(n) \, W_8^{kn}, \quad k = 0, 1, \ldots, 7$$

$$\text{Inverse } x(n) = \frac{1}{8} \sum_{k=0}^{7} X^F(k) \, W_8^{-kn}, \quad n = 0, 1, \ldots, 7$$

$$X^F(k) \rightarrow \{x_0, x_1, x_2, x_3, x_4, x_5, x_6, x_7\}$$

Level 3

$$G^F(k) \rightarrow \{x_0, x_2, x_4, x_6\} \qquad \{x_1, x_3, x_5, x_7\} \leftarrow H^F(k)$$

Level 2

$$A^F(k) \rightarrow \{x_0, x_4\} \quad \{x_2, x_6\} \leftarrow B^F(k) \quad C^F(k) \rightarrow \{x_1, x_5\} \quad \{x_3, x_7\} \leftarrow D^F(k)$$

Level 1

$$x_0 \quad x_4 \quad x_2 \quad x_6 \qquad\qquad x_1 \quad x_5 \quad x_3 \quad x_7$$

$G^F(k)$ is DFT of $\{x_0, x_2, x_4, x_6\}$ | $A^F(k)$ is DFT of $\{x_0, x_4\}$

$B^F(k)$ is DFT of $\{x_2, x_6\}$

$H^F(k)$ is DFT of $\{x_1, x_3, x_5, x_7\}$ | $C^F(k)$ is DFT of $\{x_1, x_5\}$

$D^F(k)$ is DFT of $\{x_3, x_7\}$

$$X^F(k) = \left[G^F(k)\right] + \left[W_8^k H^F(k)\right] \quad k = 0, 1, 2, 3$$
$$X^F(k+4) = \left[G^F(k)\right] - \left[W_8^k H^F(k)\right]$$
$$G^F(k) = \left[A^F(k)\right] + \left[W_4^k B^F(k)\right] \quad k = 0, 1$$
$$G^F(k+2) = \left[A^F(k)\right] - \left[W_4^k B^F(k)\right]$$
$$H^F(k) = \left[C^F(k)\right] + \left[W_4^k D^F(k)\right] \quad k = 0, 1$$
$$H^F(k+2) = \left[C^F(k)\right] - \left[W_4^k D^F(k)\right]$$

$$\begin{bmatrix} A^F(0) \\ A^F(1) \end{bmatrix} = \begin{bmatrix} 1 & 1 \\ 1 & -1 \end{bmatrix} \begin{pmatrix} x_0 \\ x_4 \end{pmatrix} \qquad \begin{bmatrix} B^F(0) \\ B^F(1) \end{bmatrix} = \begin{bmatrix} 1 & 1 \\ 1 & -1 \end{bmatrix} \begin{pmatrix} x_2 \\ x_6 \end{pmatrix}$$

$$\begin{bmatrix} C^F(0) \\ C^F(1) \end{bmatrix} = \begin{bmatrix} 1 & 1 \\ 1 & -1 \end{bmatrix} \begin{pmatrix} x_1 \\ x_5 \end{pmatrix} \qquad \begin{bmatrix} C^F(0) \\ C^F(1) \end{bmatrix} = \begin{bmatrix} 1 & 1 \\ 1 & -1 \end{bmatrix} \begin{pmatrix} x_3 \\ x_7 \end{pmatrix}$$

$G^F(0) = A^F(0) + B^F(0)$ | $G^F(1) = A^F(1) + W_4 B^F(1)$

$G^F(2) = A^F(0) - B^F(0)$ | $G^F(3) = A^F(1) - W_4 B^F(1)$

$H^F(0) = C^F(0) + D^F(0)$ | $H^F(1) = C^F(1) + W_4 D^F(1)$

$H^F(2) = C^F(0) - D^F(0)$ | $H^F(3) = C^F(1) - W_4 D^F(1)$

In matrix form

$$\begin{bmatrix} G^F(1) \\ G^F(3) \end{bmatrix} = \begin{bmatrix} 1 & 1 \\ 1 & -1 \end{bmatrix} \begin{bmatrix} 1 & 0 \\ 0 & W_4 \end{bmatrix} \begin{bmatrix} A^F(1) \\ B^F(1) \end{bmatrix}$$

Based on the radix-2 DIT/FFT for $N = 8$, the (8×8) DFT matrix shown in (2.13) with the columns rearranged in bit reversed order is as follows:

$$
\begin{matrix}
\text{Columns} \rightarrow & 0 & 4 & 2 & 6 & 1 & 5 & 3 & 7 & & \text{Rows} \downarrow
\end{matrix}
$$

$$
\begin{bmatrix} X^F(0) \\ X^F(1) \\ X^F(2) \\ X^F(3) \\ X^F(4) \\ X^F(5) \\ X^F(6) \\ X^F(7) \end{bmatrix} =
\begin{bmatrix}
1 & 1 & 1 & 1 & 1 & 1 & 1 & 1 \\
1 & -1 & W^2 & -W^2 & W & -W & W^3 & -W^3 \\
1 & 1 & -1 & -1 & W^2 & W^2 & -W^2 & -W^2 \\
1 & -1 & -W^2 & W^2 & W^3 & -W^3 & W & -W \\
1 & 1 & 1 & 1 & -1 & -1 & -1 & -1 \\
1 & -1 & W^2 & -W^2 & -W & W & -W^3 & W^3 \\
1 & 1 & -1 & -1 & -W^2 & -W^2 & W^2 & W^2 \\
1 & -1 & -W^2 & W^2 & -W^3 & W^3 & -W & W
\end{bmatrix}
\begin{bmatrix} x(0) \\ x(4) \\ x(2) \\ x(6) \\ x(1) \\ x(5) \\ x(3) \\ x(7) \end{bmatrix}
\begin{matrix} 0 \\ 1 \\ 2 \\ 3 \\ 4 \\ 5 \\ 6 \\ 7 \end{matrix}
$$

$$(\text{Here } W = W_8 \text{ and } W_8^l = W_8^{l \bmod 8})$$

(3.3)

Sparse matrix factors (SMF) of this matrix are

$$
\begin{bmatrix} [I_4] & [I_4] \\ [I_4] & -[I_4] \end{bmatrix} (\text{diag}[[I_4], W_8^0, W_8^1, W_8^2, W_8^3])
$$
$$
\times \left(\text{diag} \left[\begin{pmatrix} [I_2] & [I_2] \\ [I_2] & -[I_2] \end{pmatrix}, \begin{pmatrix} [I_2] & [I_2] \\ [I_2] & -[I_2] \end{pmatrix} \right] \right)
$$
$$
\times \left(\text{diag} \left[[I_3], W_8^2, [I_3], W_8^2 \right] \right)
$$
$$
\times \left(\text{diag} \left[\begin{pmatrix} 1 & 1 \\ 1 & -1 \end{pmatrix}, \begin{pmatrix} 1 & 1 \\ 1 & -1 \end{pmatrix}, \begin{pmatrix} 1 & 1 \\ 1 & -1 \end{pmatrix}, \begin{pmatrix} 1 & 1 \\ 1 & -1 \end{pmatrix} \right] \right)
$$

The DIT FFT flowgraph based on this decomposition is shown in Figs. 3.2 and 3.3.

3.1.1   Sparse Matrix Factors for the IFFT $N = 8$

Transform sequence in natural order.
Data sequence in BRO.

$$
\left(\text{diag} \left[\begin{pmatrix} 1 & 1 \\ 1 & -1 \end{pmatrix}, \begin{pmatrix} 1 & 1 \\ 1 & -1 \end{pmatrix}, \begin{pmatrix} 1 & 1 \\ 1 & -1 \end{pmatrix}, \begin{pmatrix} 1 & 1 \\ 1 & -1 \end{pmatrix} \right] \right)
$$
$$
\times \left(\text{diag} \left[[I_3], W_8^{-2}, [I_3], W_8^{-2} \right] \right)
$$
$$
\times \left(\text{diag} \left[\begin{pmatrix} [I_2] & [I_2] \\ [I_2] & -[I_2] \end{pmatrix}, \begin{pmatrix} [I_2] & [I_2] \\ [I_2] & -[I_2] \end{pmatrix} \right] \right)
$$
$$
\times \left(\text{diag} \left[[I_4], W_8^0, W_8^{-1}, W_8^{-2}, W_8^{-3} \right] \right) \begin{bmatrix} [I_4] & [I_4] \\ [I_4] & -[I_4] \end{bmatrix}
$$

Draw the flowgraph based on these SMF.

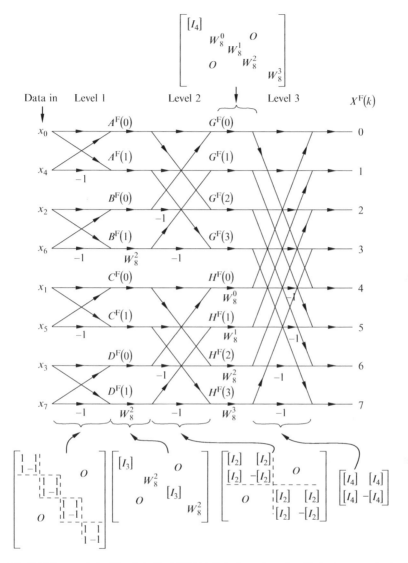

Fig. 3.2 FFT flowgraph for $N = 8$ (radix-2 DIT-FFT) is drawn based on SMF, $W_8 = \exp(-j2\pi/8)$

3.2 Fast Algorithms by Sparse Matrix Factorization

By rearranging the rows of $\left[W_N^{nk}\right]$ in bit reversed order (BRO) (Table 3.1), it can be factored into $\log_2 N$ sparse matrices i.e., $\left[W_N^{nk}\right]_{\text{BRO}} = [A_1][A_2]\cdots[A_{\log_2 N}]$, where N is an integer power of 2. (No proof of the sparse matrix factorization is given.) This is illustrated for $N = 8$.

$$\underline{X}_{\text{BRO}}^{\text{F}} = \left[W_N^{nk}\right]_{\text{BRO}} \underline{x}_{\text{NO}} \tag{3.4}$$

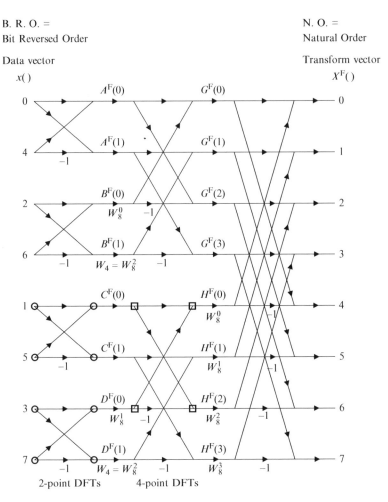

Fig. 3.3 FFT flowgraph for $N = 8$ (radix-2 DIT-FFT) (# of adds = 24, # of multiplies = 5), $W_8 = \exp(-j2\pi/8)$

Table 3.1 An example of bit reversed order for $N = 8$

Natural order	Binary	Bit reverse	Bit reversed order
0	000	000	0
1	001	100	4
2	010	010	2
3	011	110	6
4	100	001	1
5	101	101	5
6	110	011	3
7	111	111	7

Since the DFT matrix $[W_N^{nk}]$ is symmetric, it follows that

$$\underline{X}_{NO}^F = \left([W_N^{nk}]_{BRO}\right)^T \underline{x}_{BRO} \tag{3.5}$$

where $[W_N^{nk}]$ and $[W_N^{nk}]_{BRO}$ are defined in (3.6) and (3.7) for $N = 8$. $\underline{x}_{NO}$ and $\underline{x}_{BRO}$ are data vectors in natural and bit reversed orders, and X_{NO}^F and X_{BRO}^F are the DFT vectors in natural and bit reversed orders.

DFT:

Eight-point DFT

$$X^F(k) = \sum_{n=0}^{7} x(n) \exp\left[\frac{-j2\pi}{8} nk\right], \quad k = 0, 1, \ldots, 7$$

IDFT:

$$x(n) = \frac{1}{8} \sum_{k=0}^{7} X^F(k) \exp\left[\frac{j2\pi}{8} kn\right], \quad n = 0, 1, \ldots, 7$$

Eight roots of unity are equally distributed 45o apart on the unit circle.
$W_8^m = \exp(-j2\pi m/8), m = 0, 1, \ldots, 7$ are the eight roots of unity (Fig. 3.4).

Transform vector (8×8) DFT matrix   Data vector

$X^F(k)$	$n=0$	1	2	3	4	5	6	7	$x(n)$

$$
\begin{bmatrix} X^F(0) \\ X^F(1) \\ X^F(2) \\ X^F(3) \\ X^F(4) \\ X^F(5) \\ X^F(6) \\ X^F(7) \end{bmatrix}
=
\begin{bmatrix}
1 & 1 & 1 & 1 & 1 & 1 & 1 & 1 \\
1 & W & W^2 & W^3 & -1 & -W & -W^2 & -W^3 \\
1 & W^2 & -1 & -W^2 & 1 & W^2 & -1 & -W^2 \\
1 & W^3 & -W^2 & W & -1 & -W^3 & W^2 & -W \\
1 & -1 & 1 & -1 & 1 & -1 & 1 & -1 \\
1 & -W & W^2 & -W^3 & -1 & W & -W^2 & W^3 \\
1 & -W^2 & -1 & W^2 & 1 & -W^2 & -1 & W^2 \\
1 & -W^3 & -W^2 & -W & -1 & W^3 & W^2 & W
\end{bmatrix}
\begin{bmatrix} x(0) \\ x(1) \\ x(2) \\ x(3) \\ x(4) \\ x(5) \\ x(6) \\ x(7) \end{bmatrix}
$$

$\left(\text{Here } W = W_8 \text{ and } W_8^l = W_8^{l \bmod 8}\right)$

Symmetric Matrix

$$(3.6)$$

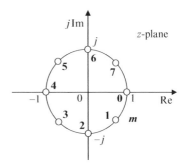

Fig. 3.4 DFT of $x(n)$ is the Z-transform at eight equally distributed points on the unit circle with center at the origin in the z-plane. $W_8^m = \exp(-j2\pi m/8)$, $m = 0, 1, \ldots, 7$ are the eight roots of unity. All the eight roots are distributed $45°$ apart

By rearranging the rows of the DFT matrix in bit reversed order (BRO) (Table 3.1), the eight-point DFT can be expressed as

$$
\begin{bmatrix} X^F(0) \\ X^F(4) \\ X^F(2) \\ X^F(6) \\ X^F(1) \\ X^F(5) \\ X^F(3) \\ X^F(7) \end{bmatrix} = \begin{bmatrix} \text{Rows of}(8 \times 8) \\ \text{DFT matrix} \\ \text{rearranged in} \\ \text{bit reversed} \\ \text{order (BRO)} \end{bmatrix} \begin{bmatrix} x(0) \\ x(1) \\ x(2) \\ x(3) \\ x(4) \\ x(5) \\ x(6) \\ x(7) \end{bmatrix}
$$

Note that the basic DFT is invariant provided $X^F(k)$ is also rearranged in BRO. $[W_8^{nk}]_{\text{BRO}}$ is the 8×8 DFT matrix with rows rearranged in bit reversed order (BRO). In this matrix $W = W_8$.

$$
[W_8^{nk}]_{\text{BRO}} = \begin{array}{c} \text{Row} \\ \begin{array}{c} 0 \\ 4 \\ 2 \\ 6 \\ 1 \\ 5 \\ 3 \\ 7 \end{array} \begin{bmatrix} 1 & 1 & 1 & 1 & 1 & 1 & 1 & 1 \\ 1 & -1 & 1 & -1 & 1 & -1 & 1 & -1 \\ 1 & W^2 & -1 & -W^2 & 1 & W^2 & -1 & -W^2 \\ 1 & -W^2 & -1 & W^2 & 1 & -W^2 & -1 & W^2 \\ 1 & W & W^2 & W^3 & -1 & -W & -W^2 & -W^3 \\ 1 & -W & W^2 & -W^3 & -1 & W & -W^2 & W^3 \\ 1 & W^3 & -W^2 & W & -1 & -W^3 & W^2 & -W \\ 1 & -W^3 & -W^2 & -W & -1 & W^3 & W^2 & W \end{bmatrix} \end{array}
\tag{3.7}
$$

DFT matrix whose rows are rearranged in BRO.

$$
\begin{bmatrix} (4 \times 4) & (4 \times 4) \\ (4 \times 4) & (4 \times 4) \end{bmatrix}
$$

Observation: $[W_8^{nk}]_{\text{BRO}}$

Note that (1) the top two (4×4) submatrices in (3.7) are the same, whereas (2) the bottom two (4×4) submatrices are negatives of each other. The row bit reversed (8×8) DFT matrix has the following sparse matrix factors (SMF). (No proof is given.)

$$
[W_8^{nk}]_{\text{BRO}} = [A_1][A_2][A_3]
\tag{3.8a}
$$

$$
[A_1] = \text{diag}\left[\begin{pmatrix} 1 & W_8^0 \\ 1 & -W_8^0 \end{pmatrix}, \begin{pmatrix} 1 & W_8^2 \\ 1 & -W_8^2 \end{pmatrix}, \begin{pmatrix} 1 & W_8^1 \\ 1 & -W_8^1 \end{pmatrix}, \begin{pmatrix} 1 & W_8^3 \\ 1 & -W_8^3 \end{pmatrix} \right]
$$

$$
= \text{diag}([\alpha], [\beta], [\gamma], [\delta])
\tag{3.8b}
$$

$$[A_2] = \text{diag}\left[\begin{pmatrix} [I_2] & W_8^0[I_2] \\ [I_2] & -W_8^0[I_2] \end{pmatrix}, \begin{pmatrix} [I_2] & W_8^2[I_2] \\ [I_2] & -W_8^2[I_2] \end{pmatrix}\right] = \text{diag}([a],[b]) \quad (3.8c)$$

$$[A_3] = \begin{bmatrix} [I_4] & [I_4] \\ [I_4] & -[I_4] \end{bmatrix} \quad (3.8d)$$

where $[I_4] = \begin{bmatrix} 1 & 0 & 0 & 0 \\ 0 & 1 & 0 & 0 \\ 0 & 0 & 1 & 0 \\ 0 & 0 & 0 & 1 \end{bmatrix}$.

$$[A_2][A_3] = \begin{bmatrix} [a] & 0 \\ 0 & [b] \end{bmatrix}\begin{bmatrix} [I_4] & [I_4] \\ [I_4] & -[I_4] \end{bmatrix} = \begin{bmatrix} [a] & [a] \\ [b] & -[b] \end{bmatrix}$$

$$[A_1][A_2][A_3] = \begin{bmatrix} [\alpha] & 0 & 0 & 0 \\ 0 & [\beta] & 0 & 0 \\ 0 & 0 & [\gamma] & 0 \\ 0 & 0 & 0 & [\delta] \end{bmatrix}\begin{bmatrix} [a] & [a] \\ [b] & -[b] \end{bmatrix}$$

$$= \begin{bmatrix} [\alpha] & [\alpha] & [\alpha] & [\alpha] \\ [\beta] & [\beta] & [\beta] & [\beta] \\ [\gamma] & -W_8^2[\gamma] & -[\gamma] & W_8^2[\gamma] \\ [\delta] & -W_8^2[\delta] & -[\delta] & W_8^2[\delta] \end{bmatrix} = \left[W_8^{nk}\right]_{\text{BRO}}$$

Notation

$$\text{diag}(a_{11}, a_{22}, \ldots, a_{nn}) = \text{diagonal matrix} = \begin{bmatrix} a_{11} & & & O \\ & a_{22} & & \\ & & \ddots & \\ O & & & a_{nn} \end{bmatrix}$$

$$\left[\underline{X}^{\text{F}}(k)\right]_{\text{BRO}} = [A_1][A_2][A_3][\underline{x}(n)] = [A_1][A_2]\left[\underline{x}^{(1)}(n)\right]$$

$$= [A_1]\left[\underline{x}^{(2)}(n)\right]$$

$$= \left[\underline{x}^{(3)}(n)\right] \quad (3.9)$$

where

$$\left[\underline{x}^{(1)}(n)\right] = [A_3][\underline{x}(n)], \left[\underline{x}^{(2)}(n)\right] = [A_2]\left[\underline{x}^{(1)}(n)\right], \quad \text{and} \quad \left[\underline{x}^{(3)}(n)\right] = [A_1]\left[\underline{x}^{(2)}(n)\right].$$

$$[A_3] = \begin{bmatrix} [I_4] & W_8^0[I_4] \\ \hline [I_4] & -W_8^0[I_4] \end{bmatrix} = \begin{bmatrix} [I_4] & [I_4] \\ \hline [I_4] & -[I_4] \end{bmatrix}$$

$$[A_2] = \begin{bmatrix} \begin{bmatrix} [I_2] & W_8^0[I_2] \\ [I_2] & -W_8^0[I_2] \end{bmatrix} & O \\ \hline O & \begin{bmatrix} [I_2] & W_8^2[I_2] \\ [I_2] & -W_8^2[I_2] \end{bmatrix} \end{bmatrix} = \begin{bmatrix} \begin{bmatrix} [I_2] & [I_2] \\ [I_2] & -[I_2] \end{bmatrix} & O \\ \hline O & \begin{bmatrix} [I_2] & W_8^2[I_2] \\ [I_2] & -W_8^2[I_2] \end{bmatrix} \end{bmatrix}$$

$$[A_1] = \begin{bmatrix} 1 & W_8^0 & & & & & & \\ 1 & -W_8^0 & & & & O & & \\ \hline & & 1 & W_8^2 & & & & \\ & & 1 & -W_8^2 & & & & \\ \hline & & & & 1 & W_8^1 & & \\ & & & & 1 & -W_8^1 & & \\ \hline & O & & & & & 1 & W_8^3 \\ & & & & & & 1 & -W_8^3 \end{bmatrix}, \qquad (0, 2, 1, 3) \text{ is BRO of } (0, 1, 2, 3)$$

$$\begin{bmatrix} [I_2] & W_8^0[I_2] \\ \hline [I_2] & -W_8^0[I_2] \end{bmatrix} = \begin{bmatrix} 1 & 0 & 1 & 0 \\ 0 & 1 & 0 & 1 \\ 1 & 0 & -1 & 0 \\ 0 & 1 & 0 & -1 \end{bmatrix}$$

$$\begin{bmatrix} [I_2] & W_8^2[I_2] \\ \hline [I_2] & -W_8^2[I_2] \end{bmatrix} = \begin{bmatrix} 1 & 0 & W_8^2 & 0 \\ 0 & 1 & 0 & W_8^2 \\ 1 & 0 & -W_8^2 & 0 \\ 0 & 1 & 0 & -W_8^2 \end{bmatrix}$$

$$\left[\underline{X}^F(k) \right]_{BRO} = [A_1][A_2][A_3][\underline{x}(n)]$$

$$[A_3][\underline{x}(n)] = \begin{bmatrix} 1 & 0 & 0 & 0 & 1 & 0 & 0 & 0 \\ 0 & 1 & 0 & 0 & 0 & 1 & 0 & 0 \\ 0 & 0 & 1 & 0 & 0 & 0 & 1 & 0 \\ 0 & 0 & 0 & 1 & 0 & 0 & 0 & 1 \\ \hline 1 & 0 & 0 & 0 & -1 & 0 & 0 & 0 \\ 0 & 1 & 0 & 0 & 0 & -1 & 0 & 0 \\ 0 & 0 & 1 & 0 & 0 & 0 & -1 & 0 \\ 0 & 0 & 0 & 1 & 0 & 0 & 0 & -1 \end{bmatrix} \begin{bmatrix} x(0) \\ x(1) \\ x(2) \\ x(3) \\ x(4) \\ x(5) \\ x(6) \\ x(7) \end{bmatrix} = \left[\underline{x}^{(1)}(n) \right]$$

The sparse matrix factorization of $[W_8^{\pi k}]_{BRO}$ as described in (3.8) is the key for the eight-point FFT. The flowgraph for eight-point FFT based on SMF is shown in Fig. 3.5. The number of multiplies can be reduced by a factor of 2 by using the butterfly described in Fig. 3.1. The FFT for $N = 8$ requires 24 adds and five multiplies. The DFT matrix whose rows are rearranged in BRO is no longer symmetric as can be observed from (3.7) for $N = 8$. The same flowgraph (Fig. 3.5) described for FFT ($N = 8$) can be used for IFFT by making the following changes:

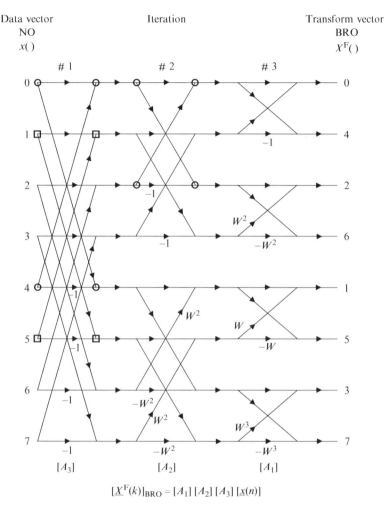

Data vector
NO
$x(\ )$

Iteration

Transform vector
BRO
$X^{F}(\ )$

$$[\underline{X}^{F}(k)]_{BRO} = [A_1]\,[A_2]\,[A_3]\,[\underline{x}(n)]$$

Fig. 3.5 Flow graph for eight-point FFT based on sparse matrix factors $[A_1][A_2][A_3]$. Here $W = W_g = \exp(-j2\pi/8)$ (NO = Natural order, BRO = Bit reversed order)

1. Reverse the direction of all arrows i.e. left by right, right by left, bottom by top and top by bottom.
2. Transform vector $\left[\underline{X}^{F}(k)\right]_{BRO}$ will be the input (on the right) in BRO.
3. Data vector $[\underline{x}(n)]$ will be the output (on the left) in natural order.
4. Replace all multiplies by their conjugates i.e., W^{nk} by W^{-nk}.
5. Add the scale factor $1/8$.

The SMF representation of IFFT for $N = 8$ is

$$[\underline{x}(n)] = \frac{1}{8}\left([A_3]^{*}\right)^{T}\left([A_2]^{*}\right)^{T}\left([A_1]^{*}\right)^{T}\left[\underline{X}^{F}(k)\right]_{BRO} \qquad (3.10)$$

It is straight forward to draw the flowgraph based on (3.10).

The sparse matrix factors for (16×16) DFT matrix with rows rearranged in bit reversed order are shown below. i.e.,

$$[\underline{X}^{\mathrm{F}}(k)]_{\mathrm{BRO}} = [B_1][B_2][B_3][B_4][\underline{x}(n)] \tag{3.11a}$$

Here $W = W_{16} = \exp(-j2\pi/16)$ and $W_{16}^{2n} = W_8^n$.

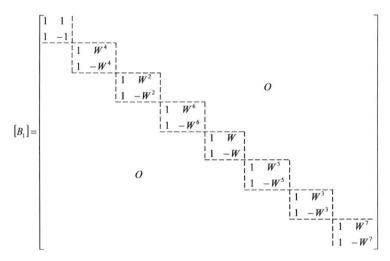

The flowgraph for FFT $N = 16$ based on (3.11a) can be easily developed. The IFFT for $N = 16$ corresponding to (3.10) is

$$[\underline{x}(n)] = \frac{1}{16}[B_4]^{*T}[B_3]^{*T}[B_2]^{*T}[B_1]^{*T}[\underline{X}^{\mathrm{F}}(k)]_{\mathrm{BRO}} \tag{3.11b}$$

The computational complexity for the DFT is compared with the FFT in tabular (Table 3.2) and graphical (Fig. 3.6) formats.

Radix-2 and radix-4 FFT algorithms with both input and output in natural order also have been developed [LA10, LA11].

Table 3.2 Numbers of multiplies and adds for the radix-2 DIT-FFT compared with brute force DFT[1]

Data size N	Brute force		Radix-2 DIT-FFT	
	Number of multiplies	Number of adds	Number of multiplies	Number of adds
8	64	56	12	24
16	256	240	32	64
32	1,024	992	80	160
64	4,096	4,032	192	384

[1]These are based on $O(N^2)$ for brute force and $O(N\log_2 N)$ for radix-2 DIT-FFT. For the later # of multiplies are reduced much more (see Fig. 3.3)

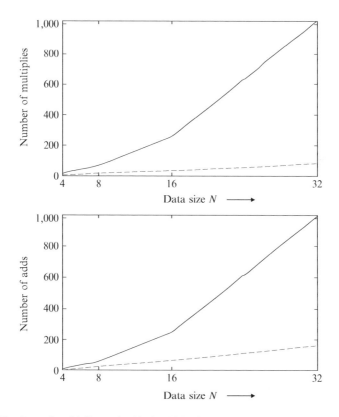

Fig. 3.6 Numbers of multiplies and adds for DFT via the brute force (*solid line*) and via the radix-2 DIT-FFT (*dashed line*)

3.3   Radix-2 DIF-FFT

Similar to radix-2 DIT-FFT (Section 3.1) radix-2 DIF-FFT can be developed as follows:

$$X^{\mathrm{F}}(k) = \sum_{n=0}^{N-1} x(n)\, W_N^{nk}, \quad k = 0, 1, \ldots, N-1, \quad N\text{-point DFT} \qquad (2.1a)$$

Here N is an integer power of two.

$$X^{\mathrm{F}}(k) = \sum_{n=0}^{(N/2)-1} x(n)\, W_N^{nk} + \sum_{n=N/2}^{N-1} x(n)\, W_N^{nk} = I + II \qquad (3.12)$$

The second summation changes to (let $n = m + N/2$)

$$\sum_{m=0}^{(N/2)-1} x\!\left(m + \tfrac{N}{2}\right) W_N^{(m+N/2)k} = \sum_{n=0}^{(N/2)-1} x\!\left(n + \tfrac{N}{2}\right) W_N^{(N/2)k}\, W_N^{nk}$$

Hence (3.12) becomes

$$X^{\mathrm{F}}(k) = \sum_{n=0}^{(N/2)-1} x(n)\, W_N^{nk} + W_N^{(N/2)k} \sum_{n=0}^{(N/2)-1} x\!\left(n + \tfrac{N}{2}\right) W_N^{nk}, \quad k = 0, 1, \ldots, N-1$$

$$X^{\mathrm{F}}(k) = \sum_{n=0}^{(N/2)-1} \left[x(n) + (-1)^k x\!\left(n + \tfrac{N}{2}\right) \right] W_N^{nk}, \qquad k = 0, 1, \ldots, N-1$$

$$(3.13)$$

Since $W_N^{N/2} = -1$. For $k =$ even integer $= 2r$, and for $k =$ odd integer $= 2r + 1$, (3.13) reduces to

$$X^{\mathrm{F}}(2r) = \sum_{n=0}^{(N/2)-1} \left[x(n) + x\!\left(n + \tfrac{N}{2}\right) \right] W_N^{2nr} \qquad (3.14a)$$

$$X^{\mathrm{F}}(2r + 1) = \sum_{n=0}^{(N/2)-1} \left[x(n) - x\!\left(n + \tfrac{N}{2}\right) \right] W_N^{n}\, W_N^{2nr} \qquad (3.14b)$$

Since $W_N^{2nr} = \exp\!\left(\dfrac{-j2\pi 2nr}{N} \right) = \exp\!\left(\dfrac{-j2\pi nr}{N/2} \right) = W_{N/2}^{nr}$

$$X^{\mathrm{F}}(2r) = \sum_{n=0}^{(N/2)-1} \left[x(n) + x\!\left(n + \tfrac{N}{2}\right) \right] W_{N/2}^{nr}$$

Fig. 3.7 Butterfly to compute (3.15) for $n = 0, 1, \ldots, \frac{N}{2} - 1$

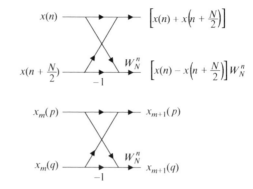

$X^F(2r)$ is $N/2$-point DFT of $\left[x(n) + x\left(n + \frac{N}{2}\right)\right]$, $\quad r, n = 0, 1, \ldots, \frac{N}{2} - 1$ (3.15a)

Similarly

$X^F(2r + 1)$ is $N/2$-point DFT of $\left[x(n) - x\left(n + \frac{N}{2}\right)\right] W_N^n$, $\quad r, n = 0, 1, \ldots, \frac{N}{2} - 1$

(3.15b)

Flowgraph for the expressions within the square brackets of (3.15) is shown in Fig. 3.7.

The N-point DFT can be executed from the two $N/2$-point DFTs as described in (3.15). This results in reduction of computational complexity as in the case of DIT-FFT algorithm. Further savings can be achieved by successively repeating this decomposition (i.e. first half of the samples and second half of the samples). The DIF-FFT algorithm is illustrated for $N = 8$.

3.3.1   DIF-FFT N = 8

DFT

$$X^F(k) = \sum_{n=0}^{7} x(n) W_8^{kn}, \quad k = 0, 1, \ldots, 7$$

IDFT

$$x(n) = \frac{1}{8} \sum_{k=0}^{7} X^F(k) W_8^{-kn}, \quad n = 0, 1, \ldots, 7$$

For $N = 8$, (3.15) becomes

$$X^{\mathrm{F}}(2r) = \sum_{n=0}^{3} \left[x(n) + x\left(n + \frac{N}{2}\right) \right] W_4^{nr}, \quad r = 0, 1, 2, 3 \qquad (3.16a)$$

$$X^{\mathrm{F}}(2r + 1) = \sum_{n=0}^{3} \left[x(n) - x\left(n + \frac{N}{2}\right) \right] W_8^{n} W_4^{nr}, \quad r = 0, 1, 2, 3 \qquad (3.16b)$$

Equation (3.16) can be explicitly described as follows:

$X^{\mathrm{F}}(2r)$, $r = 0, 1, 2, 3$ is a four-point DFT of
$$\{x(0) + x(4), x(1) + x(5), x(2) + x(6), x(3) + x(7)\}.$$

$$X^{\mathrm{F}}(0) = [x(0) + x(4)] + [x(1) + x(5)] + [x(2) + x(6)] + [x(3) + x(7)]$$
$$X^{\mathrm{F}}(2) = [x(0) + x(4)] + [x(1) + x(5)] W_4^1 + [x(2) + x(6)] W_4^2 + [x(3) + x(7)] W_4^3$$
$$X^{\mathrm{F}}(4) = [x(0) + x(4)] + [x(1) + x(5)] W_4^2 + [x(2) + x(6)] W_4^4 + [x(3) + x(7)] W_4^6$$
$$X^{\mathrm{F}}(6) = [x(0) + x(4)] + [x(1) + x(5)] W_4^3 + [x(2) + x(6)] W_4^6 + [x(3) + x(7)] W_4^9$$

$X^{\mathrm{F}}(2r + 1)$ is a four-point DFT of

$$\{x(0) - x(4), [x(1) - x(5)] W_8^1, [x(2) - x(6)] W_8^2 [x(3) - x(7)] W_8^3\}, \quad r = 0, 1, 2, 3$$
$$X^{\mathrm{F}}(1) = (x(0) - x(4)) + (x(1) - x(5)) W_8^1 + (x(2) - x(6)) W_8^2 + (x(3) - x(7)) W_8^3$$
$$X^{\mathrm{F}}(3) = (x(0) - x(4)) + \left[(x(1) - x(5)) W_8^1\right] W_4^1 + \left[(x(2) - x(6)) W_8^2\right] W_4^2$$
$$\qquad + \left[(x(3) - x(7)) W_8^3\right] W_4^3$$
$$X^{\mathrm{F}}(5) = (x(0) - x(4)) + \left[(x(1) - x(5)) W_8^1\right] W_4^2 + \left[(x(2) - x(6)) W_8^2\right] W_4^4$$
$$\qquad + \left[(x(3) - x(7)) W_8^3\right] W_4^6$$
$$X^{\mathrm{F}}(7) = (x(0) - x(4)) + \left[(x(1) - x(5)) W_8^1\right] W_4^3 + \left[(x(2) - x(6)) W_8^2\right] W_4^6$$
$$\qquad + \left[(x(3) - x(7)) W_8^3\right] W_4^9$$

Each of the four-point DFTs in (3.16) can be implemented by using two two-point DFTs. Hence the eight-point DIF-FFT is obtained in three stages i.e., I stage: two-point DFTs (Fig. 3.8), II stage: four-point DFTs (Fig. 3.9), III stage: eight-point DFT (Fig. 3.10). Compare this flowgraph with that shown in Fig. 3.5. These two have the same structure but for some minor changes in the multipliers. Figures 3.3, 3.5 and 3.10 exemplify the versatility of the various radix-2 FFT algorithms. At each stage the decomposition can be DIT based or DIF based. This leads to DIT/DIF radix-2 FFTs (same for IFFTs). The SMF representation of Fig. 3.10 is

$$\left[\underline{X}^{\mathrm{F}}(k)\right]_{\mathrm{BRO}} = \left[\tilde{A}_1\right] \left[\tilde{A}_2\right] \left[\tilde{A}_3\right] \left[\tilde{A}_4\right] \left[\tilde{A}_5\right] \left[\underline{x}(n)\right] \qquad (3.17)$$

Data sequence   Transform sequence

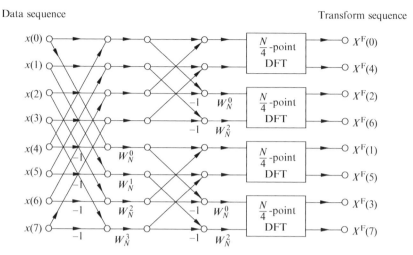

Fig. 3.8 Flow graph of decimation-in frequency decomposition of an eight-point DFT into two-point DFT computations. $W_N = W_8$

Data sequence   DFT sequence

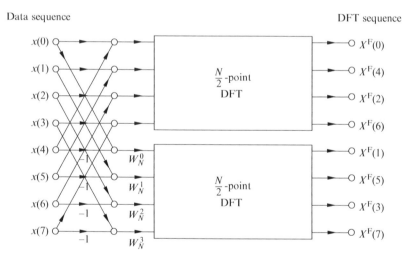

Fig. 3.9 Flow graph of decimation-in frequency decomposition of an N-point DFT into two $N/2$-point DFT computations ($N = 8$) ($W_N = W_8$)

where

$$\text{diag}\left[\begin{pmatrix} 1 & 1 \\ 1 & -1 \end{pmatrix}, \begin{pmatrix} 1 & 1 \\ 1 & -1 \end{pmatrix}, \begin{pmatrix} 1 & 1 \\ 1 & -1 \end{pmatrix}, \begin{pmatrix} 1 & 1 \\ 1 & -1 \end{pmatrix}\right] = [\tilde{A}_1]$$

Data sequence   Transform sequence

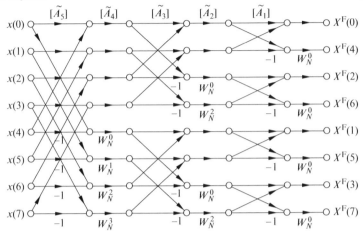

Fig. 3.10 Flow graph of complete decimation-in frequency decomposition of an eight-point DFT computation. $W_N = W_8$

$$\text{diag}\left(1, 1, 1, W_8^2, 1, 1, 1, W_8^2\right) = \left[\tilde{A}_2\right]$$

$$\text{diag}\left[\left(\begin{matrix} [I_2] & [I_2] \\ [I_2] & -[I_2] \end{matrix}\right), \left(\begin{matrix} [I_2] & [I_2] \\ [I_2] & -[I_2] \end{matrix}\right)\right] = \left[\tilde{A}_3\right]$$

$$\text{diag}\left(1, 1, 1, 1, 1, W_8^1, W_8^2, W_8^3\right) = \left[\tilde{A}_4\right]$$

$$\begin{bmatrix} [I_4] & [I_4] \\ [I_4] & -[I_4] \end{bmatrix} = \left[\tilde{A}_5\right]$$

By implementing the same changes in Fig. 3.10 as outlined for Fig. 3.5 (see just before (3.10)), eight-point DIF/IFFT can be obtained.

Radix-2 DIF-FFT algorithm can be summarized as follows (to simplify the notation, replace $X^F(k)$ by X_k^F and $x(n)$ by x_n):

$$\frac{1}{NT} = f_0 \qquad N\text{-point DFT } (N = 2^n) \qquad \qquad \xrightarrow{\quad T \quad} |\!\leftarrow$$

$$(x_0, x_1, x_2, x_3, x_4, x_5, x_6, x_7, x_8, \ldots, x_{N-4}, x_{N-3}, x_{N-2}, x_{N-1})$$

$$\left(x_0 + x_{\frac{N}{2}}, x_1 + x_{\frac{N}{2}+1}, \ldots, x_{\frac{N}{2}-1} + x_{N-1}\right) \qquad \left[x_0 - x_{\frac{N}{2}}, \left(x_1 - x_{\frac{N}{2}+1}\right)W_N^1, \left(x_2 - x_{\frac{N}{2}+2}\right)W_N^2,\right.$$

$$\frac{1}{(N/2)T} = 2f_0 \quad \text{Two } N/2\text{-point DFTs} \qquad \left. \ldots, \left(x_{\frac{N}{2}-1} - x_{N-1}\right)W_N^{\frac{N}{2}-1}\right]$$

An N-point sequence is broken down into two $N/2$-point sequences as shown above. The N-point DFT is obtained from these $N/2$-point DFTs. Call these two $N/2$-point sequences as

$$\left(y_0, y_1, y_2, y_3, \ldots, y_{\frac{N}{2}-1}\right), \quad y_i = x_i + x_{\frac{N}{2}+i}, \quad i = 0, 1, \ldots, \frac{N}{2} - 1$$

and

$$\left(z_0, z_1, z_2, z_3, \ldots, z_{\frac{N}{2}-1}\right), \quad z_i = \left(x_i - x_{\frac{N}{2}+i}\right) W_N^i, \quad i = 0, 1, \ldots, \frac{N}{2} - 1$$

Break down each of these two sequences into two $N/4$-point sequences as follows.

$$\left(y_0, y_1, y_2, y_3, \ldots, y_{\frac{N}{2}-3}, y_{\frac{N}{2}-2}, y_{\frac{N}{2}-1}\right)$$

$$\left(y_0 + y_{\frac{N}{4}}, y_1 + y_{\frac{N}{4}+1}, \ldots, y_{\frac{N}{4}-1} + y_{\frac{N}{2}-1}\right) \quad \left[y_0 - y_{\frac{N}{4}}, \left(y_1 - y_{\frac{N}{4}+1}\right) W_{\frac{N}{2}}^1, \left(y_2 - y_{\frac{N}{4}+2}\right) W_{\frac{N}{2}}^2, \right.$$

$$\frac{1}{(N/4)T} = 4f_0 \qquad N/4\text{-point DFTs} \qquad \left. \ldots, \left(y_{\frac{N}{4}-1} - y_{\frac{N}{2}-1}\right) W_{\frac{N}{2}}^{\frac{N}{4}-1}\right]$$

Similarly for $\left(z_0, z_1, z_2, z_3, \ldots, z_{(N/2)-1}\right)$. Repeat this process till two-point sequences are obtained.

3.3.2 In-Place Computations

Figure 3.7 shows the butterfly, a basic building block for DIF FFTs. Only the data in locations p and q of the mth array are involved in computing two output data in p and q of the $(m + 1)$th array. Thus only one array of N registers is necessary to implement the DFT computation if $x_{m+1}(p)$ and $x_{m+1}(q)$ are stored in the same registers as $x_m(p)$ and $x_m(q)$, respectively. This *in-place* computation results only if input and output nodes for each butterfly computation are horizontally adjacent. However, the in-place computation causes the transform sequence in bit reversed order for the DIF FFT as shown in Fig. 3.10 [A42].

3.4 Radix-3 DIT FFT

So far we have developed radix-2 DIT, DIF, and DIT/DIF algorithms. When the data sequence is of length $N = 3^l$, l an integer, radix-3 DIT FFT can be developed as follows:

Express (2.1a) as:

$$X_k^F = \sum_{r=0}^{(N/3)-1} x_{3r} W_N^{3rk} + \sum_{r=0}^{(N/3)-1} x_{3r+1} W_N^{(3r+1)k} + \sum_{r=0}^{(N/3)-1} x_{3r+2} W_N^{(3r+2)k}$$

$$W_N^{3rk} = \exp\left(\frac{-j2\pi}{N} 3rk\right) = \exp\left(\frac{-j2\pi}{N/3} rk\right) = W_{N/3}^{rk}$$

$$X_k^F = \sum_{r=0}^{(N/3)-1} x_{3r} W_{N/3}^{rk} + W_N^k \sum_{r=0}^{(N/3)-1} x_{3r+1} W_{N/3}^{rk} + W_N^{2k} \sum_{r=0}^{(N/3)-1} x_{3r+2} W_{N/3}^{rk}$$

$$X_k^F = A_k^F + W_N^k B_k^F + W_N^{2k} C_k^F \tag{3.18a}$$

where A_k^F, B_k^F, and C_k^F are $N/3$-point DFTs. These are DFTs of $(x_0, x_3, x_6, \ldots, x_{N-3})$, $(x_1, x_4, x_7, \ldots, x_{N-2})$ and $(x_2, x_5, x_8, \ldots, x_{N-1})$ respectively. Hence they are periodic with period $N/3$, resulting in

$$X_{k+N/3}^F = A_k^F + W_N^{(k+N/3)} B_k^F + W_N^{2(k+N/3)} C_k^F$$
$$= A_k^F + e^{-j2\pi/3} W_N^k B_k^F + e^{-j4\pi/3} W_N^{2k} C_k^F \tag{3.18b}$$

$$X_{k+2N/3}^F = A_k^F + W_N^{(k+2N/3)} B_k^F + W_N^{2(k+2N/3)} C_k^F$$
$$= A_k^F + e^{-j4\pi/3} W_N^k B_k^F + e^{-j2\pi/3} W_N^{2k} C_k^F \tag{3.18c}$$

$$W_N^{N/3} = \exp\left(\frac{-j2\pi}{N}\frac{N}{3}\right) = \exp(-j2\pi/3) \quad \text{and} \quad W_N^{4N/3} = W_N^N W_N^{N/3} = W_N^{N/3}$$

Equation (3.18) is valid for $k = 0, 1, \ldots, \frac{N}{3} - 1$. This process is repeated till the original sequence is split into sequences each of length 3. In matrix form (3.18) can be expressed as:

$$\begin{pmatrix} X_k^F \\ X_{k+N/3}^F \\ X_{k+2N/3}^F \end{pmatrix} = \begin{bmatrix} 1 & 1 & 1 \\ 1 & e^{-j2\pi/3} & e^{-j4\pi/3} \\ 1 & e^{-j4\pi/3} & e^{-j2\pi/3} \end{bmatrix} \begin{pmatrix} A_k^F \\ W_N^k B_k^F \\ W_N^{2k} C_k^F \end{pmatrix}, \quad k = 0, 1, \ldots, \frac{N}{3} - 1$$

Equation (3.18) is shown in flowgraph format in Fig. 3.11. Split the sequence $\{x(n)\}$ into three sequences as follows:

$$(x_0, x_3, x_6, \ldots, x_{N-3}), (x_1, x_4, x_7, \ldots, x_{N-2}), (x_2, x_5, x_8, \ldots, x_{N-1})$$
$$\rightarrow |3T| \leftarrow \qquad \rightarrow |3T| \leftarrow \qquad \rightarrow |3T| \leftarrow \qquad \text{each of length } N/3$$

Frequency resolution of X_k^F is $f_0 = \frac{1}{NT}$.

Frequency resolution of A_k^F, B_k^F, and C_k^F is $\frac{1}{(N/3)3T} = \frac{1}{NT} = f_0$.

Fig. 3.11 Flowgraph format
for (3.18)

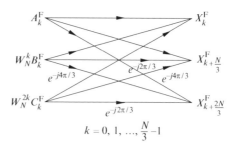

$$k = 0, 1, \ldots, \frac{N}{3} - 1$$

Frequency resolution is invariant. Time resolution goes from $3T$ to T. Hence this is called decimation in time (DIT). Split each of these sequences into three sequences each of length $N/9$ and so on till three-point sequences are obtained. This is radix-3 DIT-FFT. As in the case of radix-2 DIT-FFT, the N-point DFT ($N = 3^l$) is implemented bottoms up starting with 3-point DFTs. All the advantages outlined for radix-2 are equally valid for radix-3.

3.5 Radix-3 DIF-FFT

Similar to radix-3 DIT-FFT, when $N = 3^l$, l an integer, a radix-3 DIF-FFT algorithm can be developed.

DFT

$$X_k^F = \sum_{n=0}^{N-1} x_n W_N^{nk} \qquad k = 0, 1, \ldots, N - 1 \tag{3.19a}$$

IDFT

$$x_n = \frac{1}{N} \sum_{k=0}^{N-1} X_k^F W_N^{-nk} \qquad n = 0, 1, \ldots, N - 1 \tag{3.19b}$$

DFT

$$X_k^F = \sum_{n=0}^{(N/3)-1} x_n W_N^{nk} + \sum_{n=N/3}^{(2N/3)-1} x_n W_N^{nk} + \sum_{n=2N/3}^{N-1} x_n W_N^{nk} \tag{3.20}$$

$= I + II + III$ (these respectively represent the corresponding summation terms).

Let $n = m + N/3$ in summation *II* and let $n = m + (2N/3)$ in summation *III*. Then

$$II = \sum_{n=N/3}^{(2N/3)-1} x_n W_N^{nk} = \sum_{m=0}^{(N/3)-1} x_{m+N/3} W_N^{(m+N/3)k} = W_N^{(N/3)k} \sum_{m=0}^{(N/3)-1} x_{m+N/3} W_N^{mk}$$

(3.21a)

$$III = \sum_{n=2N/3}^{N-1} x_n W_N^{nk} = \sum_{m=0}^{(N/3)-1} x_{m+2N/3} W_N^{(m+2N/3)k} = W_N^{(2N/3)k} \sum_{m=0}^{(N/3)-1} x_{m+2N/3} W_N^{mk}$$

(3.21b)

where

$$W_N^{(N/3)k} = \exp\left(\frac{-j2\pi}{N} \frac{N}{3} k\right) = e^{-j2\pi k/3}$$

$$W_N^{(2N/3)k} = \exp\left(\frac{-j2\pi}{N} \frac{2N}{3} k\right) = e^{-j4\pi k/3}$$

Hence (3.20) becomes

$$X_k^F = \sum_{n=0}^{(N/3)-1} \left[x_n + e^{-j2\pi k/3} x_{n+N/3} + e^{-j4\pi k/3} x_{n+2N/3} \right] W_N^{nk}$$

(3.22)

In (3.22), let $k = 3m$, $k = 3m+1$, and $k = 3m+2$ for $m = 0, 1, \ldots, (N/3) - 1$ respectively.

This results in

$$X_{3m}^F = \sum_{n=0}^{(N/3)-1} \left[x_n + x_{n+N/3} + x_{n+2N/3} \right] W_{N/3}^{nm}$$

(3.23a)

This is a $N/3$-point DFT of $\left(x_n + x_{n+N/3} + x_{n+2N/3} \right), n = 0, 1, \ldots, (N/3) - 1$

$$X_{3m+1}^F = \sum_{n=0}^{(N/3)-1} \left\{ \left[x_n + \left(e^{-j2\pi/3} \right) x_{n+N/3} + \left(e^{-j4\pi/3} \right) x_{n+2N/3} \right] W_N^n \right\} W_{N/3}^{nm}$$

(3.23b)

This is a $N/3$-point DFT of $\left[x_n + \left(e^{-j2\pi/3} \right) x_{n+N/3} + \left(e^{-j4\pi/3} \right) x_{n+2N/3} \right] W_N^n$. Similarly,

$$X_{3m+2}^F = \sum_{n=0}^{(N/3)-1} \left\{ \left[x_n + \left(e^{-j4\pi/3} \right) x_{n+N/3} + \left(e^{-j2\pi/3} \right) x_{n+2N/3} \right] W_N^{2n} \right\} W_{N/3}^{nm}$$

(3.23c)

This is a $N/3$-point DFT of $\left[x_n + \left(e^{-j4\pi/3} \right) x_{n+N/3} + \left(e^{-j2\pi/3} \right) x_{n+2N/3} \right] W_N^{2n}$.

Radix-3 DIF-FFT is developed based on three $N/3$-point DFTs.

Each $N/3$-point DFT is obtained based on a linear combination of $N/3$-point data sequences. Repeat this decomposition till a three-point sequence is obtained.

The three $N/3$-point data sequences (see (3.23)) are as follows:

$$
\begin{bmatrix}
1 & 1 & 1 \\
W_N^n & (1 & e^{-j2\pi/3} & e^{-j4\pi/3}) \\
W_N^{2n} & (1 & e^{-j4\pi/3} & e^{-j2\pi/3})
\end{bmatrix}
\begin{bmatrix}
x_n \\
x_{n+N/3} \\
x_{n+2N/3}
\end{bmatrix},
\quad n = 0, 1, \ldots, \frac{N}{3} - 1
\qquad (3.24)
$$

Equation (3.24) can be expressed in flowgraph format in Fig. 3.12. Equation (3.23) represents radix-3 DIT-FFT algorithm. For example, X_{3m}^F, $m = 0, 1, \ldots, (N/3) - 1$ is a $N/3$-point DFT of $\left(x_n + x_{n+N/3} + x_{n+2N/3}\right)$. The original N samples $x_n, n = 0, 1, \ldots, N - 1$ are regrouped into $N/3$ samples as follows:

$$
\left\{ \left(x_0 + x_{N/3} + x_{2N/3}\right), \left(x_1 + x_{(N/3)+1} + x_{(2N/3)+1}\right), \ldots, \left(x_{(N/3)-1} + x_{(2N/3)-1} + x_{N-1}\right) \right\}
$$

The resolution in time of these $N/3$ samples is T. However, the resolution in frequency of $N/3$-point DFT is

$$
\frac{1}{(N/3)T} = \frac{3}{NT} = 3f_0
$$

For N-point DFT, resolution in frequency is $1/NT = f_0$.

Going from $N/3$-point to N-point DFT implies that frequency resolution is decimated by a factor of 3, i.e. from $3f_0$ to f_0 where as resolution in time remains invariant. Hence this is called DIF-FFT. Needless to say that radix-3 DIT/DIF FFTs can also be developed by appropriate decomposition at each stage. Flowgraph for radix-3 DIF-FFT for $N = 9$ is shown in Fig. 3.13. All the advantages listed for radix-2 DIT-FFT and radix-2 DIF-FFT are equally valid for radix-3 FFT algorithms.

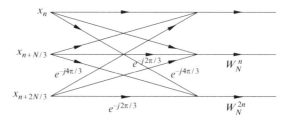

Fig. 3.12 Flowgraph format for (3.24)

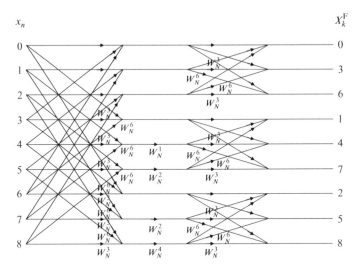

Fig. 3.13 Radix-3 DIF-FFT flowgraph for $N = 9$. Here $W_N = W_9$

3.6   FFT for N a Composite Number

When N is a composite number, mixed-radix DIT, DIF and DIT/DIF FFTs can be developed. Let

$$N = p_1 p_2 \cdots p_v = p_1 q_1, \quad q_1 = p_2 p_3 \cdots p_v = p_2 q_2, \quad q_2 = p_3 p_4 \cdots p_v$$

The process can be described for $N = 12$.
Example: $N = 12, p_1 = 3, q_1 = 4$

$$N = 12 = 3 \times 4 = 3 \times 2 \times 2, \ N = p_1 q_1$$

The decomposition of 12-point sequence for DIT-FFT is:

$$(x_0, x_1, x_2, x_3, x_4, x_5, x_6, x_7, x_8, x_9, x_{10}, x_{11})$$

$(x_0, x_3, x_6, x_9), (x_1, x_4, x_7, x_{10}), (x_2, x_5, x_8, x_{11})$ radix-3

$(x_0, x_6), (x_3, x_9), (x_1, x_7), (x_4, x_{10}), (x_2, x_8), (x_5, x_{11})$ radix-2

This is radix-3/radix-2 DIT-FFT approach. The detailed algorithm is straight forward based on radix-2 and radix-3 algorithms developed earlier.

For the DIF-FFT the decomposition is as follows:

$$(x_0, x_1, x_2, x_3, x_4, x_5, x_6, x_7, x_8, x_9, x_{10}, x_{11})$$

$$(x_0, x_1, x_2, x_3) \quad (x_4, x_5, x_6, x_7) \quad (x_8, x_9, x_{10}, x_{11}) \qquad \text{radix-3}$$

$$(x_0, x_1)(x_2, x_3) \quad (x_4, x_5)(x_6, x_7) \quad (x_8, x_9)(x_{10}, x_{11}) \qquad \text{radix-2}$$

This is radix-3 / radix-2 DIF-FFT approach. By appropriate decomposition at each stage, radix-3 / radix-2 DIT/DIF algorithms can be developed.

3.7 Radix-4 DIT-FFT [V14]

When the data sequence is of length $N = 4^n$, n an integer, radix-4 FFT can be developed based on both DIT and DIF. As before the N-point DFT of $x_n, n = 0, 1, \ldots, N - 1$ can be expressed as

$$X_k^F = \sum_{n=0}^{N-1} x_n W_N^{nk}, \qquad W_N = \exp\left(\frac{-j2\pi}{N}\right), \quad k = 0, 1, \ldots, N - 1 \qquad (3.25a)$$

Similarly the IDFT is

$$x_n = \frac{1}{N} \sum_{k=0}^{N-1} X_k^F W_N^{-nk}, \quad n = 0, 1, \ldots, N - 1 \qquad (3.25b)$$

Equation (3.25a) can be expressed as

$$X_k^F = \sum_{n=0}^{N/4-1} x_{4n} W_N^{4nk} + \sum_{n=0}^{N/4-1} x_{4n+1} W_N^{(4n+1)k} + \sum_{n=0}^{N/4-1} x_{4n+2} W_N^{(4n+2)k} + \sum_{n=0}^{N/4-1} x_{4n+3} W_N^{(4n+3)k}$$

$$(3.26a)$$

$$X_k^F = \left(\sum_{n=0}^{N/4-1} x_{4n} W_N^{4nk}\right) + W_N^k\left(\sum_{n=0}^{N/4-1} x_{4n+1} W_N^{4nk}\right) + W_N^{2k}\left(\sum_{n=0}^{N/4-1} x_{4n+2} W_N^{4nk}\right)$$

$$+ W_N^{3k}\left(\sum_{n=0}^{N/4-1} x_{4n+3} W_N^{4nk}\right)$$

$$(3.26b)$$

This can be expressed as

$$X_k^F = A_k^F + W_N^k B_k^F + W_N^{2k} C_k^F + W_N^{3k} D_k^F, \quad k = 0, 1, \ldots, N - 1 \qquad (3.27a)$$

where $A_k^F, B_k^F, C_k^F,$ and D_k^F, are $N/4$-point DFTs. Therefore they are periodic with period $N/4$. Hence

$$X_{k+\frac{N}{4}}^F = A_k^F + W_N^{k+\frac{N}{4}} B_k^F + W_N^{2\left(k+\frac{N}{4}\right)} C_k^F + W_N^{3\left(k+\frac{N}{4}\right)} D_k^F$$

$$= A_k^F - jW_N^k B_k^F - W_N^{2k} C_k^F + jW_N^{3k} D_k^F, \quad k = 0, 1, \ldots, \frac{N}{4} - 1$$

$$(3.27b)$$

Similarly

$$X^F_{k+\frac{N}{2}} = A^F_k - W^k_N B^F_k + W^{2k}_N C^F_k - W^{3k}_N D^F_k, \quad k = 0, 1, \ldots, \tfrac{N}{4} - 1 \qquad (3.27c)$$

$$X^F_{k+\frac{3N}{4}} = A^F_k + jW^k_N B^F_k - W^{2k}_N C^F_k - jW^{3k}_N D^F_k, \quad k = 0, 1, \ldots, \tfrac{N}{4} - 1 \qquad (3.27d)$$

Equation (3.27) can be expressed in matrix form as

$$\begin{bmatrix} X^F_k \\ X^F_{k+N/4} \\ X^F_{k+N/2} \\ X^F_{k+3N/4} \end{bmatrix} = \begin{bmatrix} 1 & 1 & 1 & 1 \\ 1 & -j & -1 & j \\ 1 & -1 & 1 & -1 \\ 1 & j & -1 & -j \end{bmatrix} \begin{bmatrix} A^F_k \\ W^k_N B^F_k \\ W^{2k}_N C^F_k \\ W^{3k}_N D^F_k \end{bmatrix}, \quad k = 0, 1, \ldots, \tfrac{N}{4} - 1 \quad (3.28)$$

This is also shown in flowgraph format in Fig. 3.14. This requires three multiplies and 12 adds.

The breakdown of the original N-point sequence (sampling interval is T) into four $N/4$-point sequences (this is an iterative process) is shown below:

$$\{x_0, x_1, x_2, x_3, x_4, x_5, \ldots, x_{N-3}, x_{N-2}, x_{N-1}\}$$

$\qquad\qquad A^F_k \qquad\qquad\qquad\qquad B^F_k \qquad\qquad\qquad\qquad C^F_k \qquad\qquad\qquad\qquad D^F_k$

$\{x_0, x_4, x_8, x_{12}, \ldots\} \; \{x_1, x_5, x_9, x_{13}, \ldots\} \; \{x_2, x_6, x_{10}, x_{14}, \ldots\} \; \{x_3, x_7, x_{11}, x_{15}, \ldots\}$

Four $N/4$ - point DFTs, $(N = 4^n)$

$$\{x_{4n}\}, \; \{x_{4n+1}\}, \; \{x_{4n+2}\}, \; \{x_{4n+3}\}, \quad n = 0, 1, \ldots, \tfrac{N}{4} - 1$$

$$A^F_k$$

$$\{x_0, x_4, x_8, x_{12}, x_{16}, x_{20}, x_{24}, x_{28}, x_{32}, x_{36}, x_{40}, x_{44}, x_{48}, x_{52}, x_{56} \ldots\}$$

$$\longrightarrow |4T| \longleftarrow$$

$\qquad\qquad E^F_k \qquad\qquad\qquad\qquad F^F_k \qquad\qquad\qquad\qquad G^F_k \qquad\qquad\qquad\qquad H^F_k$

$\{x_0, x_{16}, x_{32}, x_{48}, \ldots\} \; \{x_4, x_{20}, x_{36}, x_{52}, \ldots\} \; \{x_8, x_{24}, x_{40}, x_{56}, \ldots\} \{x_{12}, x_{28}, x_{44}, x_{60}, \ldots\}$

$$\longrightarrow |16T| \longleftarrow$$

A^F_k is $N/4$-point DFT. E^F_k, F^F_k, G^F_k, and H^F_k, are $N/16$-point DFTs. Obtain A^F_k from E^F_k, F^F_k, G^F_k, and H^F_k. Repeat this process for the remaining $N/4$-point DFTs to obtain B^F_k, C^F_k, and D^F_k. This is similar to radix-3 DIT-FFT (see Section 3.4).

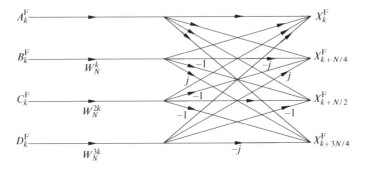

Fig. 3.14 N-point DFT X^F_k from four $N/4$-point DFTs

Radix-4 DIT-DFT can be illustrated for $N = 64$.

$$X_k^{\mathrm{F}}$$

$$\{x_0, x_1, x_2, x_3, x_4, x_5, x_6, x_7, x_8, x_9, x_{10}, x_{11}, x_{12}, \ldots, x_{55}, x_{56}, x_{57}, x_{58}, x_{59}, x_{60}, x_{61}x_{62}, x_{63}\}$$

I Stage

$$A_k^{\mathrm{F}} \qquad\qquad B_k^{\mathrm{F}} \qquad\qquad C_k^{\mathrm{F}} \qquad\qquad D_k^{\mathrm{F}}$$

$$\{x_0, x_4, x_8, x_{12}, \ldots, x_{60}\}\{x_1, x_5, x_9, x_{13}, \ldots, x_{61}\}\{x_2, x_6, x_{10}, x_{14}, \ldots, x_{62}\}\{x_3, x_7, x_{11}, x_{15}, \ldots, x_{63}\}$$

II Stage

$$A_k$$
$$\{x_0, x_4, x_8, x_{12}, \ldots, x_{60}\}$$

$$E_k^{\mathrm{F}} \qquad\qquad F_k^{\mathrm{F}}$$
$$\{x_0, x_{16}, x_{32}, x_{48}\}, \quad \{x_4, x_{20}, x_{36}, x_{52}\}$$
$$G_k^{\mathrm{F}} \qquad\qquad H_k^{\mathrm{F}}$$
$$\{x_8, x_{24}, x_{40}, x_{56}\}, \quad \{x_{12}, x_{28}, x_{44}, x_{60}\}$$

$$B_k^{\mathrm{F}}$$
$$\{x_1, x_5, x_9, x_{13}, \ldots, x_{61}\}$$

$$I_k^{\mathrm{F}} \qquad\qquad J_k^{\mathrm{F}}$$
$$\{x_1, x_{17}, x_{33}, x_{49}\}, \quad \{x_5, x_{21}, x_{37}, x_{53}\}$$

$$K_k^{\mathrm{F}} \qquad\qquad L_k^{\mathrm{F}}$$
$$\{x_9, x_{25}, x_{41}, x_{57}\}, \quad \{x_{13}, x_{29}, x_{45}, x_{61}\}$$

$$C_k^{\mathrm{F}}$$
$$\{x_2, x_6, x_{10}, x_{14}, \ldots, x_{62}\}$$

$$M_k^{\mathrm{F}} \qquad\qquad N_k^{\mathrm{F}}$$
$$\{x_2, x_{18}, x_{34}, x_{50}\}, \quad \{x_6, x_{22}, x_{38}, x_{54}\}$$
$$O_k^{\mathrm{F}} \qquad\qquad P_k^{\mathrm{F}}$$
$$\{x_{10}, x_{26}, x_{42}, x_{58}\}, \quad \{x_{14}, x_{30}, x_{46}, x_{62}\}$$

$$D_k^{\mathrm{F}}$$
$$\{x_3, x_7, x_{11}, x_{15}, \ldots, x_{63}\}$$

$$Q_k^{\mathrm{F}} \qquad\qquad R_k^{\mathrm{F}}$$
$$\{x_3, x_{19}, x_{35}, x_{51}\}, \quad \{x_7, x_{23}, x_{39}, x_{55}\}$$
$$S_k^{\mathrm{F}} \qquad\qquad T_k^{\mathrm{F}}$$
$$\{x_{11}, x_{27}, x_{43}, x_{59}\}, \quad \{x_{15}, x_{31}, x_{47}, x_{63}\}$$

$(N = 64)$ $\qquad\qquad\qquad\qquad\qquad\qquad X_k^{\mathrm{F}}$

$(N = 16)$ $\qquad A_k^{\mathrm{F}} \qquad\qquad B_k^{\mathrm{F}} \qquad\qquad C_k^{\mathrm{F}} \qquad\qquad D_k^{\mathrm{F}}$

$(N = 4)$ $\quad E_k^{\mathrm{F}}\ F_k^{\mathrm{F}}\ G_k^{\mathrm{F}}\ H_k^{\mathrm{F}} \quad I_k^{\mathrm{F}}\ J_k^{\mathrm{F}}\ K_k^{\mathrm{F}}\ L_k^{\mathrm{F}} \quad M_k^{\mathrm{F}}\ N_k^{\mathrm{F}}\ O_k^{\mathrm{F}}\ P_k^{\mathrm{F}} \quad Q_k^{\mathrm{F}}\ R_k^{\mathrm{F}}\ S_k^{\mathrm{F}}\ T_k^{\mathrm{F}}$

$E_k^{\mathrm{F}}, F_k^{\mathrm{F}}, \ldots, T_k^{\mathrm{F}}$ are four-point DFTs. Hence no further reduction.

N-point DFT ($N = 4^m$, $m = $ integer)

$$\{x_0, x_1, x_2, x_3, x_4, x_5, \ldots, x_{N-3}, x_{N-2}, x_{N-1}\}$$
$$\rightarrow |T| \leftarrow$$

$$f_0 = \frac{1}{NT} = \text{ resolution in the frequency domain}$$

$$\{x_0, x_4, x_8, x_{12}, \ldots\} \quad \{x_1, x_5, x_9, x_{13}, \ldots\} \quad \{x_2, x_6, x_{10}, x_{14}, \ldots\} \quad \{x_3, x_7, x_{11}, x_{15}, \ldots\}$$
$$\rightarrow |4T| \leftarrow \qquad \rightarrow |4T| \leftarrow \qquad \rightarrow |4T| \leftarrow \qquad \rightarrow |4T| \leftarrow$$

Resolution in the frequency domain $= \dfrac{1}{4T\frac{N}{4}} = \dfrac{1}{NT} = f_0$.

In going backward to get the N-point DFT, frequency resolution is same but time resolution is decimated by a factor of 4 at each stage. Hence this is DIT-FFT.

Four-point DFT can be described as follows:

$$X_k^F = \sum_{n=0}^{3} x_n W_4^{nk}, \quad k = 0, 1, 2, 3 \tag{3.29}$$

$$W_4 = e^{-j2\pi/4} = e^{-j\pi/2} = -j, \quad W_4^0 = 1, \quad W_4^2 = -1, \quad W_4^3 = j$$

Row

$$n, k = 0, 1, 2, 3, \quad [W_4^{nk}] = \begin{bmatrix} 1 & 1 & 1 & 1 \\ 1 & -j & -1 & j \\ 1 & -1 & 1 & -1 \\ 1 & j & -1 & -j \end{bmatrix} \begin{matrix} 0 \\ 1 \\ 2 \\ 3 \end{matrix}$$

(4×4) DFT matrix

Rearrange the rows in bit reversed order

$$[W_4^{nk}]_{\text{BRO}} = \begin{bmatrix} 1 & 1 & 1 & 1 \\ 1 & -1 & 1 & -1 \\ 1 & -j & -1 & j \\ 1 & j & -1 & -j \end{bmatrix} = \begin{bmatrix} \begin{array}{cc|cc} 1 & 1 & & \\ 1 & -1 & & O \\ \hline & & 1 & -j \\ O & & 1 & j \end{array} \end{bmatrix} \begin{bmatrix} [I_2] & [I_2] \\ [I_2] & -[I_2] \end{bmatrix}$$

The flowgraph for four-point DFT is shown in Fig. 3.15.
Example: Radix-4 DIT-FFT for $N = 16$

$$X_k^F$$
$$|$$
$$\{x_0, x_1, x_2, x_3, x_4, x_5, x_6, x_7, x_8, x_9, x_{10}, x_{11}, x_{12}, x_{13}, x_{14}, x_{15}\}$$
$$|$$

$$\boxed{A_k^F} \qquad \boxed{B_k^F} \qquad \boxed{C_k^F} \qquad \boxed{D_k^F}$$
$$\{x_0, x_4, x_8, x_{12}\} \quad \{x_1, x_5, x_9, x_{13}\} \quad \{x_2, x_6, x_{10}, x_{14}\} \quad \{x_3, x_7, x_{11}, x_{15}\}$$

Fig. 3.15 Four-point DIT-
FFT flowgraph

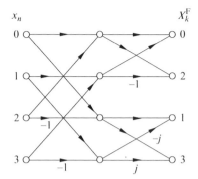

These are four-point DFTs.

$$X_k^F = \sum_{n=0}^{3} x_{4n} W_4^{nk} + W_{16}^k \left(\sum_{n=0}^{3} x_{4n+1} W_4^{nk} \right) + W_{16}^{2k} \left(\sum_{n=0}^{3} x_{4n+2} W_4^{nk} \right)$$

$$+ W_{16}^{3k} \left(\sum_{n=0}^{3} x_{4n+3} W_4^{nk} \right)$$

$$(3.30a)$$

$$X_k^F = A_k^F + W_{16}^k B_k^F + W_{16}^{2k} C_k^F + W_{16}^{3k} D_k^F, \quad k = 0, 1, 2, 3 \qquad (3.30b)$$

In matrix form (3.30) can be expressed as

$$\begin{bmatrix} X_k^F \\ X_{k+\frac{N}{4}}^F \\ X_{k+\frac{N}{2}}^F \\ X_{k+\frac{3N}{4}}^F \end{bmatrix} = \begin{bmatrix} 1 & 1 & 1 & 1 \\ 1 & -j & -1 & j \\ 1 & -1 & 1 & -1 \\ 1 & j & -1 & -j \end{bmatrix} \begin{bmatrix} A_k^F \\ W_{16}^k B_k^F \\ W_{16}^{2k} C_k^F \\ W_{16}^{3k} D_k^F \end{bmatrix}, \quad k = 0, 1, 2, 3 \qquad (3.31)$$

where

$$A_k^F = \sum_{n=0}^{3} x_{4n} W_4^{nk}, \quad k = 0, 1, 2, 3$$

$$\begin{bmatrix} A_0^F \\ A_1^F \\ A_2^F \\ A_3^F \end{bmatrix} = \begin{bmatrix} 1 & 1 & 1 & 1 \\ 1 & -j & -1 & j \\ 1 & -1 & 1 & -1 \\ 1 & j & -1 & -j \end{bmatrix} \begin{bmatrix} x_0 \\ x_4 \\ x_8 \\ x_{12} \end{bmatrix} \qquad (3.32)$$

This is represented in flowgraph format (Fig. 3.16). Similarly B_k^F, C_k^F and D_k^F are four-point DFTs of $\{x_1, x_5, x_9, x_{13}\}$, $\{x_2, x_6, x_{10}, x_{14}\}$ and $\{x_3, x_7, x_{11}, x_{15}\}$ respectively (see Fig. 3.17).

Fig. 3.16 Flowgraph for
four-point DFT of
$\{x_0, x_4, x_8, x_{12}\}$

Input data   Transform data

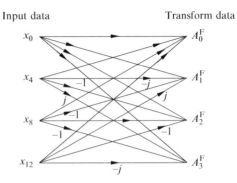

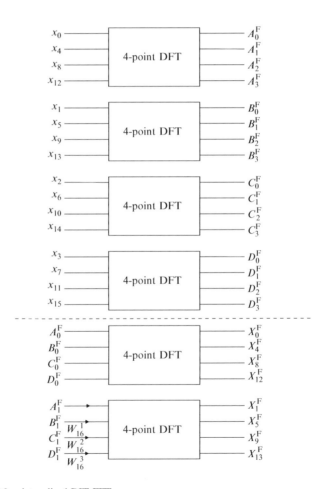

Fig. 3.17 16-point radix-4 DIT-FFT

3.8   Radix-4 DIF-FFT

Similar to radix-4 DIT-FFT, when $N = 4^n$, a radix-4 DIF-FFT can be developed.

$$
\begin{aligned}
X_k^{\mathrm{F}} &= \sum_{n=0}^{N-1} x_n W_N^{nk}, \quad k = 0, 1, \ldots, N-1 \\
&= \sum_{n=0}^{\frac{N}{4}-1} x_n W_N^{nk} + \sum_{n=\frac{N}{4}}^{\frac{N}{2}-1} x_n W_N^{nk} + \sum_{n=\frac{N}{2}}^{\frac{3N}{4}-1} x_n W_N^{nk} + \sum_{n=\frac{3N}{4}}^{N-1} x_n W_N^{nk}
\end{aligned}
\tag{3.33}
$$

$= I + II + III + IV$ represent respectively the four summation terms.
   Let $n = m + \frac{N}{4}$ in II, $n = m + \frac{N}{2}$ in III, $n = m + \frac{3N}{4}$ in IV.
   Then II, III, and IV become

$$
\sum_{n=\frac{N}{4}}^{\frac{N}{2}-1} x_n W_N^{nk} = \sum_{n=0}^{\frac{N}{4}-1} x_{m+\frac{N}{4}} W_N^{\left(m+\frac{N}{4}\right)k} = W_N^{\frac{N}{4}k} \sum_{n=0}^{\frac{N}{4}-1} x_{n+\frac{N}{4}} W_N^{nk} = (-j)^k \sum_{n=0}^{\frac{N}{4}-1} x_{n+\frac{N}{4}} W_N^{nk} \tag{3.34a}
$$

$$
\sum_{n=\frac{N}{2}}^{\frac{3N}{4}-1} x_n W_N^{nk} = \sum_{m=0}^{\frac{N}{4}-1} x_{m+\frac{N}{2}} W_N^{\left(m+\frac{N}{2}\right)k} = W_N^{\frac{N}{2}k} \sum_{n=0}^{\frac{N}{4}-1} x_{n+\frac{N}{2}} W_N^{nk} = (-1)^k \sum_{n=0}^{\frac{N}{4}-1} x_{n+\frac{N}{2}} W_N^{nk}
$$

$$
\tag{3.34b}
$$

Similarly

$$
\sum_{n=\frac{3N}{4}}^{N-1} x_n W_N^{nk} = (j)^k \sum_{n=0}^{\frac{N}{4}-1} x_{n+\frac{3N}{4}} W_N^{nk} \tag{3.34c}
$$

Substitute (3.34) in (3.33) to get

$$
X_k^{\mathrm{F}} = \sum_{n=0}^{\frac{N}{4}-1} \left[x_n + (-j)^k x_{n+\frac{N}{4}} + (-1)^k x_{n+\frac{N}{2}} + (j)^k x_{n+\frac{3N}{4}} \right] W_N^{nk} \tag{3.35}
$$

Let $k = 4m$ in (3.35), ($m = 0, 1, \ldots, \frac{N}{4} - 1$). Then

$$
X_{4m}^{\mathrm{F}} = \sum_{n=0}^{\frac{N}{4}-1} \left[x_n + x_{n+\frac{N}{4}} + x_{n+\frac{N}{2}} + x_{n+\frac{3N}{4}} \right] W_{\frac{N}{4}}^{nm} \tag{3.36a}
$$

This is a $N/4$-point DFT of $\left[x_n + x_{n+\frac{N}{4}} + x_{n+\frac{N}{2}} + x_{n+\frac{3N}{4}} \right]$.
Let $k = 4m + 1$ in (3.35).

$$X^{\mathrm{F}}_{4m+1} = \sum_{n=0}^{\frac{N}{4}-1} x_n W_N^{n(4m+1)} + \sum_{n=\frac{N}{4}}^{\frac{N}{2}-1} x_n W_N^{n(4m+1)} + \sum_{n=\frac{N}{2}}^{\frac{3N}{4}-1} x_n W_N^{n(4m+1)} + \sum_{n=\frac{3N}{4}}^{N-1} x_n W_N^{n(4m+1)}$$

$$= V + VI + VII + VIII$$

(3.36b)

V, VI, VII, and VIII represent respectively the four summation terms.
Let $n = m + \frac{N}{4}$ in VI, $n = m + \frac{N}{2}$ in $\dot{V}II$, $n = m + \frac{3N}{4}$ in VIII.
Then (3.36b) becomes

$$X^{\mathrm{F}}_{4m+1} = \sum_{n=0}^{\frac{N}{4}-1} \left(\left[x_n - jx_{n+\frac{N}{4}} - x_{n+\frac{N}{2}} + jx_{n+\frac{3N}{4}} \right] W_N^n \right) W_{\frac{N}{4}}^{nm}$$

(3.36c)

This is again a $N/4$-point DFT of $\left(\left[x_n - jx_{n+\frac{N}{4}} - x_{n+\frac{N}{2}} + jx_{n+\frac{3N}{4}} \right] W_N^n \right)$.
Let $k = 4m + 2$ in (3.35). As before

$$X^{\mathrm{F}}_{4m+2} = \sum_{n=0}^{\frac{N}{4}-1} \left(\left[x_n - x_{n+\frac{N}{4}} + x_{n+\frac{N}{2}} - x_{n+\frac{3N}{4}} \right] W_N^{2n} \right) W_{\frac{N}{4}}^{nm}$$

(3.36d)

This is a $N/4$-point DFT of $\left(\left[x_n - x_{n+\frac{N}{4}} + x_{n+\frac{N}{2}} - x_{n+\frac{3N}{4}} \right] W_N^{2n} \right)$.
Let $k = 4m + 3$ in (3.35). As before

$$X^{\mathrm{F}}_{4m+3} = \sum_{n=0}^{\frac{N}{4}-1} \left(\left[x_n + jx_{n+\frac{N}{4}} - x_{n+\frac{N}{2}} - jx_{n+\frac{3N}{4}} \right] W_N^{3n} \right) W_{\frac{N}{4}}^{nm}$$

(3.36e)

This is a $N/4$-point DFT of $\left(\left[x_n + jx_{n+\frac{N}{4}} - x_{n+\frac{N}{2}} - jx_{n+\frac{3N}{4}} \right] W_N^{3n} \right)$.
  In conclusion, a radix-4 N-point DIF-FFT is developed based on four $N/4$-point DFTs. Each $N/4$-point DFT is obtained based on a linear combination of the original N-point data sequence. The N-point DFT

$$X^{\mathrm{F}}_k = \sum_{n=0}^{N-1} x_n W_N^{nk}, \quad k = 0, 1, \ldots, N-1$$

is decomposed into

$$X^{\mathrm{F}}_{4m} = \sum_{n=0}^{\frac{N}{4}-1} \left[x_n + x_{n+\frac{N}{4}} + x_{n+\frac{N}{2}} + x_{n+\frac{3N}{4}} \right] W_{\frac{N}{4}}^{nm}$$

(3.37a)

$$X^{\mathrm{F}}_{4m+1} = \sum_{n=0}^{\frac{N}{4}-1} \left(\left[x_n - jx_{n+\frac{N}{4}} - x_{n+\frac{N}{2}} + jx_{n+\frac{3N}{4}} \right] W_N^n \right) W_{\frac{N}{4}}^{nm}$$

(3.37b)

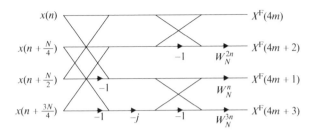

Fig. 3.18 Butterfly to compute (3.37) for $m, n = 0, 1, \ldots, \frac{N}{4} - 1$ where $N = 4$

$$X^F_{4m+2} = \sum_{n=0}^{\frac{N}{4}-1} \left(\left[x_n - x_{n+\frac{N}{4}} + x_{n+\frac{N}{2}} - x_{n+\frac{3N}{4}} \right] W_N^{2n} \right) W_{\frac{N}{4}}^{nm} \qquad (3.37c)$$

$$X^F_{4m+3} = \sum_{n=0}^{\frac{N}{4}-1} \left(\left[x_n + jx_{n+\frac{N}{4}} - x_{n+\frac{N}{2}} - jx_{n+\frac{3N}{4}} \right] W_N^{3n} \right) W_{\frac{N}{4}}^{nm} \qquad (3.37d)$$

$$m = 0, 1, \ldots, \frac{N}{4} - 1$$

All these are $N/4$-point DFTs (Fig. 3.18).
The four $N/4$-point data sequences are formed as follows:

$$\begin{bmatrix} & 1 & 1 & 1 & 1 \\ W_N^n & (1 & -j & -1 & j) \\ W_N^{2n} & (1 & -1 & 1 & -1) \\ W_N^{3n} & (1 & j & -1 & -j) \end{bmatrix} \begin{bmatrix} x_n \\ x_{n+N/4} \\ x_{n+N/2} \\ x_{n+3N/4} \end{bmatrix}, \quad n = 0, 1, \ldots, \frac{N}{4} - 1 \qquad (3.38)$$

The N-point DFT is obtained from four $N/4$-point DFTs. Each of the $N/4$-point data sequences is formed as described above. This process is repeated till (at the end) four-point data sequences are obtained (Figs. 3.19 and 3.20). Radix-4 DIT-DIF-FFT can be obtained by mixing DIT-DIF algorithms at any stage (or even in each stage). Similar to this, radix-6 and radix-8 DIT-FFT, DIF-FFT and DIT-DIF-FFT algorithms can be developed.

A high speed FFT processor for OFDM systems has been developed based on radix-4 DIF FFT. The processor can operate at 42 MHz and implement a 256-point complex FFT in 6 μs [O8].

3.9 Split-Radix FFT Algorithm

We have developed radix-2 and radix-4 DIT and DIF algorithms. Combination of these two algorithms results in what is called a split-radix FFT (SRFFT) algorithm [SR1, A42], which has a number of advantages including the lowest number of

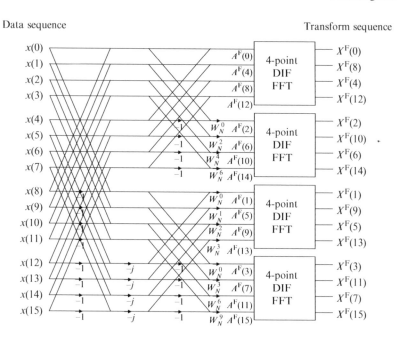

Fig. 3.19 Flowgraph of 16-point DIF FFT is developed from (3.37). See Fig. 3.18 for the flowgraph of four-point DIF FFT (see also [A42])

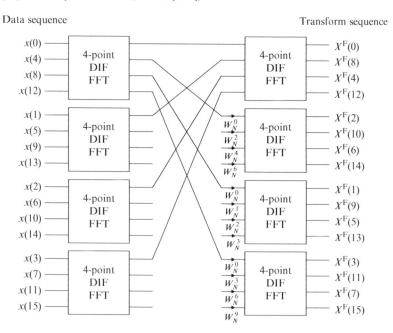

Fig. 3.20 Flowgraph of 16-point DIF FFT. Only one of four butterflies is shown to minimize confusion

multiplies/adds among the 2^m algorithms. The algorithm is based on observation that at each stage a radix-4 is better for the odd DFT coefficients and a radix-2 is better for the even DFT coefficients. The algorithm is described as follows:

The N-point DFT for $N = 2^n$ can be decomposed into

$X^{\mathrm{F}}(2k), k = 0, 1, \ldots, \frac{N}{2} - 1,\ X^{\mathrm{F}}(4k+1)$ and $X^{\mathrm{F}}(4k+3), k = 0, 1, \ldots, \frac{N}{4} - 1$, i.e.,

$$X^{\mathrm{F}}(2k) = \sum_{n=0}^{(N/2)-1} x(n)\, W_N^{2nk} + \sum_{n=N/2}^{N-1} x(n)\, W_N^{2nk}$$

$$(3.39)$$

By changing the variable in the second summation from n to m as $n = m + \frac{N}{2}$, (3.39) can be expressed as

$$X^{\mathrm{F}}(2k) = \sum_{n=0}^{(N/2)-1} \left[x(n) + x\left(n + \frac{N}{2}\right) \right] W_{N/2}^{nk}, \quad k = 0, 1, \ldots, \frac{N}{2} - 1 \qquad (3.40a)$$

This is a $N/2$-point DIF-FFT. Similarly based on the radix-4 DIF-FFT,

$$X^{\mathrm{F}}(4k+1) = \sum_{n=0}^{(N/4)-1} \left(\left[x(n) - jx\left(n + \frac{N}{4}\right) - x\left(n + \frac{N}{2}\right) + jx\left(n + \frac{3N}{4}\right) \right] W_N^n \right) W_{N/4}^{nk}$$

$$(3.40b)$$

and

$$X^{\mathrm{F}}(4k+3) = \sum_{n=0}^{(N/4)-1} \left(\left[x(n) + jx\left(n + \frac{N}{4}\right) - x\left(n + \frac{N}{2}\right) - jx\left(n + \frac{3N}{4}\right) \right] W_N^{3n} \right) W_{N/4}^{nk}$$

$$k = 0, 1, \ldots, \frac{N}{4} - 1$$

$$(3.40c)$$

These two are radix-4 $N/4$-point DIF-FFTs. By successive use of these decompositions at each step, the N-point DFT can be obtained. The computational complexity of this split-radix algorithm is lower than either radix-2 or radix-4 algorithms [SR1].

Yeh and Jen [O9] have developed a SRFFT pipeline architecture for implementing the SRFFT algorithm. It is also suitable for VLSI implementation. The objective is to design high performance (high speed) and low power FFTs for applications in OFDM which is adopted by the European digital radio/audio broadcasting (DVB-T/DAB) standards. Other applications of OFDM include wireless local area networks (WLANs), HYPERLAN/2 systems, fourth-generation cellular phone and new WLAN systems. Details on the pipeline architecture design for implementing the SRFFT are described in [O9] along with the references cited at the end.

A new technique called decomposition partial transmit sequence (D-PTS) which reduces the multiplicative complexity while reducing the peak-to-average power ratio (PAPR) of an OFDM signal is developed in [O19]. To generate the PTS, multiple transforms are computed within the FFT. Details of the low complexity IFFT based PAPR reduction are described in [O19].

3.10 Fast Fourier and BIFORE Transforms by Matrix Partitioning

Matrix factoring techniques have been developed for implementing the fast Fourier and fast BIFORE transforms (FFT, FBT). Matrix partitioning methods can be similarly used for developing FFT and FBT algorithms. Several versions [T1, T2, T3, T4, T5, T6, T8] of the matrix factoring methods for computing the FFT and FBT have been developed. The intent, here, is to develop matrix partitioning techniques, which illustrate the computational savings, in evaluating the Fourier and BIFORE (BInary FOurier REpresentation) coefficients.

3.10.1 Matrix Partitioning

BIFORE or Hadamard transform [T6, T7] (BT or HT) of an N-periodic sequence, $\underline{x}(n) = \{x(0), x(1), \ldots, x(N-1)\}^T$, where $x(n)$ and superscript T denote column vector and transpose respectively, is defined as

$$\underline{B}_x(k) = \frac{1}{N}[H(g)]\,\underline{x}(n) \qquad (3.41)$$

$B_x(k)$ for $k = 0, 1, \ldots, N-1$ are the BIFORE coefficients, $g = \log_2 N$, and $[H(g)]$ is an $(N \times N)$ Hadamard matrix [B6, p. 156] which can be successively generated as follows:

$$[H(0)] = [1],\ [H(1)] = \begin{bmatrix} 1 & 1 \\ 1 & -1 \end{bmatrix}\ [H(2)] = \begin{bmatrix} [H(1)] & [H(1)] \\ [H(1)] & -[H(1)] \end{bmatrix}, \ldots,$$

$$[H(g)] = \begin{bmatrix} [H(g-1)] & [H(g-1)] \\ [H(g-1)] & -[H(g-1)] \end{bmatrix} \qquad (3.42)$$

The matrix partitioning method can be best illustrated with an example. Consider $N = 8$. Using (3.41) and (3.42), $\underline{B}_x(k)$, $[H(g)]$, and $\underline{x}(n)$ are partitioned as follows:

$$
\begin{bmatrix} B_x(0) \\ B_x(1) \\ B_x(2) \\ B_x(3) \\ \hline B_x(4) \\ B_x(5) \\ B_x(6) \\ B_x(7) \end{bmatrix}
= \frac{1}{8}
\left[
\begin{array}{cc|cc}
\begin{bmatrix} 1 & 1 \\ 1 & -1 \end{bmatrix} & \begin{bmatrix} 1 & 1 \\ 1 & -1 \end{bmatrix} & \begin{bmatrix} 1 & 1 \\ 1 & -1 \end{bmatrix} & \begin{bmatrix} 1 & 1 \\ 1 & -1 \end{bmatrix} \\
\begin{bmatrix} 1 & 1 \\ 1 & -1 \end{bmatrix} & -1\begin{bmatrix} 1 & 1 \\ 1 & -1 \end{bmatrix} & \begin{bmatrix} 1 & 1 \\ 1 & -1 \end{bmatrix} & -1\begin{bmatrix} 1 & 1 \\ 1 & -1 \end{bmatrix} \\
\hline
\begin{bmatrix} 1 & 1 \\ 1 & -1 \end{bmatrix} & \begin{bmatrix} 1 & 1 \\ 1 & -1 \end{bmatrix} & \begin{bmatrix} 1 & 1 \\ 1 & -1 \end{bmatrix} & \begin{bmatrix} 1 & 1 \\ 1 & -1 \end{bmatrix} \\
\begin{bmatrix} 1 & 1 \\ 1 & -1 \end{bmatrix} & -1\begin{bmatrix} 1 & 1 \\ 1 & -1 \end{bmatrix} & \begin{bmatrix} 1 & 1 \\ 1 & -1 \end{bmatrix} & -1\begin{bmatrix} 1 & 1 \\ 1 & -1 \end{bmatrix}
\end{array}
\right]
\begin{bmatrix} x(0) \\ x(1) \\ x(2) \\ x(3) \\ x(4) \\ x(5) \\ x(6) \\ x(7) \end{bmatrix}
\tag{3.43}
$$

The structure of (3.43) suggests repeated application of matrix partitioning to obtain the following set of equations:

$$
\begin{bmatrix} B_x(0) \\ B_x(1) \\ B_x(2) \\ B_x(3) \end{bmatrix}
= \frac{1}{8}
\left[
\begin{array}{cc}
\begin{bmatrix} 1 & 1 \\ 1 & -1 \end{bmatrix} & -1\begin{bmatrix} 1 & 1 \\ 1 & -1 \end{bmatrix} \\
\begin{bmatrix} 1 & 1 \\ 1 & -1 \end{bmatrix} & -1\begin{bmatrix} 1 & 1 \\ 1 & -1 \end{bmatrix}
\end{array}
\right]
\begin{bmatrix} x(0)+x(4) \\ x(1)+x(5) \\ x(2)+x(6) \\ x(3)+x(7) \end{bmatrix}
$$

$$
= \frac{1}{8}
\left[
\begin{array}{cc}
\begin{bmatrix} 1 & 1 \\ 1 & -1 \end{bmatrix} & -1\begin{bmatrix} 1 & 1 \\ 1 & -1 \end{bmatrix} \\
\begin{bmatrix} 1 & 1 \\ 1 & -1 \end{bmatrix} & -1\begin{bmatrix} 1 & 1 \\ 1 & -1 \end{bmatrix}
\end{array}
\right]
\begin{bmatrix} x_1(0) \\ x_1(1) \\ x_1(2) \\ x_1(3) \end{bmatrix}
\tag{3.44a}
$$

$$
\begin{bmatrix} B_x(4) \\ B_x(5) \\ B_x(6) \\ B_x(7) \end{bmatrix}
= \frac{1}{8}
\left[
\begin{array}{cc}
\begin{bmatrix} 1 & 1 \\ 1 & -1 \end{bmatrix} & -1\begin{bmatrix} 1 & 1 \\ 1 & -1 \end{bmatrix} \\
\begin{bmatrix} 1 & 1 \\ 1 & -1 \end{bmatrix} & -1\begin{bmatrix} 1 & 1 \\ 1 & -1 \end{bmatrix}
\end{array}
\right]
\begin{bmatrix} x(0)-x(4) \\ x(1)-x(5) \\ x(2)-x(6) \\ x(3)-x(7) \end{bmatrix}
$$

$$
= \frac{1}{8}
\left[
\begin{array}{cc}
\begin{bmatrix} 1 & 1 \\ 1 & -1 \end{bmatrix} & -1\begin{bmatrix} 1 & 1 \\ 1 & -1 \end{bmatrix} \\
\begin{bmatrix} 1 & 1 \\ 1 & -1 \end{bmatrix} & -1\begin{bmatrix} 1 & 1 \\ 1 & -1 \end{bmatrix}
\end{array}
\right]
\begin{bmatrix} x_1(4) \\ x_1(5) \\ x_1(6) \\ x_1(7) \end{bmatrix}
\tag{3.44b}
$$

$$
\begin{bmatrix} B_x(0) \\ B_x(1) \end{bmatrix}
= \frac{1}{8}\begin{bmatrix} 1 & 1 \\ 1 & -1 \end{bmatrix}
\begin{bmatrix} x_1(0)+x_1(2) \\ x_1(1)+x_1(3) \end{bmatrix}
= \frac{1}{8}\begin{bmatrix} 1 & 1 \\ 1 & -1 \end{bmatrix}
\begin{bmatrix} x_2(0) \\ x_2(1) \end{bmatrix}
\tag{3.45a}
$$

$$
\begin{bmatrix} B_x(2) \\ B_x(3) \end{bmatrix}
= \frac{1}{8}\begin{bmatrix} 1 & 1 \\ 1 & -1 \end{bmatrix}
\begin{bmatrix} x_1(0)-x_1(2) \\ x_1(1)-x_1(3) \end{bmatrix}
= \frac{1}{8}\begin{bmatrix} 1 & 1 \\ 1 & -1 \end{bmatrix}
\begin{bmatrix} x_2(2) \\ x_2(3) \end{bmatrix}
\tag{3.45b}
$$

$$
\begin{bmatrix} B_x(4) \\ B_x(5) \end{bmatrix}
= \frac{1}{8}\begin{bmatrix} 1 & 1 \\ 1 & -1 \end{bmatrix}
\begin{bmatrix} x_1(4)+x_1(6) \\ x_1(5)+x_1(7) \end{bmatrix}
= \frac{1}{8}\begin{bmatrix} 1 & 1 \\ 1 & -1 \end{bmatrix}
\begin{bmatrix} x_2(4) \\ x_2(5) \end{bmatrix}
\tag{3.46a}
$$

$$
\begin{bmatrix} B_x(6) \\ B_x(7) \end{bmatrix}
= \frac{1}{8}\begin{bmatrix} 1 & 1 \\ 1 & -1 \end{bmatrix}
\begin{bmatrix} x_1(4)-x_1(6) \\ x_1(5)-x_1(7) \end{bmatrix}
= \frac{1}{8}\begin{bmatrix} 1 & 1 \\ 1 & -1 \end{bmatrix}
\begin{bmatrix} x_2(6) \\ x_2(7) \end{bmatrix}
\tag{3.46b}
$$

$$B_x(0) = \frac{1}{8}\{x_2(0) + x_2(1)\} = \frac{1}{8}x_3(0), \quad B_x(1) = \frac{1}{8}\{x_2(0) - x_2(1)\} = \frac{1}{8}x_3(1)$$

$$B_x(2) = \frac{1}{8}\{x_2(2) + x_2(3)\} = \frac{1}{8}x_3(2), \quad B_x(3) = \frac{1}{8}\{x_2(2) - x_2(3)\} = \frac{1}{8}x_3(3)$$

$$B_x(4) = \frac{1}{8}\{x_2(4) + x_2(5)\} = \frac{1}{8}x_3(4), \quad B_x(5) = \frac{1}{8}\{x_2(4) - x_2(5)\} = \frac{1}{8}x_3(5)$$

$$B_x(6) = \frac{1}{8}\{x_2(6) + x_2(7)\} = \frac{1}{8}x_3(6), \quad B_x(7) = \frac{1}{8}\{x_2(6) - x_2(7)\} = \frac{1}{8}x_3(7)$$

$$(3.47)$$

The above sequence of computations is schematically shown in Fig. 3.21. Apart from the 1/8 constant multiplier, the total number of arithmetic operations (real additions and subtractions) required for computing the BT is $8 \times 3 = 24$, or in general, $N\log_2 N$, compared to N^2 operations as implied by (3.41).

3.10.2 DFT Algorithm

Matrix partitioning can also be extended to DFT. As is well known [T2], the DFT of $\underline{x}(k)$ is

$$X^F(k) = \sum_{n=0}^{N-1} x(n) W_N^{nk}, \quad k = 0, 1, \ldots, N-1 \qquad (2.1a)$$

where $W_N = e^{-j2\pi/N}$ and $j = \sqrt{-1}$. Using the properties, $W^{Nl+r} = W^r$ and $W_N^{(N/2)+r} = -W^r$, DFT for $N = 8$ can be expressed as (here $W = W_8$)

$$
\begin{bmatrix} X^F(0) \\ X^F(1) \\ X^F(2) \\ X^F(3) \\ X^F(4) \\ X^F(5) \\ X^F(6) \\ X^F(7) \end{bmatrix}
=
\begin{matrix} E \\ O \\ E \\ O \\ E \\ O \\ E \\ O \end{matrix}
\begin{bmatrix}
1 & 1 & 1 & 1 & 1 & 1 & 1 & 1 \\
1 & W & W^2 & W^3 & -1 & -W & -W^2 & -W^3 \\
1 & W^2 & -1 & -W^2 & 1 & W^2 & -1 & -W^2 \\
1 & W^3 & -W^2 & -W^5 & -1 & -W^3 & W^2 & -W \\
1 & -1 & 1 & -1 & 1 & -1 & 1 & -1 \\
1 & -W & W^2 & -W^3 & -1 & W & -W^2 & W^3 \\
1 & -W^2 & -1 & W^2 & 1 & -W^2 & -1 & W^2 \\
1 & -W^3 & -W^2 & W^5 & -1 & W^3 & W^2 & W
\end{bmatrix}
\begin{bmatrix} x(0) \\ x(1) \\ x(2) \\ x(3) \\ x(4) \\ x(5) \\ x(6) \\ x(7) \end{bmatrix}
$$

$$(3.48)$$

Here E means even vector and O means odd vector at the midpoint. This, however, holds no advantage regarding computational savings. The trick is to rearrange (3.48), based on the bit reversal of $\{X^F(k)\}$, i.e.,

$$
\begin{bmatrix}
X^{\mathrm{F}}(0)\\
X^{\mathrm{F}}(4)\\
X^{\mathrm{F}}(2)\\
X^{\mathrm{F}}(6)\\
\hline
X^{\mathrm{F}}(1)\\
X^{\mathrm{F}}(5)\\
X^{\mathrm{F}}(3)\\
X^{\mathrm{F}}(7)
\end{bmatrix}
=
\left[
\begin{array}{cc|cc}
\begin{bmatrix}1 & 1\\ 1 & -1\end{bmatrix} & 1\begin{bmatrix}1 & 1\\ 1 & -1\end{bmatrix} & \begin{bmatrix}1 & 1\\ 1 & -1\end{bmatrix} & 1\begin{bmatrix}1 & 1\\ 1 & -1\end{bmatrix}\\[2mm]
\begin{bmatrix}1 & W^2\\ 1 & -W^2\end{bmatrix} & -1\begin{bmatrix}1 & W^2\\ 1 & -W^2\end{bmatrix} & \begin{bmatrix}1 & W^2\\ 1 & -W^2\end{bmatrix} & -1\begin{bmatrix}1 & W^2\\ 1 & -W^2\end{bmatrix}\\[2mm]
\hline
\begin{bmatrix}1 & W\\ 1 & -W\end{bmatrix} & W^2\begin{bmatrix}1 & W\\ 1 & -W\end{bmatrix} & -1\begin{bmatrix}1 & W\\ 1 & -W\end{bmatrix} & -W^2\begin{bmatrix}1 & W\\ 1 & -W\end{bmatrix}\\[2mm]
\begin{bmatrix}1 & W^3\\ 1 & -W^3\end{bmatrix} & -W^2\begin{bmatrix}1 & W^3\\ 1 & -W^3\end{bmatrix} & -1\begin{bmatrix}1 & W^3\\ 1 & -W^3\end{bmatrix} & W^2\begin{bmatrix}1 & W^3\\ 1 & -W^3\end{bmatrix}
\end{array}
\right]
\begin{bmatrix}
x(0)\\
x(1)\\
x(2)\\
x(3)\\
x(4)\\
x(5)\\
x(6)\\
x(7)
\end{bmatrix}
\tag{3.49}
$$

Observe the similarities in the symmetry structure and partitioning of the square matrices in (3.43) and (3.49). The difference is in the "weighting factors", ± 1 and $\pm W_N^2$ in FFT, whereas ± 1 only in FBT. Repetition of the process indicated by (3.44) through (3.47) on (3.49) results in a corresponding flow graph for FFT (Fig. 3.21). The total number of arithmetic operations for DFT computation, which, however, involves complex multiplications and additions, is $N\log_2 N$.

Analogous to factoring of transform matrices, matrix partitioning also can yield computational savings in evaluating BT and DFT. These techniques, both factoring and partitioning, can be extended to other transforms [T4] (Table 3.3).

The sparse matrix factors based on Fig. 3.21 are as follows:

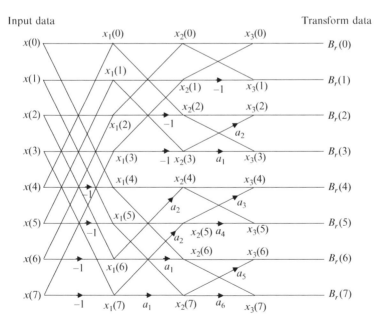

Fig. 3.21 Flow graph for fast implementation of BT, CBT and DFT, $N = 8$. For the DFT the output is in bit reversed order (BRO) (see Table 3.3)

Table 3.3 Values of $a_1, a_2, \ldots, a_6$ shown in Fig. 3.21. Here $W = W_8$

Multiplier	BT $= [G_0(3)]$	CBT $= [G_1(3)]$	DFT
a_1	-1	j	j
a_2	1	$-j$	$-j$
a_3	1	1	W^1
a_4	-1	-1	W^2
a_5	1	1	W^3
a_6	-1	-1	W^4

CBT = complex BIFORE transform

3.10.3 BT (BIFORE Transform)

$$[G_0(3)] = \begin{bmatrix} 1 & 1 & & & & & & \\ 1 & -1 & & & & O & & \\ & & 1 & 1 & & & & \\ & & 1 & -1 & & & & \\ & & & & 1 & 1 & & \\ & O & & & 1 & -1 & & \\ & & & & & & 1 & 1 \\ & & & & & & 1 & -1 \end{bmatrix} \begin{bmatrix} [I_2] & [I_2] & & \\ [I_2] & -[I_2] & & O \\ & & [I_2] & [I_2] \\ O & & [I_2] & -[I_2] \end{bmatrix} \begin{bmatrix} [I_4] & [I_4] \\ [I_4] & -[I_4] \end{bmatrix} \qquad (3.50)$$

3.10.4 CBT (Complex BIFORE Transform)

$$[G_1(3)] = \begin{bmatrix} 1 & 1 & & & & & & \\ 1 & -1 & & & & O & & \\ & & 1 & -j & & & & \\ & & 1 & j & & & & \\ & & & & 1 & 1 & & \\ & O & & & 1 & -1 & & \\ & & & & & & 1 & 1 \\ & & & & & & 1 & -1 \end{bmatrix} \begin{bmatrix} [I_2] & [I_2] & & \\ [I_2] & -[I_2] & & O \\ & & [I_2] & -j[I_2] \\ O & & [I_2] & j[I_2] \end{bmatrix} \begin{bmatrix} [I_4] & [I_4] \\ [I_4] & -[I_4] \end{bmatrix} \qquad (3.51)$$

3.10.5 DFT (Sparse Matrix Factorization)

Note that here $W = W_8$.

(8×8) DFT matrix with rows rearranged in bit reversed order (BRO)

$$= \begin{bmatrix} 1 & 1 & & & & & & \\ 1 & -1 & & & & O & & \\ & & 1 & -j & & & & \\ & & 1 & j & & & & \\ & & & & 1 & W^1 & & \\ & O & & & 1 & W^5 & & \\ & & & & & & 1 & W^3 \\ & & & & & & 1 & W^7 \end{bmatrix} \begin{bmatrix} [I_2] & [I_2] & & \\ [I_2] & -[I_2] & & O \\ & & [I_2] & -j[I_2] \\ O & & [I_2] & j[I_2] \end{bmatrix} \begin{bmatrix} [I_4] & [I_4] \\ [I_4] & -[I_4] \end{bmatrix} \qquad (3.52)$$

3.11   The Winograd Fourier Transform Algorithm [V26]

Relative to the fast Fourier transform (FFT), Winograd Fourier transform algorithm (WFTA) significantly reduces the number of multiplications; it does not increase the number of additions in many cases.

Winograd originally developed a fast algorithm now called WFTA. WFTA can be best explained by some examples.

3.11.1   Five-Point DFT (Fig. 3.22)

$$u = -2\pi/5 \qquad j = \sqrt{-1}$$
$$s_1 = x(1) + x(4) \quad s_2 = x(1) - x(4) \quad s_3 = x(3) + x(2) \quad s_4 = x(3) - x(2)$$
$$s_5 = x(1) + x(3) \quad s_6 = x(1) - x(3) \quad s_7 = s_2 + s_4 \qquad s_8 = s_5 + x(0)$$
$$a_0 = 1 \qquad\qquad a_1 = (\cos u + \cos 2u)/2 - 1 \quad a_2 = (\cos u - \cos 2u)/2$$
$$a_3 = j(\sin u + \sin 2u)/2 \quad a_4 = j\sin 2u \qquad\qquad a_5 = j(\sin u - \sin 2u)$$
$$m_0 = a_0 \cdot s_8 \qquad m_1 = a_1 \cdot s_5 \qquad m_2 = a_2 \cdot s_6 \qquad m_3 = a_3 \cdot s_2$$
$$m_4 = a_4 \cdot s_7 \qquad m_5 = a_5 \cdot s_4$$
$$s_9 = m_0 + m_1 \qquad s_{10} = s_9 + m_2 \qquad s_{11} = s_9 - m_2 \qquad s_{12} = m_3 - m_4$$
$$s_{13} = m_4 + m_5 \quad s_{14} = s_{10} + s_{12} \quad s_{15} = s_{10} - s_{12} \quad s_{16} = s_{11} + s_{13}$$
$$s_{17} = s_{11} - s_{13}$$
$$X^F(0) = m_0 \quad X^F(1) = s_{14} \quad X^F(2) = s_{16} \quad X^F(3) = s_{17} \quad X^F(4) = s_{15}$$

$$(3.53)$$

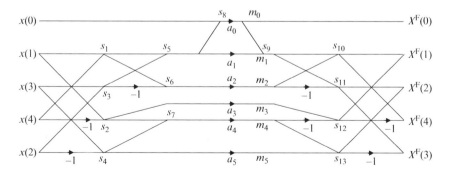

Fig. 3.22 Flow graph of a five-point DFT computed by the WFTA

3.11.2 Seven-Point DFT (Fig. 3.23)

$u = -2\pi/7 \qquad j = \sqrt{-1}$

$s_1 = x(1) + x(6) \quad s_2 = x(1) - x(6) \quad s_3 = x(4) + x(3) \quad s_4 = x(4) - x(3)$

$s_5 = x(2) + x(5) \quad s_6 = x(2) - x(5) \quad s_7 = s_1 + s_3 \qquad\quad s_8 = s_7 + s_5$

$s_9 = s_8 + x(0) \qquad s_{10} = s_1 - s_3 \qquad\quad s_{11} = s_3 - s_5 \qquad\quad s_{12} = s_5 - s_1$

$s_{13} = s_2 + s_4 \qquad s_{14} = s_{13} + s_6 \qquad s_{15} = s_2 - s_4 \qquad\quad s_{16} = s_4 - s_6$

$s_{17} = s_6 - s_2$

$a_0 = 1 \quad a_1 = (\cos u + \cos 2u + \cos 3u)/3 - 1 \quad a_2 = (2\cos u - \cos 2u - \cos 3u)/3$

$a_3 = (\cos u - 2\cos 2u + \cos 3u)/3 \qquad\qquad a_4 = (\cos u + \cos 2u - 2\cos 3u)/3$

$a_5 = j(\sin u + \sin 2u - \sin 3u)/3 \qquad\qquad a_6 = j(2\sin u - \sin 2u + \sin 3u)/3$

$a_7 = j(\sin u - 2\sin 2u - \sin 3u)/3 \qquad\qquad a_8 = j(\sin u + \sin 2u + 2\sin 3u)/3$

$m_0 = a_0 \cdot s_9 \qquad m_1 = a_1 \cdot s_8 \qquad m_2 = a_2 \cdot s_{10} \qquad m_3 = a_3 \cdot s_{11}$

$m_4 = a_4 \cdot s_{12} \qquad m_5 = a_5 \cdot s_{14} \qquad m_6 = a_6 \cdot s_{15} \qquad m_7 = a_7 \cdot s_{16}$

$m_8 = a_8 \cdot s_{17}$

$s_{18} = m_0 + m_1 \qquad s_{19} = s_{18} + m_2 \qquad s_{20} = s_{19} + m_3 \qquad s_{21} = s_{18} - m_2$

$s_{22} = s_{21} - m_4 \qquad s_{23} = s_{18} - m_3 \qquad s_{24} = s_{23} + m_4 \qquad s_{25} = m_5 + m_6$

$s_{26} = s_{25} + m_7 \qquad s_{27} = m_5 - m_6 \qquad s_{28} = s_{27} - m_8 \qquad s_{29} = m_5 - m_7$

$s_{30} = s_{29} + m_8 \qquad s_{31} = s_{20} + s_{26} \qquad s_{32} = s_{20} - s_{26} \qquad s_{33} = s_{22} + s_{28}$

$s_{34} = s_{22} - s_{28} \qquad s_{35} = s_{24} + s_{30} \qquad s_{36} = s_{24} - s_{30}$

$X^F(0) = m_0 \qquad X^F(1) = s_{31} \qquad X^F(2) = s_{33} \qquad X^F(3) = s_{36}$

$X^F(4) = s_{35} \qquad X^F(5) = s_{34} \qquad X^F(6) = s_{32}$

$$(3.54)$$

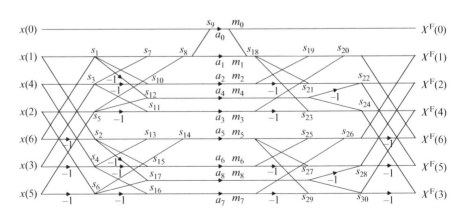

Fig. 3.23 Flow graph of a seven-point DFT computed by the WFTA

3.11.3 Nine-Point DFT (Fig. 3.24)

$$u = -2\pi/9 \qquad j = \sqrt{-1}$$
$$s_1 = x(1) + x(8) \quad s_2 = x(1) - x(8) \quad s_3 = x(7) + x(2) \quad s_4 = x(7) - x(2)$$
$$s_5 = x(3) + x(6) \quad s_6 = x(3) - x(6) \quad s_7 = x(4) + x(5) \quad s_8 = x(4) - x(5)$$
$$s_9 = s_1 + s_3 \qquad s_{10} = s_9 + s_7 \qquad s_{11} = s_{10} + s_5 \qquad s_{12} = s_{11} + x(0)$$
$$s_{13} = s_2 + s_4 \qquad s_{14} = s_{13} + s_8 \qquad s_{15} = s_1 - s_3 \qquad s_{16} = s_3 - s_7$$
$$s_{17} = s_7 - s_1 \qquad s_{18} = s_2 - s_4 \qquad s_{19} = s_4 - s_8 \qquad s_{20} = s_8 - s_2$$

$$a_0 = 1 \quad a_1 = -\tfrac{1}{2} \quad a_2 = j\sin 3u \quad a_3 = \cos 3u - 1 \quad a_4 = j\sin 3u$$
$$a_5 = (2\cos u - \cos 2u - \cos 4u)/3 \quad a_6 = (\cos u + \cos 2u - 2\cos 4u)/3$$
$$a_7 = (\cos u - 2\cos 2u + \cos 4u)/3 \quad a_8 = j(2\sin u + \sin 2u - \sin 4u)/3$$
$$a_9 = j(\sin u - \sin 2u - 2\sin 4u)/3 \quad a_{10} = j(\sin u + 2\sin 2u + \sin 4u)/3$$

$$m_0 = a_0 \cdot s_{12} \qquad m_1 = a_1 \cdot s_{10} \qquad m_2 = a_2 \cdot s_{14} \qquad m_3 = a_3 \cdot s_5$$
$$m_4 = a_4 \cdot s_6 \qquad m_5 = a_5 \cdot s_{15} \qquad m_6 = a_6 \cdot s_{16} \qquad m_7 = a_7 \cdot s_{17}$$
$$m_8 = a_8 \cdot s_{18} \qquad m_9 = a_9 \cdot s_{19} \qquad m_{10} = a_{10} \cdot s_{20}$$

$$s_{21} = m_1 + m_1 \qquad s_{22} = s_{21} + m_1 \quad s_{23} = s_{22} + m_0 \qquad s_{24} = s_{23} + m_2$$
$$s_{25} = s_{23} - m_2 \qquad s_{26} = m_0 + m_3 \quad s_{27} = s_{26} + s_{21} \qquad s_{28} = s_{27} + m_5$$
$$s_{29} = s_{28} + m_6 \qquad s_{30} = s_{27} - m_6 \quad s_{31} = s_{30} + m_7 \qquad s_{32} = s_{27} - m_5$$
$$s_{33} = s_{32} - m_7 \qquad s_{34} = m_4 + m_8 \quad s_{35} = s_{34} + m_9 \qquad s_{36} = m_4 - m_9$$
$$s_{37} = s_{36} + m_{10} \quad s_{38} = m_4 - m_8 \quad s_{39} = s_{38} - m_{10} \qquad s_{40} = s_{29} + s_{35}$$
$$s_{41} = s_{29} - s_{35} \qquad s_{42} = s_{31} + s_{37} \quad s_{43} = s_{31} - s_{37} \qquad s_{44} = s_{33} + s_{39}$$
$$s_{45} = s_{33} - s_{39}$$

$$X^F(0) = m_0 \quad X^F(1) = s_{40} \quad X^F(2) = s_{43} \quad X^F(3) = s_{24} \quad X^F(4) = s_{44}$$
$$X^F(5) = s_{45} \quad X^F(6) = s_{25} \quad X^F(7) = s_{42} \quad X^F(8) = s_{41}$$

$$(3.55)$$

Winograd DFT is computationally more efficient than the radix-3 DIF FFT
(Table 3.4).

3.11.4 DFT Algorithms for Real-Valued Input Data

There has been significant literature discussing efficient algorithms for real valued
FFTs [B29, A38]. Basically, the algorithms fall into two categories: *prime factor
algorithms* (PFA) and *common factor algorithms* (CFA). PFAs are used when the
transform length N can be decomposed into two or more relatively prime factors.
An example of PFA is the Winograd Fourier transform algorithm (WFTA)

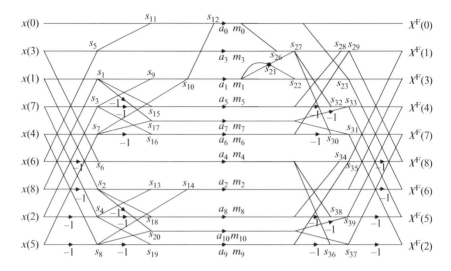

Fig. 3.24 Flow graph of a nine-point DFT computed by the WFTA

Table 3.4 Performance of WFTAs

| DFT size N | WFTA | | Radix-3 DIF FFT | |
	Number of real multiplies	Number of real adds	Number of real multiplies	Number of real adds
5	5	17	–	–
7	8	36	–	–
9	10	45	64	74

(Section 3.11). See [T1] for an application. CFAs are used when the factors are not relatively prime. Examples of CFA are the traditional Cooley-Tukey FFT algorithms and the split radix algorithms. Both PFAs and CFAs are based on permuting the input data into a two dimensional matrix so that the corresponding two-dimensional DFT is a separable transform with minimum number of twiddle factors. Especially for PFAs, the row and column DFTs are independent and are not related by any twiddle factors.

In our case, the length 15 can be decomposed into five and three, which are mutually prime. Hence, PFAs are applicable in our case. Two PFA algorithms are available: the PFA algorithms developed by Burrus et al. [B29] and the WFTA. The WFTA algorithm uses Winograd short-N DFT modules for prime lengths N as building blocks. WFTA algorithms typically minimize the number of multiplications (by using a *nesting* procedure) at the expense of a slight increase in additions [B1, B29]. However, for short lengths (including 15), the WFTA algorithm uses less number of multiplications and the same number of additions as PFA. Hence, we will be using the WFTA algorithm for our 15-point FFT. Our survey of the existing literature shows that the 15-point real WFTA is the least complex among the available algorithms [B1, B29].

3.11.5   Winograd Short-N DFT Modules

Winograd short-N DFT modules are the building blocks for constructing the WFTA for longer lengths. The short-N modules are defined for prime lengths. Specifically, we need three-point and five-point DFT modules for the 15-point transform.

The Winograd DFT modules are based on a fast cyclic convolution algorithm for prime lengths using the theoretically minimum number of multiplications [B1, B29, A36]. This optimum convolution algorithm can be mapped to DFT using Rader's technique [B29] to give very efficient DFT modules for prime lengths.

In mathematical terms, Winograd's algorithm achieves a canonical decomposition of the DFT matrix as shown below. Using the matrix notation in [A35]

$$[D_N] = [S_N][C_N][T_N] \tag{3.56}$$

where

$[D_N]$ $(= [F])$ is the $N \times N$ DFT matrix,
$[S_N]$ is an $N \times J$ matrix having 0, 1, -1 only as elements,
$[C_N]$ is a $J \times J$ *diagonal* matrix,
$[T_N]$ is a $J \times N$ matrix having 0, 1, -1 only as elements.

That is, $[S_N]$ and $[T_N]$ are just addition matrices and $[C_N]$ is the multiplier matrix. Moreover, the elements of $[C_N]$ matrix are either purely real or purely imaginary, so for real input data, we will have just one real multiplication for each element of $[C_N]$. Hence, the number of multiplications will be J. Winograd algorithm is powerful because, for small and prime N (such as 3, 5, 7, 11), J is very close to N. That is, we need only about N multiplications instead of N^2 multiplications required in brute force approach. For example, for $N = 3$, $[S_3]$, $[C_3]$ and $[T_3]$ can be derived from [A37] as follows.

$$[S_3] = \begin{bmatrix} 1 & 0 & 0 \\ 1 & 1 & 1 \\ 1 & 1 & -1 \end{bmatrix} \quad [C_3] = \text{diag}\left[1, \cos\left(\tfrac{2\pi}{3}\right), -1, -j\sin\left(\tfrac{2\pi}{3}\right)\right] \quad [T_3] = \begin{bmatrix} 1 & 1 & 1 \\ 0 & 1 & 1 \\ 0 & 1 & -1 \end{bmatrix}$$

$$\tag{3.57}$$

Note that the $[S_N]$ and $[T_N]$ matrices can be factorized into sparse matrices to minimize the number of additions. For example $[S_3]$ and $[T_3]$ in (3.57) can be factorized as

$$[S_3] = \begin{bmatrix} 1 & 0 & 0 \\ 0 & 1 & 1 \\ 0 & 1 & -1 \end{bmatrix} \begin{bmatrix} 1 & 0 & 0 \\ 1 & 1 & 0 \\ 0 & 0 & 1 \end{bmatrix} \quad [T_3] = \begin{bmatrix} 1 & 1 & 0 \\ 0 & 1 & 0 \\ 0 & 0 & 1 \end{bmatrix} \begin{bmatrix} 1 & 0 & 0 \\ 0 & 1 & 1 \\ 0 & 1 & -1 \end{bmatrix} \tag{3.58}$$

For $N = 4$, $[S_4], [C_4]$ and $[T_4]$ can be derived from [A37] as follows.

$$[S_4] = \begin{bmatrix} 1 & 0 & 0 & 0 \\ 0 & 0 & 1 & 1 \\ 0 & 1 & 0 & 0 \\ 0 & 0 & 1 & -1 \end{bmatrix} \qquad [C_4] = \mathrm{diag}\left[1, 1, 1, \ j\sin\left(-\frac{\pi}{2}\right)\right]$$

$$[T_4] = \begin{bmatrix} 1 & 0 & 1 & 0 \\ 1 & 0 & -1 & 0 \\ 0 & 1 & 0 & 0 \\ 0 & 0 & 0 & 1 \end{bmatrix} \begin{bmatrix} 1 & 0 & 1 & 0 \\ 1 & 0 & -1 & 0 \\ 0 & 1 & 0 & 1 \\ 0 & 1 & 0 & -1 \end{bmatrix} = \begin{bmatrix} 1 & 1 & 1 & 1 \\ 1 & -1 & 1 & -1 \\ 1 & 0 & -1 & 0 \\ 0 & 1 & 0 & -1 \end{bmatrix}$$

(3.59)

For $N = 5$, $[S_5], [C_5]$ and $[T_5]$ can be derived from (3.53) and Fig. 3.22 as follows.

$$[S_5] = \begin{bmatrix} 1 & 0 & 0 & 0 & 0 \\ 0 & 1 & 0 & 0 & 0 \\ 0 & 0 & 1 & 0 & 1 \\ 0 & 0 & 1 & 0 & -1 \\ 0 & 0 & 0 & 1 & 0 \end{bmatrix} \begin{bmatrix} 1 & 0 & 0 & 0 & 0 \\ 0 & 1 & 0 & 1 & 0 \\ 0 & 0 & 1 & 0 & 0 \\ 0 & 1 & 0 & -1 & 0 \\ 0 & 0 & 0 & 0 & 1 \end{bmatrix} \begin{bmatrix} 1 & 0 & 0 & 0 & 0 \\ 0 & 1 & 1 & 0 & 0 \\ 0 & 1 & -1 & 0 & 0 \\ 0 & 0 & 0 & 1 & 0 \\ 0 & 0 & 0 & 0 & 1 \end{bmatrix} \begin{bmatrix} 1 & 0 & 0 & 0 & 0 \\ 1 & 1 & 0 & 0 & 0 \\ 0 & 0 & 1 & 0 & 0 \\ 0 & 0 & 1 & -1 & 0 \\ 0 & 0 & 0 & 1 & 1 \end{bmatrix}$$

$$[C_5] = \mathrm{diag}\left[1, \ \frac{\cos(u) + \cos(2u)}{2} - 1, \ \frac{\cos(u) - \cos(2u)}{2}, \right.$$

$$\left. j(\sin u + \sin 2u), \ j\sin(2u), \ j(\sin u - \sin 2u)\right] \qquad u = \frac{-2\pi}{5}$$

$$[T_5] = \begin{bmatrix} 1 & 1 & 0 & 0 & 0 & 0 \\ 0 & 1 & 0 & 0 & 0 & 0 \\ 0 & 0 & 1 & 0 & 0 & 0 \\ 0 & 0 & 0 & 1 & 0 & 0 \\ 0 & 0 & 0 & 0 & 1 & 0 \\ 0 & 0 & 0 & 0 & 0 & 1 \end{bmatrix} \begin{bmatrix} 1 & 0 & 0 & 0 & 0 \\ 0 & 1 & 1 & 0 & 0 \\ 0 & 1 & -1 & 0 & 0 \\ 0 & 0 & 0 & 0 & 1 \\ 0 & 0 & 0 & 1 & 1 \\ 0 & 0 & 0 & 1 & 0 \end{bmatrix} \begin{bmatrix} 1 & 0 & 0 & 0 & 0 \\ 0 & 1 & 0 & 0 & 1 \\ 0 & 0 & 1 & 1 & 0 \\ 0 & 0 & -1 & 1 & 0 \\ 0 & 1 & 0 & 0 & -1 \end{bmatrix}$$

(3.60)

3.11.6 Prime Factor Map Indexing

The idea behind mapping a one-dimensional array into a two-dimensional array is to divide the bigger problem into several smaller problems and solve each of the smaller problems effectively [B29]. The mappings are based on modulo integer arithmetic to exploit the periodicity property of the complex exponentials in DFT. Let $N = N_1 N_2$. Let N_1 and N_2 be co-prime. Let n, n_1, n_2, k, k_1, k_2 be the corresponding index variables in time and frequency domains. The mappings from n_1, n_2, k_1, k_2 to n and k can be defined as follows.

$$n = \langle K_1 n_1 + K_2 n_2 \rangle_N \quad n_1 = 0, 1, \ldots, N_1 - 1 \quad n_2 = 0, 1, \ldots, N_2 - 1$$
$$k = \langle K_3 k_1 + K_4 k_2 \rangle_N \quad k_1 = 0, 1, \ldots, N_1 - 1 \quad k_2 = 0, 1, \ldots, N_2 - 1 \quad (3.61)$$
$$n, k = 0, 1, \ldots, N - 1$$

where $\langle \cdot \rangle_N$ represents modulo-N operation.

There always exist integers K_1, K_2, K_3 and K_4 such that the above mapping is unique (i.e., all values of n and k between 0 and $N - 1$ can be generated by using all combinations of n_1, n_2, and k_1, k_2). The mapping is called a *prime factor map* (PFM) if

$$K_1 = aN_2 \quad \text{and} \quad K_2 = bN_1$$
$$K_3 = cN_2 \quad \text{and} \quad K_4 = bN_1 \quad (3.62)$$

A solution that satisfies the PFM and also decouples the row and column DFTs is

$$K_1 = N_2 \quad \text{and} \quad K_2 = N_1$$
$$K_3 = N_2 \langle N_2^{-1} \rangle_{N_1} \quad \text{and} \quad K_4 = N_1 \langle N_1^{-1} \rangle_{N_2} \quad (3.63)$$

where $\langle N_1^{-1} \rangle_{N_2}$ is defined as the smallest natural number that satisfies $\langle N_1^{-1} N_1 \rangle_{N_2} = 1$. For our case $N_1 = 3$ and $N_2 = 5$.

$$\langle N_1^{-1} \rangle_{N_2} = 2 \qquad \text{since } N_1 \times 2 = 1 \times N_2 + 1$$
$$\langle N_2^{-1} \rangle_{N_1} = 2 \qquad \text{since } N_2 \times 2 = 3 \times N_1 + 1$$

Thus $K_3 = N_2 \langle N_2^{-1} \rangle_{N_1} = 10$ and $K_4 = N_1 \langle N_1^{-1} \rangle_{N_2} = 6$.

For another example, let $N_1 = 3$ and $N_2 = 4$.

$$\langle N_1^{-1} \rangle_{N_2} = 3 \qquad \text{since } N_1 \times 3 = 2 \times N_2 + 1$$
$$\langle N_2^{-1} \rangle_{N_1} = 1 \qquad \text{since } N_2 \times 1 = 1 \times N_1 + 1$$

Thus $K_3 = N_2 \langle N_2^{-1} \rangle_{N_1} = 4$ and $K_4 = N_1 \langle N_1^{-1} \rangle_{N_2} = 9$. The DFT then becomes

$$\hat{X}^F(k_1, k_2) = \sum_{n_1=0}^{N_1-1} \sum_{n_2=0}^{N_2-1} \hat{x}(n_1, n_2) W_{N_1}^{n_1 k_1} W_{N_2}^{n_2 k_2} \quad (3.64)$$

where

$$\hat{x}(n_1, n_2) = x(\langle K_1 n_1 + K_2 n_2 \rangle_N)$$
$$\hat{X}^F(k_1, k_2) = X^F(\langle K_3 k_1 + K_4 k_2 \rangle_N) \quad (3.65)$$

That is, the DFT is first applied along the columns and then along the rows.

An alternative indexing method. In case $N = N_1 \times N_2$ where N_1 and N_2 are relatively prime, then a group z_N is isomorphic to a group $z_{N_1} \times z_{N_2}$ (z in z_n represents integers or *Zahlen* in German). The DFT matrix can be partitioned into blocks of $N_2 \times N_2$ cyclic matrices, and such that the blocks form an $N_1 \times N_1$ cyclic matrix [A36].

For example, since $15 = 3 \times 5$ we have the isomorphism

$0 \to 0 \times (1, 1) = (0, 0)$	$1 \to 1 \times (1, 1) = (1, 1)$	$2 \to 2 \times (1, 1) = (2, 2)$
$3 \to 3 \times (1, 1) = (0, 3)$	$4 \to 4 \times (1, 1) = (1, 4)$	$5 \to 5 \times (1, 1) = (2, 0)$
$6 \to 6 \times (1, 1) = (0, 1)$	$7 \to 7 \times (1, 1) = (1, 2)$	$8 \to 8 \times (1, 1) = (2, 3)$
$9 \to 9 \times (1, 1) = (0, 4)$	$10 \to 10 \times (1, 1) = (1, 0)$	$11 \to 11 \times (1, 1) = (2, 1)$
$12 \to 12 \times (1, 1) = (0, 2)$	$13 \to 13 \times (1, 1) = (1, 3)$	$14 \to 14 \times (1, 1) = (2, 4)$

Put this in lexicographical order to get the rearrangement: 0, 6, 12, 3, 9, 10, 1, 7, 13, 4, 5, 11, 2, 8, 14. Here (1,1) is a generator of group $z_{N_1} \times z_{N_2}$.

$$4 \to 4 \times (1, 1) = (4 \times 1, 4 \times 1) = (4 \bmod 3, 4 \bmod 5) = (1, 4)$$
$$6 \to 6 \times (1, 1) = (6 \times 1, 6 \times 1) = (6 \bmod 3, 6 \bmod 5) = (0, 1)$$

Matrix Interpretation. If we fix a value for n_1 and vary the values for n_2 from 0 to $N_2 - 1$ in the above mapping and do this for all values of n_1 from 0 to $N_1 - 1$, we obtain a permuted sequence of values of n from 0 to $N - 1$. Let $[P_i]$ be the input permutation matrix describing these steps. Similarly, we define $[P_o]$ to be the output permutation matrix. Then it can be shown that [A35]

$$[P_o]\,\underline{X}^{\mathrm{F}} = ([D_{N_1}] \otimes [D_{N_2}])[P_i]\,\underline{x} \qquad \text{(see Fig. 3.25)} \tag{3.66}$$

where $\otimes$ denotes Kronecker product (see Appendix E). That is, a two-dimensional $N_1 \times N_2$ DFT can be viewed as a Kronecker product of the row and column DFT matrices $[D_{N_1}]$ and $[D_{N_2}]$. This matrix representation and the corresponding permutation matrices for $N = 15$ are illustrated in the MATLAB code of Appendix H.1 (Fig. 3.25).

3.11.7 Winograd Fourier Transform Algorithm (WFTA)

WFTA takes forward the matrix interpretation of the PFA given above. If we have short-N DFT modules for lengths N_1 and N_2, then

$$[D_{N_1}] \otimes [D_{N_2}] = ([S_{N_1}][C_{N_1}][T_{N_1}]) \otimes ([S_{N_2}][C_{N_2}][T_{N_2}]) \tag{3.67}$$

Using the properties of Kronecker products, it can be shown that [A35, A36]

$$\begin{aligned} [D_{N_1}] \otimes [D_{N_2}] &= ([S_{N_1}] \otimes [S_{N_2}])([C_{N_1}] \otimes [C_{N_2}])([T_{N_1}] \otimes [T_{N_2}]) \\ &= [S_N][C_N][T_N] \end{aligned} \tag{3.68}$$

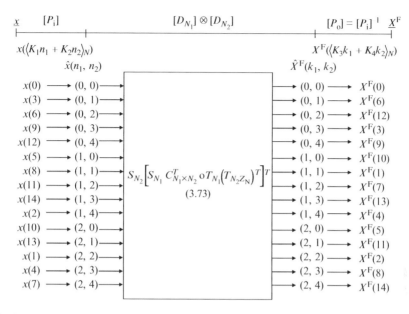

Fig. 3.25 Permutations of input and output sequences in the WFTA for $N = 3 \times 5$. The symbol $\circ$ denotes element-by-element product of two matrices of the same size

This last step in (3.68) is called the *nesting* procedure in WFTA [A35]. Because the multiplications are nested, this algorithm generally yields the lowest multiplication count. The above equations can be further simplified as follows. Suppose

$$\hat{\underline{x}} = [P_i]\underline{x} = \left(\hat{x}_0, \hat{x}_1, \ldots, \hat{x}_{N_2-1}, \Big| \hat{x}_{N_2}, \hat{x}_{N_2+1}, \ldots, \hat{x}_{2N_2-1}, \Big| \hat{x}_{2N_2}, \hat{x}_{2N_2+1}, \ldots, \hat{x}_{N_1 N_2-1}\right)^T \quad (3.69)$$

Define

$$[z_N] \equiv \begin{bmatrix} \hat{x}_0 & \hat{x}_1 & \cdots & \hat{x}_{N_2-1} \\ \hat{x}_{N_2} & \hat{x}_{N_2+1} & \cdots & \hat{x}_{2N_2-1} \\ \vdots & \vdots & \ddots & \vdots \\ \hat{x}_{(N_1-1)N_2} & \hat{x}_{(N_1-1)N_2+1} & \cdots & \hat{x}_{N_1 N_2-1} \end{bmatrix} \quad (3.70)$$

Similarly, define $[Z_N]$ for the output coefficients. $[P_i]$ and $[P_o]$ are defined in (3.66). The matrices $[z_N]$ and $[Z_N]$ map into vectors $\hat{\underline{x}}$ and $\widehat{\underline{X}}^F$ by row ordering (see Appendix E.3). Then it can be shown that [A35, A37]

$$\left([T_{N_1}] \otimes [T_{N_2}]\right)\underline{x} \Rightarrow \left([T_{N_2}]([T_{N_1}][z_N])^T\right)^T \quad (3.71)$$

$$\left([T_{N_1}] \otimes [T_{N_2}]\right)\underline{x} \Rightarrow [T_{N_1}][z_N][T_{N_2}]^T. \quad \text{(The right-hand side is same as (E.11))} \quad (3.72)$$

$$[Z_N] = [S_{N_2}]\left([S_{N_1}][C_{N_1 \times N_2}]^T \circ [T_{N_1}]([T_{N_2}][Z_N])^T\right)^T \qquad (3.73)$$

One transposition operation has canceled out another in (3.73). $[C_{N_1 \times N_2}]$ is defined as

$$
\begin{aligned}
([C_{N_1}] \otimes [C_{N_2}])\underline{x} \Rightarrow \left([C_{N_2}]([C_{N_1}][z_N])^T\right)^T &= [C_{N_1}][z_N][C_{N_2}]^T \\
&= [C_{N_1}][z_N][C_{N_2}] = [C_{N_1 \times N_2}] \circ [z_N] \qquad \text{see (F.12)}
\end{aligned}
\qquad (3.74)
$$

where $[C_{N_1}]$ and $[C_{N_2}]$ are diagonal matrices, and

$$
\begin{aligned}
C_{N_1 \times N_2}(n,m) = C_{N_1}(n,n)C_{N_2}(m,m) \quad & n = 0,1,\ldots,M_{N_1}-1 \quad (M_{N_1}=3) \\
& m = 0,1,\ldots,M_{N_2}-1 \quad (M_{N_2}=6)
\end{aligned}
\qquad (3.75)
$$

$$([C_{N_1}][z_N][C_{N_2}])^T = [C_{N_2}][z_N]^T[C_{N_1}] = [C_{N_1 \times N_2}] \circ [z_N]^T = [C_{N_1 \times N_2}]^T \circ [z_N]^T \quad (3.76)$$

M_{N_1} and M_{N_2} are the lengths of the diagonals (i.e., the number of multiplications) of $[C_{N_1}]$ and $[C_{N_2}]$ respectively. The symbol $\circ$ denotes element-by-element product of two matrices of the same size. The superscript T denotes matrix transpose.

Applications. To develop five-point discrete cosine transform (DCT), we can simply use Winograd's five-point DFT module combined with Heideman's mapping [A40, A41].

3.12   Sparse Factorization of the DFT Matrix

3.12.1   *Sparse Factorization of the DFT Matrix Using Complex Rotations*

3.12.1.1   Preliminary

Property. Given an $N \times N$ DFT matrix $[F]$, we can find the sparse factorization of the matrix using only row permutations and complex rotations.

3.12.1.1   Preliminary

Since rotations are being considered, and a matrix can be reduced to upper triangular form (triangularization) by applying a sequence of complex rotations to the matrix, each involving four elements of the matrix, we need to consider rotations of 2×2 matrices only. (In an upper triangular matrix all the elements below the diagonal are zero.) In addition, the elements of the DFT matrix are complex so that the Givens rotations [B27] must also be complex. We, therefore, first prove that a sequence of *complex* Givens rotations will result in an orthogonal matrix (not a unitary matrix!).

Proof.   Consider the 2×2 matrix given by:

$$[R] = \begin{bmatrix} \cos \gamma & \sin \gamma \\ -\sin \gamma & \cos \gamma \end{bmatrix} \tag{3.77}$$

where γ is complex.

For $[R]$ to be orthogonal, $[R]^T[R] = [I]$. Note that $[R]$ is not required to be unitary, only orthogonal. Thus we have:

$$\begin{aligned} [R]^T[R] &= \begin{bmatrix} \cos \gamma & -\sin \gamma \\ \sin \gamma & \cos \gamma \end{bmatrix} \begin{bmatrix} \cos \gamma & \sin \gamma \\ -\sin \gamma & \cos \gamma \end{bmatrix} \\ &= \begin{bmatrix} \cos^2 \gamma + \sin^2 \gamma & 0 \\ 0 & \cos^2 \gamma + \sin^2 \gamma \end{bmatrix} = \begin{bmatrix} 1 & 0 \\ 0 & 1 \end{bmatrix} \end{aligned} \tag{3.78}$$

For (3.78) to be true, we need the identity

$$\cos^2 \gamma + \sin^2 \gamma = 1 \quad \text{for complex } \gamma$$

This is easily shown as follows. Let $\gamma = \alpha + j\beta$ where α and β are real. Using the compound angle formulae for the sine and cosine functions, we can easily show:

$$\cos^2 \gamma + \sin^2 \gamma = (\cos^2 \alpha + \sin^2 \alpha)\cosh^2 \beta - (\cos^2 \alpha + \sin^2 \alpha)\sinh^2 \beta = 1 \quad (3.79)$$

where $\cosh \beta$ and $\sinh \beta$ are defined as

$$\cos(j\theta) = \cosh \theta \qquad \sin(j\theta) = j \sinh \theta \tag{3.80}$$

$$\cosh \theta = (e^0 + e^{-0})/2 \qquad \sinh \theta = (e^0 - e^{-0})/2 \tag{3.81}$$

$$\cosh^2 \theta - \sinh^2 \theta = 1 \tag{3.82}$$

Thus the complex Givens rotation $[R]$ is orthogonal.

3.12.1.2   Analysis

Let $[R] = \begin{bmatrix} \cos \gamma & \sin \gamma \\ -\sin \gamma & \cos \gamma \end{bmatrix}$, and let $[A] = \begin{bmatrix} a_{11} & a_{12} \\ a_{21} & a_{22} \end{bmatrix}$, where the elements of the matrices are complex. We require a rotation matrix to null such that $([R][A])_{21} = 0$. This leads to the following equation:

$$a_{11} \sin \gamma - a_{21} \cos \gamma = 0 \tag{3.83}$$

giving $\tan \gamma = a_{21}/a_{11}$ provided $a_{11} \neq 0$.

1. If $a_{11} = 0$, $\gamma = \pi/2$.
2. If $a_{11} \neq 0$, we have to solve for the complex angle γ in the equation:

$$\tan \gamma = a_{21}/a_{11} = z_3$$

Let $u = e^{j\gamma}$, so that we can express the sine and cosine functions as:

$$\sin \gamma = \frac{1}{2j}(u - u^{-1}) \quad \text{and} \quad \cos \gamma = \frac{1}{2}(u + u^{-1})$$

These give the equation:

$$\tan \gamma = \frac{1}{j} \frac{u - u^{-1}}{u + u^{-1}} \tag{3.84}$$

Equation (3.84) can be solved using complex analysis. The solution for u is

$$u^2 = \frac{1 + jz_2}{1 - jz_3} \quad \text{if } z_3 \neq -j$$

1. If $z_3 \neq -j$, let $u^2 = z_4 = re^{j\phi}$. It is then seen that there are two solutions:

$$u_1 = \sqrt{r}e^{j\phi/2} \quad \text{and} \quad u_2 = \sqrt{r}e^{j(\phi+2\pi)/2} = -u_1$$

The solution for u is unique if we constrain the argument of the complex number to be inside $(0, \pi)$. Let this u be $\rho e^{j\theta}$ and recall that $u = e^{j\gamma}$. Thus we have

$$\gamma = \theta - j \ln \rho$$

2. If $z_3 = -j$, we simply perform a row permutation on the 2×2 matrix, thus interchanging the positions of a_{11} and a_{21}. The resulting z_3 will be equal to j.

We have thus shown that there exist *complex* rotations that can reduce (triangularize) an $N \times N$ DFT matrix to upper triangular form, when row permutations are allowed.

3.12.2   Sparse Factorization of the DFT Matrix Using Unitary Matrices

In the previous summary titled "sparse factorization of the DFT matrix using complex rotations," we have examined that the DFT matrix can be reduced to upper triangular form (triangularization) using complex Givens rotations and row

permutations. It was indeed shown that an orthogonal matrix exists that will reduce
the DFT matrix to upper triangular form. The previous summary examines the case
of 2×2 matrices, which are sufficient. The equation is

$$[F] = [Q][R] \tag{3.85}$$

where $[Q]$ is the transpose of a sequence of complex rotations given by

$$[Q] = ([Q_n][Q_{n-1}] \cdots [Q_1])^T$$

in which $[Q_m][Q_m]^T = [I]$ showing the orthogonality of each of the complex rotations.

The reduced matrix $[R]$ is upper triangular. Since it is known that $[F]$ is a unitary
matrix, the fact that $[Q]$ is orthogonal implies no particular symmetry for the upper
triangular matrix $[R]$. However, it is easy to see that it cannot be orthogonal
otherwise it would make $[F]$ orthogonal as well. Since $[R]$ is not orthogonal, it
can be seen it cannot be factorized into a sequence of complex rotations, each of
which is orthogonal.

The above observation prompts the question which is the subject of this summary. The question is, can the orthogonality condition of the complex rotation be
generalized to a unitary condition so that the DFT matrix is factored into a sequence
of unitary matrices, each of which may imply a complex rotation of some kind?

To examine this question, we consider first the well-known QR decomposition
[B27] of the DFT matrix.

3.12.2.1   The QR decomposition of the DFT matrix [F]

It is known that for a unitary matrix $[F]$, its $[Q]$ and $[R]$ factors are respectively
unitary and diagonal. Thus $[F] = [Q][R]$ such that $[F]^{*T}[F] = [I]$ and $[Q]^{*T}[Q] = [I]$
where * denotes conjugate, T denotes transpose, and the matrix $[R]$ is diagonal. We
show this for the 2×2 case. Suppose that $[R]$ is the upper triangular matrix from the
factorization of the unitary matrix $[F]$ into $[Q][R]$ where $[Q]$ is also a unitary matrix.
Let

$$[R] = \begin{bmatrix} g_{11} & g_{12} \\ 0 & g_{22} \end{bmatrix} \tag{3.86}$$

where the elements are complex.

Since $[R]^{*T}[R] = [I]$, we have the following equations for the elements of $[R]$:

$$[R]^{*T}[R] = \begin{bmatrix} g_{11}^* & 0 \\ g_{12}^* & g_{22}^* \end{bmatrix} \begin{bmatrix} g_{11} & g_{12} \\ 0 & g_{22} \end{bmatrix} = [I] \tag{3.87}$$

where the superscript * denotes the complex conjugate. From (3.87), we get

$$g_{11}^* g_{11} = 1, \quad g_{11}^* g_{12} = 0 \quad \text{and} \quad g_{12}^* g_{12} + g_{22}^* g_{22} = 1 \tag{3.88}$$

Express g_{11} and g_{12} in terms of their real and imaginary parts as $g_{11} = x_1 + jy_1$ and $g_{12} = x_2 + jy_2$. Then we have the equations for the components for g_{12}:

$$g_{11}^* g_{12} = (x_1 - jy_1)(x_2 + jy_2) = 0$$

Hence

$$x_1 x_2 + y_1 y_2 = 0 \quad \text{and} \quad x_1 x_2 + y_1 y_2 = 0$$

Noting that the determinant of the coefficient matrix $[R]$ of (3.85) is the square amplitude of g_{11}, which is one, there is only the trivial solution for x_2 and y_2, meaning that g_{12} is zero. Thus the matrix $[R]$ is *diagonal*.

3.12.2.2 The unitary matrices in the form of rotations of some kind

Let us first examine the complex "rotation" matrix of the form:

$$[q] = \begin{bmatrix} \cos \gamma & \sin \gamma \\ -\sin^* \gamma & \cos^* \gamma \end{bmatrix}$$

From this definition of the rotation matrix, we can examine the elements of the product $[q]^{*T}[q]$.

$$\left([q]^{*T}[q]\right)_{11} = \cos \gamma \cos^* \gamma + \sin \gamma \sin^* \gamma$$

$$\left([q]^{*T}[q]\right)_{12} = \cos \gamma \cos^* \gamma - \sin \gamma \sin^* \gamma = 0$$

$$\left([q]^{*T}[q]\right)_{21} = \left([q]^{*T}[q]\right)_{12} \quad \text{and} \quad \left([q]^{*T}[q]\right)_{11} = \left([q]^{*T}[q]\right)_{22}$$

If the matrix $[q]$ is scaled by the square root of the value $([q]^{*T}[q])_{11}$, the resulting matrix will be a unitary matrix. Thus we now define the unitary *rotation* matrix as

$$[Q] = \frac{1}{\sqrt{k}} \begin{bmatrix} \cos \gamma & \sin \gamma \\ -\sin^* \gamma & \cos^* \gamma \end{bmatrix} \tag{3.89}$$

where

$$k = \cos \gamma \cos^* \gamma + \sin \gamma \sin^* \gamma = \cosh^2 \beta + \sinh^2 \beta$$

Suppose this matrix $[Q]$ is applied to a 2×2 unitary matrix given by

$$[F] = \begin{bmatrix} a_{11} & a_{12} \\ a_{21} & a_{22} \end{bmatrix}$$

Then γ satisfies the following equation:

$$([Q][F])_{21} = 0, \quad \text{or} \quad \frac{1}{\sqrt{k}}(-a_{11}\sin^*\gamma + a_{21}\cos^*\gamma) = 0 \qquad (3.90)$$

Hence we see that the development for the solution of the complex angle is the same as before (see (3.83)). The only difference is that now the complex conjugate appears in the equation.

We have shown in this summary that it is possible to completely factor an $N \times N$ DFT matrix into unitary "rotation" matrices, and a diagonal matrix. The unitary matrix is made up of elements in the 2×2 matrix defined by

$$[Q] = \frac{1}{\sqrt{k}} \begin{bmatrix} \cos \gamma & \sin \gamma \\ -\sin^*\gamma & \cos^*\gamma \end{bmatrix}$$

where $k = \cos \gamma \cos^*\gamma + \sin \gamma \sin^*\gamma$.

Example 3.1   For $N = 2$

$$[F] = \sqrt{\frac{1}{2}}\begin{bmatrix} 1 & 1 \\ 1 & e^{-j\pi} \end{bmatrix} = \sqrt{\frac{1}{2}}\begin{bmatrix} 1 & 1 \\ 1 & -1 \end{bmatrix} \quad [Q] = \sqrt{\frac{1}{2}}\begin{bmatrix} 1 & 1 \\ 1 & -1 \end{bmatrix} \quad [R] = \begin{bmatrix} 1 & 0 \\ 0 & 1 \end{bmatrix}$$

For $N = 3$

$$[F] = \sqrt{\frac{1}{3}}\begin{bmatrix} 1 & 1 & 1 \\ 1 & W^1 & W^2 \\ 1 & W^2 & W^4 \end{bmatrix} \quad [Q] = \sqrt{\frac{1}{3}}\begin{bmatrix} -1 & 1 & 1 \\ -1 & W^1 & W^2 \\ -1 & W^2 & W^4 \end{bmatrix} \quad [R] = \begin{bmatrix} -1 & 0 & 0 \\ 0 & 1 & 0 \\ 0 & 0 & 1 \end{bmatrix}$$

Here the MATLAB function QR is used.

3.13   Unified Discrete Fourier–Hartley Transform

Oraintara [I-29] has developed the theory and structure of the unified discrete Fourier–Hartley transforms (UDFHT). He has shown that with simple modifications, UDFHT can be used to implement DFT, DHT (discrete Hartley transform [I-28, I-31]), DCT and DST (discrete sine transform [B23]) of types I-IV (Table 3.5). He has thus unified all these transforms and has derived efficient algorithms by utilizing the FFT techniques. An added advantage is that the FFT based hardware/software can be utilized with pre/post processing steps to implement the family of DCTs, DHTs and DSTs. The basis functions of this family of discrete transforms are listed in Table 3.5.

Table 3.5 The basis functions of the DFT, DHT, DCT and DST of types I–IV where μ_k and λ_n are normalization factors and $0 \leq k, n \leq N - 1$. [1-29] © 2002 IEEE

Type	DFT (F_{kn})	DHT (H_{kn})	DCT (C_{kn})	DST (S_{kn})
I	$\exp\left[\dfrac{-j2\pi kn}{N}\right]$	$\cos\left[\dfrac{2\pi kn}{N}\right] + \sin\left[\dfrac{2\pi kn}{N}\right]$	$\lambda_n \mu_k \cos\left[\dfrac{\pi kn}{N-1}\right]$	$\cos\left[\dfrac{\pi(k+1)(n+1)}{N+1}\right]$
II	$\exp\left[\dfrac{-j2\pi k(n+0.5)}{N}\right]$	$\cos\left[\dfrac{2\pi k(n+0.5)}{N}\right] + \sin\left[\dfrac{2\pi k(n+0.5)}{N}\right]$	$\mu_k \cos\left[\dfrac{\pi k(n+0.5)}{N}\right]$	$\mu_k \sin\left[\dfrac{\pi(k+1)(n+0.5)}{N}\right]$
III	$\exp\left[\dfrac{-j2\pi(k+0.5)n}{N}\right]$	$\cos\left[\dfrac{2\pi(k+0.5)n}{N}\right] + \sin\left[\dfrac{2\pi(k+0.5)n}{N}\right]$	$\lambda_n \cos\left[\dfrac{\pi(k+0.5)n}{N}\right]$	$\lambda_n \sin\left[\dfrac{\pi(k+0.5)(n+1)}{N}\right]$
IV	$\exp\left[\dfrac{-j2\pi(k+0.5)(n+0.5)}{N}\right]$	$\cos\left[\dfrac{2\pi(k+0.5)(n+0.5)}{N}\right] + \sin\left[\dfrac{2\pi(k+0.5)(n+0.5)}{N}\right]$	$\cos\left[\dfrac{\pi(k+0.5)(n+0.5)}{N}\right]$	$\sin\left[\dfrac{\pi(k+0.5)(n+0.5)}{N}\right]$

The UDFHT is defined as an $(N \times N)$ matrix $[T]$ whose (k, n) elements are given by

$$
\begin{aligned}
T_{kn} &= \mu_k \lambda_n \left\{ A e^{-j2\pi(k+k_0)(n+n_0)/N} + B e^{j2\pi(k+k_0)(n+n_0)/N} \right\} \\
&= \mu_k \lambda_n \left\{ (A+B) \cos\left[\tfrac{2\pi}{N}(k+k_0)(n+n_0)\right] - j(A-B) \sin\left[\tfrac{2\pi}{N}(k+k_0)(n+n_0)\right] \right\}
\end{aligned}
$$
(3.91)

where A and B are constants and μ_k and λ_n are normalization factors so that the basis functions of $[T]$ have the same norm, $\sqrt{N}$.

The UDFHT is orthogonal if any one of the following is true:

1. $AB = 0$, or
2. $n_0 = p/2$, $k_0 = q/2$ and $\phi_A - \phi_B = (2r-1)\pi/2$, or
3. $n_0 = 0$ or $1/2$, $k_0 = (2q-1)/4$, and $\phi_A - \phi_B = r\pi$, or
4. $k_0 = 0$ or $1/2$, $n_0 = (2p-1)/4$, and $\phi_A - \phi_B = r\pi$,

(3.92)

where p, q, r are some integers, and ϕ_A and ϕ_B are the phases of A and B, respectively.

Table 3.6 lists the values of A, B, k_0 and n_0 for DFT/DHT of types I–IV.

As stated earlier the family of DCTs/DSTs (types I–IV) can also be implemented via UDFHT. However, besides the choice of A, B, k_0 and n_0 (Table 3.7), this implementation requires permuting data or transform sequence with possible sign changes.

Table 3.6 DFT and DHT of types I–IV via UDFHT. [I-29] © 2002 IEEE

	A	B	k_0	n_0
DFT-I	1	0	0	0
DFT-II	1	0	0	1/2
DFT-III	1	0	1/2	0
DFT-IV	1	0	1/2	1/2
DHT-I	$(1+j)/2$	$(1-j)/2$	0	0
DHT-II	$(1+j)/2$	$(1-j)/2$	0	1/2
DHT-III	$(1+j)/2$	$(1-j)/2$	1/2	0
DHT-IV	$(1+j)/2$	$(1-j)/2$	1/2	1/2

Table 3.7 DCT and DST of types II–IV via UDFHT. [I-29] © 2002 IEEE

	A	B	k_0	n_0
DCT-II	1/2	1/2	0	1/4
DCT-III	1/2	1/2	1/4	0
DCT-IV	1/2	1/2	1/4	1/2
DST-II	$j/2$	$-j/2$	0	1/4
DST-III	$j/2$	$-j/2$	1/4	0
DST-IV	$j/2$	$-j/2$	1/4	1/2

For example, using Table 3.7 for DCT-III, leads to

$$T_{kn} = \lambda_n \cos\left[\frac{2\pi}{N}\left(k + \frac{1}{4}\right)n\right] = \lambda_n \cos\left[\frac{\pi}{N}\left(2k + \frac{1}{2}\right)n\right] \tag{3.93}$$

For $0 \leq k \leq N/2$,

$$T_{kn} = C^{III}_{2k,n} \quad \text{and} \quad T_{N-1-k,n} = C^{III}_{2k+1,n} \tag{3.94}$$

Let $[P_i]$ be an $(N \times N)$ permutation matrix

$$[P_i] = \begin{bmatrix} 1 & 0 & 0 & \cdots & 0 & 0 & 0 & 0 \\ 0 & 0 & 0 & \cdots & 0 & 0 & 0 & (-1)^i \\ 0 & 1 & 0 & \cdots & 0 & 0 & 0 & 0 \\ 0 & 0 & 0 & \cdots & 0 & 0 & (-1)^i & 0 \\ & \vdots & & & \vdots & & \vdots & & \vdots \end{bmatrix} \tag{3.95}$$

Then it is easy to show that $[C^{III}] = [P_0][T]$ where $[C^{III}]$ is the $(N \times N)$ DCT matrix of type III. Similarly from Table 3.7 $[C^{IV}] = [P_1][T]$ where $[C^{IV}]$ is the $(N \times N)$ DCT matrix of type IV. Generalizing, the family of DCTs and DSTs can be implemented via UDFHT as follows:

$$[C^{II}] = [T]^T[P_0]^T, \quad [C^{III}] = [P_0][T], \quad [C^{IV}] = [P_1][T]$$
$$[S^{II}] = [T]^T[P_1]^T, \quad [S^{III}] = [P_1][T], \quad [S^{IV}] = [P_0][T] \tag{3.96}$$

It is shown here that UDHFT can be used to obtain DCT-III (see (3.91), (3.93)–(3.96) and Table 3.7).

$$T(k) = \sum_{n=0}^{N-1} \lambda_n \cos\left[\frac{\pi}{N}(2k + \frac{1}{2})n\right] \Leftrightarrow C^{III}(k) = \sum_{n=0}^{N-1} \lambda_n \cos\left[\frac{\pi}{N}(k + \frac{1}{2})n\right] \tag{3.97}$$

$$\text{where } \lambda_p = \begin{cases} 1/\sqrt{N} & p = 0 \\ \sqrt{2/N} & p \neq 0 \end{cases} \quad \text{and} \quad k = 0, 1, \ldots, N-1.$$

Proof. From (3.97) for $N = 8$,

$$T(0) = C^{III}(0), \ T(1) = C^{III}(2), \ T(2) = C^{III}(4) \text{ and } T(3) = C^{III}(6) \tag{3.98}$$

Thus the first equation in (3.94) is true. To show that the second equation in (3.94) is true, let $k = N - 1 - m$ and put it in (3.97) to obtain

$$T(N - 1 - m) = \sum_{n=0}^{N-1} \lambda_n \cos\left[\frac{\pi}{N}\left(2N - 2m - 2 + \frac{1}{2}\right)n\right] \tag{3.99a}$$

Index m in (3.99a) can be changed to k, and $\cos(2\pi - a) = \cos(a)$. Thus

$$T(N - 1 - k) = \sum_{n=0}^{N-1} \lambda_n \cos\left[\frac{\pi}{N}\left((2k + 1) + \frac{1}{2}\right)n\right]$$

$$= C^{III}(2k + 1) \qquad 0 \leq k < N/2 \tag{3.99b}$$

Equations (3.98) and (3.99) can be expressed in matrix form as

$$
\begin{bmatrix}
C^{III}(0) \\
C^{III}(1) \\
C^{III}(2) \\
C^{III}(3) \\
C^{III}(4) \\
C^{III}(5) \\
C^{III}(6) \\
C^{III}(7)
\end{bmatrix}
=
\begin{bmatrix}
1 & 0 & 0 & 0 & 0 & 0 & 0 & 0 \\
0 & 0 & 0 & 0 & 0 & 0 & 0 & 1 \\
0 & 1 & 0 & 0 & 0 & 0 & 0 & 0 \\
0 & 0 & 0 & 0 & 0 & 0 & 1 & 0 \\
0 & 0 & 1 & 0 & 0 & 0 & 0 & 0 \\
0 & 0 & 0 & 0 & 0 & 1 & 0 & 0 \\
0 & 0 & 0 & 1 & 0 & 0 & 0 & 0 \\
0 & 0 & 0 & 0 & 1 & 0 & 0 & 0
\end{bmatrix}
\begin{bmatrix}
T(0) \\
T(1) \\
T(2) \\
T(3) \\
T(4) \\
T(5) \\
T(6) \\
T(7)
\end{bmatrix}
\tag{3.100}
$$

The property in (3.97) can be illustrated in a pictorial fashion referring to Fig. D.1b. Symmetrically extend $C^{III}(k)$, whose basis function is $\cos\left[\frac{\pi}{N}\left(k + \frac{1}{2}\right)n\right]$ and then decimate the resulting sequence to obtain $T(k)$, whose basis function is $\cos\left[\frac{\pi}{N}\left(2k + \frac{1}{2}\right)n\right]$. Then $C^{III}(N)$ is replaced by $C^{III}(N + 1)$ as $C^{III}(N) = C^{III}(N + 1)$ and other coefficients are obtained similarly. This property for $C^{I}(k), S^{I}(k), C^{II}(k)$ (Fig. D.1c), and $S^{II}(k)$ will be used in Appendix D.3, where $C^{I}(k)$ is a coefficient of DCT type-I and so on.

3.13.1 Fast Structure for UDFHT

The UDFHT can be calculated from the DFT algorithms. Consider only $n_0 = p/2$ where p is an integer. Fast structures for DCT-II and DST-II can be obtained from the transposes of DCT-III and DST-III, respectively. Let

$$\hat{n} = (-n - 2n_0) \bmod N, \quad \text{i.e.} \quad \hat{n} = s_n N - n - 2n_0 \tag{3.101}$$

and s_n is an integer such that $0 \leq \hat{n} \leq N - 1$. Notice that $s_{\hat{n}} = s_n$ where $n = s_{\hat{n}} N - \hat{n} - 2n_0$.

$$\frac{X(k)}{\mu_k} = \sum_{n=0}^{N-1} T_{kn} x(n)$$

$$= \sum_{n=0}^{N-1} \lambda_n \left[A e^{-j\frac{2\pi}{N}(k+k_0)(n+n_0)} + B e^{j\frac{2\pi}{N}(k+k_0)(n+n_0)} \right] x(n)$$

$$= \sum_{n=0}^{N-1} \left[\lambda_n A x(n) + \lambda_{\hat{n}} B e^{j2\pi s_n k_0} x(\hat{n}) \right] e^{j\frac{2\pi}{N}(k+k_0)(n+n_0)}$$

$$= W^{(k+k_0)n_0} \sum_{n=0}^{N-1} \left[\lambda_n A x(n) + \lambda_{\hat{n}} B W^{-s_0 k_0 N} x(\hat{n}) \right] W^{k_0 n} W^{kn} \quad (3.102)$$

The UDFHT can be expressed in vector-matrix form as

$$\underline{X} = [T] \underline{x}$$

where $\underline{X}$ is the N-point transform vector, $\underline{x}$ is the N-point data vector and $[T]$ is the $(N \times N)$ UDFHT matrix.

The UDFHT can be implemented via DFT-I with pre and post processing as shown in Fig. 3.26 for $N = 8$ and $n_0 = 2$. Note from (3.101) $\hat{n} = 1 \times N - n - 2 \times n_0 = 4$ and thus $s_0 = 1$, for $n = 0$. From Fig. 3.26, it is clear that, if Mult(N) is the number of (complex) multiplications used in the operation of N-point FFT, the total number of multiplications used in the UDFHT is approximately

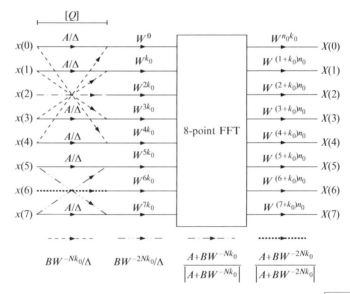

Fig. 3.26 Fast structure for non-normalized UDFHT for $N = 8$, $n_0 = 2$, $\Delta = \sqrt{|A|^2 + |B|^2}$ and $W = e^{-j2\pi/N}$. [I-29] © 2002 IEEE

Mult(N) + $4N$ where the $4N$ extra multiplications are from the butterflies in $[Q]$ and the pre- and post-multiplications $\left(W^{k_0 n}, W^{(k+k_0)n_0}\right)$ of the FFT block.

The orthogonality of the UDFHT can also be seen in Fig. 3.26. Since pre- and post-multiplying of the FFT block are orthogonal operations and the FFT itself is an orthogonal matrix, the entire UDFHT is orthogonal if and only if the pre-processing $[Q]$ whose (k, n) element is given by

$$Q_{kn} = \begin{cases} A/\Delta & \text{if } k = n \text{ and } k \neq \hat{n} \\ BW^{-s_n k_0 N}/\Delta & \text{if } k \neq n \text{ and } k = \hat{n} \\ \left(A + BW^{-s_n k_0 N}\right)/\left|A + BW^{-s_n k_0 N}\right| & \text{if } k = n = \hat{n} \\ 0 & \text{otherwise} \end{cases} \qquad (3.103)$$

where $\Delta = \sqrt{|A|^2 + |B|^2}$ and $\hat{n} = s_n N - n - 2 n_0$, is orthogonal, which implies that

$$AB = 0 \quad \text{or} \quad \cos\left(2\pi s_n k_0 - \phi_A + \phi_B\right) = 0 \qquad (3.104)$$

Using (3.91), (3.97)–(3.103) and Table 3.7, the flowgraph for eight-point DCT-III via FFT is shown in Fig. 3.27. Here $A = B = 1/2$, $k_0 = 1/4$ and $n_0 = 0$. Thus from (3.101) $\hat{n} = 1 \times N - n - 2 \times n_0 = 7$ for $n = 1$. Here the $(N \times N)$ DCT-III matrix $[C^{\text{III}}] = [P_0][T]$, and $[T]$ is defined by (3.102). Note that the input and output of the FFT block are to be in natural order. Note that the postprocessor implements $[P_0]$ defined by (3.100).

Oraintara [I-29] has generalized UDFHT called GDFHT by considering A and B to be variables as a function of n. From this, he has developed fast structures for MLT (modulated lapped transforms) followed by a flowgraph for efficient implementation of MDCT (modified DCT) via 12-point FFT.

Potipantong et al. [I-30] have proposed a novel architecture for a 256-point UDFHT using the existing FFT IP-core combined with pre and post processing

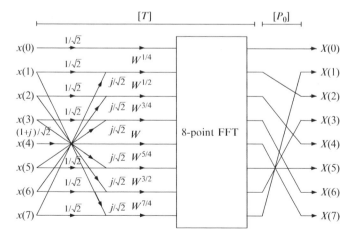

Fig. 3.27 An example of an eight-point DCT-III via FFT. Here $W = e^{-j2\pi/8}$. [I-29] © 2002 IEEE

operations. This architecture can be used to implement the family of discrete transforms DFT/DHT/DCT/DST, and is based on Xilinx Virtex-II Pro FPGA. The 256-point UDFHT can be implemented in 9.25 μs at 100 MHz operating clock frequency.

Wahid et al. [LA4] have proposed a hybrid architecture to compute three eight-point transforms – the DCT, DFT and discrete wavelet transform (DCT-DFT-DWT) to be implemented on a single Altera FPGA with a maximum operational frequency of 118 MHz. The transformation matrices are first decomposed into multiple sub-matrices; the common structures of the sub-matrices are then identified and shared among them.

The prime length DFT can be computed by first changing it into convolution. Then the convolution is computed by using FFTs. The Bluestein's FFT algorithm [A2, IP3] and the Rader prime algorithm [A34] are two different ways of transforming a DFT into a convolution.

3.14   Bluestein's FFT Algorithm

The Bluestein's FFT algorithm [A2, IP3], also called the chirp z-transform algorithm, is applicable for computing prime length DFT in $O(N\log_2 N)$ operations. The algorithm is as follows:

Multiply and divide DFT by $W_N^{(k^2+n^2)/2}$

$$X^{\mathrm{F}}(k) = \sum_{n=0}^{N-1} x(n) W_N^{kn} \quad k = 0, 1, \ldots, N-1 \tag{2.1a}$$

to obtain

$$
\begin{aligned}
X^{\mathrm{F}}(k) &= \sum_{n=0}^{N-1} x(n) W_N^{kn} \, W_N^{-(k^2+n^2)/2} \, W_N^{(k^2+n^2)/2} \\
&= W_N^{k^2/2} \sum_{n=0}^{N-1} W_N^{-(k-n)^2/2} \left(x(n) \, W_N^{n^2/2} \right) \quad (\text{i.e., } 2kn = k^2 + n^2 - (k-n)^2) \\
&= W_N^{k^2/2} \sum_{n=0}^{N-1} w(k-n) \, \hat{x}(n) \\
&= W_N^{k^2/2} \{ w(n) * \hat{x}(n) \} \quad k = 0, 1, \ldots, N-1
\end{aligned}
\tag{3.105}
$$

where $\hat{x}(n) = x(n) W_N^{n^2/2}$, $n = 0, 1, \ldots, N-1$ and $w(n) = W_N^{-n^2/2}$, $n = -N+1$, $n = -N+2, \ldots, -1, 0, 1, 2, \ldots, N-1$ (see Fig. 3.28).

Fig. 3.28 DFT computation
using Bluestein's algorithm.
The block in the middle is
implemented using one FFT
and one IFFT

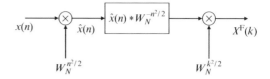

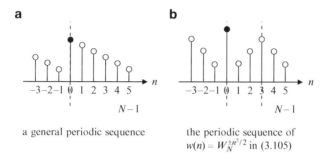

a general periodic sequence

the periodic sequence of
$w(n) = W_N^{+n^2/2}$ in (3.105)

Fig. 3.29 Periodic sequences with period N *even* where $N = 6$

For N *even*, a convolution in (3.105) can be regarded as an N-point circular
convolution of complex sequences since $W_N^{-(n+N)^2/2} = W_N^{-n^2/2}$ (Fig. 3.29b). *Circular convolution of two periodic sequences in time/spatial domain is equivalent
to multiplication in the DFT domain* (see (2.24)). Similarly, the convolution
in (3.105) can be computed with a pair of FFTs when the DFT of $w(n)$ is
pre-computed.

For N *odd*, however $W_N^{-(n+N)^2/2} = -W_N^{-n^2/2}$, so that the convolution in (3.105)
corresponds to a circular convolution of size $2N$ where $\{x(n)\}$ is extended by
adding N zeros at the end. Thus the N output samples of even index are the DFT
of the input sequence since adding zeros to $\{x(n)\}$ at the end results in interpolation
in the frequency domain (see Section 2.7).

For N even, a circular convolution in (3.105) can be best explained by examples
for $N = 4$ and 6 as

$$
\begin{pmatrix} X^F(0)/W_{2N}^{k^2} \\ X^F(1)/W_{2N}^{k^2} \\ X^F(2)/W_{2N}^{k^2} \\ X^F(3)/W_{2N}^{k^2} \end{pmatrix} = \begin{pmatrix} X^F(0) \\ X^F(1)/a \\ -X^F(2) \\ X^F(3)/a \end{pmatrix} = [w]\,\underline{\hat{x}} = \begin{pmatrix} 1 & a & -1 & a \\ a & 1 & a & -1 \\ -1 & a & 1 & a \\ a & -1 & a & 1 \end{pmatrix} \begin{pmatrix} \hat{x}(0) \\ \hat{x}(1) \\ \hat{x}(2) \\ \hat{x}(3) \end{pmatrix}
$$

(3.106a)

$$
\begin{pmatrix} X^{\mathrm{F}}(0) \\ X^{\mathrm{F}}(1)/b \\ jX^{\mathrm{F}}(2)/b \\ -jX^{\mathrm{F}}(3) \\ jX^{\mathrm{F}}(4)/b \\ X^{\mathrm{F}}(5)/b \end{pmatrix} = \begin{pmatrix} 1 & b & -jb & j & -jb & b \\ b & 1 & b & -jb & j & -jb \\ -jb & b & 1 & b & -jb & j \\ j & -jb & b & 1 & b & -jb \\ -jb & j & -jb & b & 1 & b \\ b & -jb & j & -jb & b & 1 \end{pmatrix} \begin{pmatrix} \hat{x}(0) \\ \hat{x}(1) \\ \hat{x}(2) \\ \hat{x}(3) \\ \hat{x}(4) \\ \hat{x}(5) \end{pmatrix} \qquad (3.106b)
$$

where a and b are $\exp(j\pi/N)$ with $N = 4$ and 6 respectively.

Similarly, let

$$
Y^{\mathrm{F}}(k) = \sum_{n=0}^{N-1} x(n) W_N^{-kn} \quad k = 0, 1, \ldots, N-1 \qquad (3.107)
$$

$$
Y^{\mathrm{F}}(k) = W_N^{k^2/2} \sum_{n=0}^{N-1} W_N^{-(k+n)^2/2} \left(x(n) W_N^{n^2/2} \right) \qquad (3.108)
$$

$$
= W_N^{k^2/2} \sum_{n=0}^{N-1} w(k+n)\,\hat{x}(n) \quad k = 0, 1, \ldots, N-1 \quad (N \text{ even})
$$

Then vector $\left[\underline{Y}^{\mathrm{F}}(k) \right]$ is $\left[\underline{X}^{\mathrm{F}}(k) \right]^* = [X^{\mathrm{F}}(0), X^{\mathrm{F}}(N-1), X^{\mathrm{F}}(N-2), \ldots, X^{\mathrm{F}}(2),$ $X^{\mathrm{F}}(1)]^T$, where $X^{\mathrm{F}}(k)$ is defined in (2.1a) and (3.105). Equation (3.107) corresponds to the IDFT (2.1b), after including the factor $1/N$.

3.15   Rader Prime Algorithm [A34]

An integer modulo N is indicated as

$$
((n)) = n \bmod N \qquad (3.109)
$$

If N is an integer there is a number g, such that there is a one-on-one mapping of the integers $n = 1, 2, \ldots, N-1$ to the integers $m = 1, 2, \ldots, N-1$, defined as

$$
m = ((g^n)) \qquad (3.110)
$$

For example, let $N = 7$, and $g = 3$. The mapping of n onto m is:

n	1	2	3	4	5	6
m	3	2	6	4	5	1

g is called a *primitive root* of N. $X^{\mathrm{F}}(0)$ is directly computed as

$$
X^{\mathrm{F}}(0) = \sum_{n=0}^{N-1} x(n) \qquad (3.111)
$$

$x(0)$ is not to be multiplied and is added last into the summation.

$$X^F(0) - x(0) = \sum_{n=1}^{N-1} x(n) \exp\left(-j\frac{2\pi nk}{N}\right) \quad k = 1, 2, \ldots, N-1 \qquad (3.112)$$

The terms in the summation are permuted, and the order of the equations are changed via transformations.

$$n \to ((g^n)) \qquad k \to ((g^k)) \qquad\qquad (3.113)$$

Note $((g^{N-1})) = ((g^0))$ and $((g^N)) = ((g^1))$.

$$X^F((g^k)) - x(0) = \sum_{n=0}^{N-2} x((g^n)) \exp\left(-j\frac{2\pi}{N}((g^{n+k}))\right) \quad k = 1, 2, \ldots, N-1$$

$$(3.114)$$

Equation (3.114) is the circular correlation of the sequence $\{x((g^n))\}$ and the sequence $\{\exp[-j(2\pi/N)((g^{n+k}))]\}$. Since N is prime, $N-1$ is composite. Circular correlation functions can be computed by using DFTs.

$$X^F((g^k)) - x(0) = \text{IDFT}\left\{\text{DFT}[x((g^{-n}))] \circ \text{DFT}\left[\exp\left(\frac{-j2\pi}{N}((g^n))\right)\right]\right\} \quad (3.115)$$

$$k, n = 1, 2, \ldots, N-1$$

where $\circ$ denotes the element-by-element multiplication of the two sequences. DFT and IDFT are implemented by FFT and IFFT. See the MATLAB code below.

```
N = 7;
a = [0 1 2 3 4 5 6];                    % Input to be permuted
p_a = [a(2) a(6) a(5) a(7) a(3) a(4)] ; % 1 5 4 6 2 3 = ((g^(-n)))
n = [3 2 6 4 5 1];
b = exp(-j*2*pi*n/N)
c = ifft(fft(p_a).*fft(b)) + a(1);      % Convolution
out=[sum(a) c(6) c(2) c(1) c(4) c(5) c(3)]  % Output comes after a permutation
fft(a)                                  % This should be equal to the output.
```

3.16 Summary

Followed by the definitions and properties of the DFT (Chapter 2), this chapter has covered the gamut of the fast algorithms. Radix-2, 3, 4, mixed-radix, split-radix, DIT and DIF FFTs are developed. For an application of radix-8 FFT, the reader is referred to [O18, O5] (see also Fig. 8.21). Other fast algorithms such as WFTA, prime factor FFT, UDFHT, etc. are also described in detail.

The overall objective of these algorithms is significant reduction in computational complexity, roundoff/truncation errors and memory requirements. Needless to say all these algorithms are equally applicable to IDFT. The various radices (radix 2, 3, 4, etc.) have specific advantages over particular platforms, architectures, etc., e.g. radix-16 FFT [V2]. Thus a designer can choose among these algorithms, one that best meets the specific requirements of an application customized by a DSP, microprocessor or VLSI chip.

The following chapter extends the FFT to the integer FFT. The integer FFT (Chapter 4), although recently developed, has gained prominence in terms of applications.

3.17   Problems

3.1 By rearranging the DFT matrix rows in bit reversed order, it can be factored into sparse matrices leading to a fast and efficient algorithm. This is illustrated for $N = 8$ in (3.3). (i.e., both sparse matrix factors and the flowgraph)
(a) Extend this to IDFT ($N = 8$).
(b) Extend this to DFT ($N = 16$).
(c) Extend this to IDFT ($N = 16$).

Draw the flowgraphs. Indicate the numbers of multiplies and adds required for the above. Obtain the sparse matrix factors, from the flowgraphs.

3.2 The 2D-DFT can be implemented by row/column 1D-DFT technique (see Chapter 5). Using the FFT approach of Problem 3.1, indicate the numbers of multiplications and additions required for
(a) (8×8) 2D-DFT
(b) (8×8) 2D-IDFT
(c) (16×16) 2D-DFT
(d) (16×16) 2D-IDFT

3.3 See Fig. 3.4. The sparse matrix factors (SMF) and the DFT matrix whose rows are rearranged in bit reversed order (BRO) for $N = 8$ are given in Section 3.2 i.e.,

[DFT] Rows rearranged in BRO $= [A_1][A_2][A_3]$

Note that this matrix is not symmetric. Show that this relationship is correct i.e., carry out the matrix multiplication $[A_1][A_2][A_3]$. What are the SMF for the inverse DFT for $N = 8$? Specifically obtain the SMF for [(DFT) rows rearranged in BRO]$^{-1}$. Unitary property of a matrix is invariant to the rearrangement of its rows or columns.

3.4 Repeat Problem 3.3 for ($N = 16$). See the SMF shown in (3.7).

3.5 Based on the flowgraphs for DFT $N = 8$ and 16 and from the results of Problems 3.3 and 3.4, draw the flowgraphs for the IDFT, $N = 8$ and 16.

3.6 Write down the SMF of the DFT matrix (rows rearranged in BRO) for $N = 32$ based on the SMF for $N = 8$ and 16. What are the corresponding SMF for the IDFT?

3.7   Develop a radix-3 DIF-FFT algorithm for $N = 27$. Draw the flowgraph and write down the SMF. What are the numbers of multiplies and adds required for implementing this algorithm.

3.8   Repeat Problem 3.7 for radix-3 DIT-FFT.

3.9   Repeat Problem 3.7 for $N = 9$.

3.10 Repeat Problems 3.7 and 3.8 for radix-4 for ($N = 16$).

3.11 Develop a radix-4 DIT-FFT algorithm for $N = 64$. Draw the flowgraph.

3.12 Using (3.33), develop a radix-4 DIF-FFT algorithm for $N = 64$. Draw the flowgraph.

3.13 Use $X^{\mathrm{F}}(k) = \sum_{n=0}^{N-1} x(n) W_N^{nk}$, $k = 0, 1, \ldots, N-1$ and $x(n) = \dfrac{1}{N} \sum_{k=0}^{N-1} X^{\mathrm{F}}(k) W_N^{-nk}$, $0, 1, \ldots, N-1$ as the DFT and IDFT denoted by $x(n) \Leftrightarrow X^{\mathrm{F}}(k)$. (Here $W_N = \exp(-j2\pi/N)$, and $j = \sqrt{-1}$.) $x(n)$ is the data sequence and $X^{\mathrm{F}}(k)$ is the transform sequence. Develop the following fast algorithms for $N = 16$:
(a) Radix-2 DIT-FFT
(b) Radix-2 DIF-FFT
(c) Radix-2 DIT/DIF-FFT
(d) Radix-2 DIF/DIT-FFT
(e) Radix-4 DIT-FFT
(f) Radix-4 DIF-FFT
(g) Split-radix DIT-FFT (see [SR1])
(h) Split-radix DIF-FFT (see [SR1])
(i) Draw the flow graphs for all the algorithms part (a) thru part (h).
(j) Compare the computational complexity of all these algorithms (number of adds and number of multiplies).
(k) Write down the sparse matrix factors based on part (i).
(l) Check for in-place property for all these algorithms.
(m) Are all these algorithms valid both for DFT and IDFT? What modifications, if any, are required for the IDFT?
(n) Why are the DIT and DIF so called?

See [SR1] for Problems 3.14 and 3.15.

3.14 The *split-radix* algorithm for DIF-FFT is developed and is described by (3.40). This is a combination of radix-2 and radix-4 algorithms.
(a) Starting with the DFT definition, derive (3.40).
(b) Show the flowgraph for $N = 16$.
(c) Obtain sparse matrix factors.
(d) Derive inverse transform and repeat part (b) and part (c).
(e) Compare multiplies/adds with radix-2 and radix-4 algorithms.
(f) Derive a corresponding split-radix algorithm for DIT-FFT for $N = 16$.

3.15 Start with the definition of DCT (k, N, x), (10) in [SR1]. Derive all the steps in detail leading to (22) in [SR1] from (10) in [SR-1].

3.16 (a) Derive a split-radix FFT for $N = 8$. The lifting scheme version of its flowgraph is shown in Fig. 4.4.
(b) Obtain sparse matrix factors from the flowgraph (Fig. 4.4).

(c) Derive inverse transform for $N = 8$ and draw the flowgraph.

3.17 Implement eight-point DCT-III via FFT (see Fig. 3.27) and DCT-III separately, using MATLAB. Show their transform coefficients are the same for an eight-point input sequence.

3.18 Show that (3.82) and then (3.79) are true.

3.19 Show similar equations with (3.106) for $N = 3$ and $N = 5$.

(Section 3.11: 3.20–3.21)

3.20 Find permutation matrices for the input and output sequences for the Winograd Fourier transform algorithm (WFTA) of the following sizes.

(a) $N = 2 \times 3$ (b) $N = 3 \times 4$ (c) $N = 3 \times 7$ (d) $N = 4 \times 5$ (e) $N = 5 \times 7$

3.21 Repeat Problem 3.20 using isomorphism with a generator of $(1,1)$.

(Section 3.13 UDFHT)

3.22 Show that similar to the proof of DCT-III shown in (3.97) thru (3.100), UDFHT can be used to obtain the following.

(a) DCT-II   (b) DCT-IV   (c) DST-II   (d) DST-III   (e) DST-IV

(Section 3.15: 3.23–3.24)

3.23 For the Rader prime algorithm [A34], create mapping tables.

(a) $N = 5$ and $g = 2$ (b) $N = 11$ and $g = 2$ (c) $N = 13$ and $g = 2$

3.24 Implement the Rader prime algorithm in MATLAB with the numbers given in Problem 3.23.

3.25 In [T8], Agaian and Caglayan have developed recursive fast orthogonal mapping algorithms based FFTs. Develop similar algorithms and flow graphs as follows:
(a) Walsh-Hadamard transform based FFT for $N = 16$ (see Fig. 1 in [T8]).
(b) Haar transform based FFT for $N = 8$.
(c) Combination of Walsh-Hadamard and Haar transforms based FFT for $N = 8$ (see Fig. 2 in [T8]).
(d) Lowest arithmetic complexity FFT algorithm for $N = 8$ (see Fig. 3 in [T8]).

3.18 Projects

3.1 Modify the code in Appendix H.1 to implement the Winograd Fourier transform algorithm (WFTA) with $N = 3 \times 4$.

Chapter 4
Integer Fast Fourier Transform

4.1 Introduction

Since the floating-point operation is very expensive, numbers are quantized to a fixed number of bits. The number of bits at each internal node in the implementation of FFT is fixed to a certain number of bits. Denote this number as N_n. The most significant bits (MSBs) of the result after each operation is kept up to N_n bits, and the tail is truncated. Thus this conventional fixed-point arithmetic affects the invertibility of the DFT because DFT coefficients are quantized.

Integer fast Fourier transform (IntFFT) is an integer approximation of the DFT [I-6]. The transform can be implemented by using only bit shifts and additions but no multiplications. Unlike the fixed-point FFT (FxpFFT), IntFFT is power adaptable and reversible. IntFFT has the same accuracy as the FxpFFT when the transform coefficients are quantized to a certain number of bits. Complexity of IntFFT is much lower than that of FxpFFT, as the former requires only integer arithmetic.

Since the DFT has the orthogonality property, the DFT is invertible. The inverse is just the complex conjugate transpose. Fixed-point arithmetic is often used to implement the DFT in hardware. Direct quantization of the coefficients affects the invertibility of the transform. The IntFFT guarantees the invertibility/perfect-reconstruction property of the DFT while the coefficients can be quantized to finite-length binary numbers.

Lifting factorization can replace the 2×2 orthogonal matrices appearing in fast structures to compute the DFT of input with length of $N = 2^n$ for n an integer such as split-radix, radix-2 and radix-4. The resulting transforms or IntFFTs are invertible, even though the lifting coefficients are quantized and power-adaptable, that is, different quantization step sizes can be used to quantize the lifting coefficients.

K.R. Rao et al., *Fast Fourier Transform: Algorithms and Applications*,
Signals and Communication Technology,
DOI 10.1007/978-1-4020-6629-0_4, © Springer Science+Business Media B.V. 2010

4.2 The Lifting Scheme

The lifting scheme is used to construct wavelets and perfect reconstruction (PR) filter banks [I-1, I-4, I-6]. Biorthogonal filter banks having integer coefficients can be easily implemented and can be used as integer-to-integer transform.

The two-channel system in Fig. 4.1 shows the lifting scheme. The first branch is operated by A_0 and called *dual lifting* whereas the second branch is operated by A_1 and is called *lifting*. We can see that the system is PR for any choices of A_0 and A_1. It should be noted that A_0 and A_1 can be nonlinear operations like rounding or flooring operations, etc. Flooring a value means rounding it to the nearest integer less than or equal to the value.

4.3 Algorithms

Integer fast Fourier transform algorithm approximates the twiddle factor multiplication [I-6, I-7]. Let $x = x_r + jx_i$ be a complex number. The multiplication of x with a twiddle factor $W_N^{-k} = \exp\left(\frac{j2\pi k}{N}\right) = e^{j\theta} = \cos\theta + j\sin\theta$, is the complex number, $y = W_N^{-k}x$ and can be represented as

$$y = (1,\, j)\begin{bmatrix} \cos\theta & -\sin\theta \\ \sin\theta & \cos\theta \end{bmatrix}\begin{bmatrix} x_r \\ x_i \end{bmatrix} = (1,\, j)[R_0]\begin{bmatrix} x_r \\ x_i \end{bmatrix} \tag{4.1}$$

where

$$[R_0] = \begin{bmatrix} \cos\theta & -\sin\theta \\ \sin\theta & \cos\theta \end{bmatrix} \tag{4.2}$$

The main difficulty in constructing a multiplier-less or integer transform by using the sum-of-powers-of-two (SOPOT) representation of $[R_0]$ is that *once the entries of* $[R_0]$ *are rounded to the SOPOT numbers, the entries of its inverse cannot be represented by the SOPOTs* (IntFFT covered in the first half of this chapter

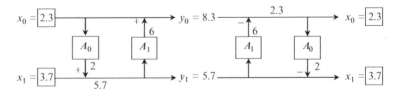

Fig. 4.1 Lifting scheme guarantees perfect reconstruction for any choices of A_0 and A_1. Perfect reconstruction means the final output equals to the input. Here A_0 and A_1 are rounding operators. [I-6] © 2002 IEEE

resolves this difficulty). In other words, if $\cos\theta$ and $\sin\theta$ in (4.1) are quantized and represented as α and β in terms of SOPOT coefficients, then an approximation of $[R_0]$ and its inverse can be represented as

$$[\tilde{R}_0] = \begin{bmatrix} \alpha & -\beta \\ \beta & \alpha \end{bmatrix} \tag{4.3}$$

$$[\tilde{R}_0]^{-1} = \frac{1}{\sqrt{\alpha^2 + \beta^2}} \begin{bmatrix} \alpha & \beta \\ -\beta & \alpha \end{bmatrix} \tag{4.4}$$

As α and β are SOPOT coefficients, the term $\sqrt{\alpha^2 + \beta^2}$ cannot in general be represented as SOPOT coefficient. The basic idea of the integer or multiplier-less transform is to decompose $[R_0]$ into three lifting steps.

If $\det([A]) = 1$ and $c \neq 0$ [1-4],

$$[A] = \begin{bmatrix} a & b \\ c & d \end{bmatrix} = \begin{bmatrix} 1 & (a-1)/c \\ 0 & 1 \end{bmatrix} \begin{bmatrix} 1 & 0 \\ c & 1 \end{bmatrix} \begin{bmatrix} 1 & (d-1)/c \\ 0 & 1 \end{bmatrix} \tag{4.5}$$

From (4.5), $[R_0]$ is decomposed as

$$\begin{aligned} [R_0] &= \begin{bmatrix} 1 & \dfrac{\cos\theta - 1}{\sin\theta} \\ 0 & 1 \end{bmatrix} \begin{bmatrix} 1 & 0 \\ \sin\theta & 1 \end{bmatrix} \begin{bmatrix} 1 & \dfrac{\cos\theta - 1}{\sin\theta} \\ 0 & 1 \end{bmatrix} = [R_1][R_2][R_3] \\[2mm] &= \begin{bmatrix} 1 & -\tan\left(\dfrac{\theta}{2}\right) \\ 0 & 1 \end{bmatrix} \begin{bmatrix} 1 & 0 \\ \sin\theta & 1 \end{bmatrix} \begin{bmatrix} 1 & -\tan\left(\dfrac{\theta}{2}\right) \\ 0 & 1 \end{bmatrix} \end{aligned} \tag{4.6}$$

$$\begin{aligned} [R_0]^{-1} &= [R_3]^{-1}[R_2]^{-1}[R_1]^{-1} \\[2mm] &= \begin{bmatrix} 1 & -\dfrac{\cos\theta - 1}{\sin\theta} \\ 0 & 1 \end{bmatrix} \begin{bmatrix} 1 & 0 \\ -\sin\theta & 1 \end{bmatrix} \begin{bmatrix} 1 & -\dfrac{\cos\theta - 1}{\sin\theta} \\ 0 & 1 \end{bmatrix} \end{aligned} \tag{4.7}$$

The coefficients in the factorization of (4.6) can be quantized to SOPOT coefficients to form

$$[R_0] \approx [S_0] = \begin{bmatrix} 1 & \alpha_0 \\ 0 & 1 \end{bmatrix} \begin{bmatrix} 1 & 0 \\ \beta_0 & 1 \end{bmatrix} \begin{bmatrix} 1 & \alpha_0 \\ 0 & 1 \end{bmatrix} \tag{4.8}$$

where α_0 and β_0 are respectively SOPOT approximations to $(\cos\theta - 1)/\sin\theta$ and $\sin\theta$ having the form

$$\alpha_0 = \sum_{k=1}^{t} a_k \, 2^{b_k} \tag{4.9}$$

where $a_k \in \{-1, 1\}$, $b_k \in \{-r, \ldots, -1, 0, 1, \ldots, r\}$, r is the range of the coefficients and t is the number of terms used in each coefficient. The variable t is usually limited so that the twiddle factor multiplication can be implemented with limited number of addition and shift operations. The integer FFT converges to the DFT when t increases.

The lifting structure has two advantages over the butterfly structure. First, the number of real multiplications is reduced from four to three, although the number of additions is increased from two to three (see Figs. 4.2 and 4.3). Second, the structure allows for quantization of the lifting coefficients and the quantization does not affect the PR property. To be specific, instead of quantizing the elements of $[R_0]$ in (4.2) directly, the lifting coefficients, s and $(s-1)/c$ are quantized and therefore, the inversion also consists of three lifting steps with the same lifting coefficients but with opposite signs.

Example 4.1 In case of the twiddle factor, W_8^1, $\theta = -\pi/4$, $(\cos\theta - 1)/\sin\theta = \sqrt{2} - 1$ and $\sin\theta = -1/\sqrt{2}$. If we round these numbers respectively to the right-hand one digit of the decimal point, then $\alpha_0 = 0.4$ and $\beta_0 = 0.7$.

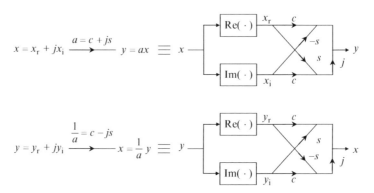

Fig. 4.2 Butterfly structure for implementing a complex multiplication above and its inverse below where $s = \sin\theta$ and $c = \cos\theta$. [I-6] © 2002 IEEE

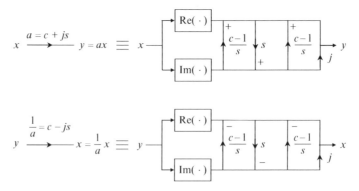

Fig. 4.3 Lifting structure for implementing a complex multiplication above and its inverse below where $s = \sin\theta$ and $c = \cos\theta$. [I-6] © 2002 IEEE

$$[S_0] = \begin{bmatrix} 1 & 0.4 \\ 0 & 1 \end{bmatrix} \begin{bmatrix} 1 & 0 \\ -0.7 & 1 \end{bmatrix} \begin{bmatrix} 1 & 0.4 \\ 0 & 1 \end{bmatrix} = \begin{bmatrix} 0.72 & 0.688 \\ -0.7 & 0.72 \end{bmatrix} \quad (4.10a)$$

$$[S_0]^{-1} = \begin{bmatrix} 1 & -0.4 \\ 0 & 1 \end{bmatrix} \begin{bmatrix} 1 & 0 \\ 0.7 & 1 \end{bmatrix} \begin{bmatrix} 1 & -0.4 \\ 0 & 1 \end{bmatrix} = \begin{bmatrix} 0.72 & -0.688 \\ 0.7 & 0.72 \end{bmatrix} \quad (4.10b)$$

The lifting scheme defined in (4.8) and (4.10) works for any numbers (real and complex) of α_0 and β_0. $[S_0]$ is no more orthogonal ($[S_0]^{-1} \neq [S_0]^T$), but developing its inverse transform is as easy as for the case of the orthogonal transform as the entries of $[S_0]$ and $[S_0]^{-1}$ are the same with different signs except 1s on the diagonal (Figs. 4.2 and 4.3), while both schemes guarantee perfect inverse or perfect reconstruction as $[S_0]^{-1}[S_0] = [I]$ (biorthogonal) and $[R_0]^T[R_0] = [I]$ (orthogonal, see [4.2]), respectively.

In summary, in implementing a complex number multiplication, a twiddle factor in matrix form has a butterfly structure and if we round the coefficients, its inverse is computationally complex, but if we decompose the twiddle factor into a lifting structure, the twiddle factor has a perfect inverse even if we round the coefficients. Once the coefficients are rounded in the lifting structure, the twiddle factor may have either a lifting or butterfly structures for perfect inverse but the lifting structure has one less multiplication.

An eight-point integer FFT based on the split-radix structure is developed in [I-6]. Figure 4.4 shows the lattice structure of the integer FFT, where the twiddle factors W_8^1 and W_8^3 are implemented using the factorization. Another integer FFT based on the radix-2 decimation-in-frequency is covered in [I-7]. At the expense of precision, we can develop computationally effective integer FFT algorithms.

When an angle is in I and IV quadrants, (4.6) is used. If $\theta \in (-\pi, -\pi/2) \cup (\pi/2, \pi)$, then $|(\cos\theta - 1)/\sin\theta| > 1$ as $\cos\theta < 0$ for II and III quadrants. Thus the absolute values of the lifting coefficients need to be controlled to be less than or equal to one by replacing R_0 by $-[R_{0+\pi}]$ as follows:

$$[R_0] = -[R_{0+\pi}] = -\begin{bmatrix} -\cos\theta & \sin\theta \\ -\sin\theta & -\cos\theta \end{bmatrix}$$
$$= -\begin{bmatrix} 1 & (c+1)/s \\ 0 & 1 \end{bmatrix} \begin{bmatrix} 1 & 0 \\ -s & 1 \end{bmatrix} \begin{bmatrix} 1 & (c+1)/s \\ 0 & 1 \end{bmatrix} \quad (4.11)$$

When an angle is in I and II quadrants, we can have another choice of lifting factorization as follows:

$$[R_0] = \begin{bmatrix} \cos\theta & -\sin\theta \\ \sin\theta & \cos\theta \end{bmatrix} = \begin{bmatrix} 0 & 1 \\ 1 & 0 \end{bmatrix} \begin{bmatrix} \sin\theta & -\cos\theta \\ \cos\theta & \sin\theta \end{bmatrix} \begin{bmatrix} 1 & 0 \\ 0 & -1 \end{bmatrix}$$
$$= \begin{bmatrix} 0 & 1 \\ 1 & 0 \end{bmatrix} \begin{bmatrix} 1 & (s-1)/c \\ 0 & 1 \end{bmatrix} \begin{bmatrix} 1 & 0 \\ c & 1 \end{bmatrix} \begin{bmatrix} 1 & (s-1)/c \\ 0 & 1 \end{bmatrix} \begin{bmatrix} 1 & 0 \\ 0 & -1 \end{bmatrix} \quad (4.12)$$

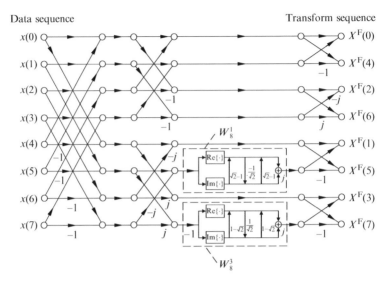

Fig. 4.4 Lattice structure of eight-point integer FFT using split-radix structure (see also [I-9]). For example, the twiddle factors are quantized/rounded off in order to be represented as 16-bit numbers (N_c bits). Multiplication results are again uniformly quantized to N_n bits. N_n the number of bits required to represent internal nodes is determined only by FFT of size N and N_i, the number of bits required to represent input signal. [I-6] © 2002 IEEE

Table 4.1 For each value of θ, only two out of four possible lifting factorizations have all their lifting coefficients falling between -1 and 1. [I-6] © 2002 IEEE

Quadrant	Range of θ	Lifting factorization
I	$(0, \pi/2)$	(4.6) and (4.13)
II	$(\pi/2, \pi)$	(4.11) and (4.13)
III	$(-\pi, -\pi/2)$	(4.11) and (4.15)
IV	$(-\pi/2, 0)$	(4.6) and (4.15)

However, if $\theta \in (-\pi, 0)$, then $\sin\theta < 0$ for III and IV quadrants and $(\sin\theta - 1)/\cos\theta$ will be greater than one. Thus $[R_0]$ should be replaced by $-[R_{0+\pi}]$ as follows (Table 4.1):

$$
[R_0] = -\begin{bmatrix} -\cos\theta & \sin\theta \\ -\sin\theta & -\cos\theta \end{bmatrix} = -\begin{bmatrix} 0 & 1 \\ 1 & 0 \end{bmatrix}\begin{bmatrix} -\sin\theta & \cos\theta \\ -\cos\theta & -\sin\theta \end{bmatrix}\begin{bmatrix} 1 & 0 \\ 0 & -1 \end{bmatrix}
$$

$$
= -\begin{bmatrix} 0 & 1 \\ 1 & 0 \end{bmatrix}\begin{bmatrix} 1 & (s+1)/c \\ 0 & 1 \end{bmatrix}\begin{bmatrix} 1 & 0 \\ -c & 1 \end{bmatrix}\begin{bmatrix} 1 & (s+1)/c \\ 0 & 1 \end{bmatrix}\begin{bmatrix} 1 & 0 \\ 0 & -1 \end{bmatrix}
$$

$$(4.13)$$

For example, suppose we are given the twiddle factor, $W_8^3 = e^{-j6\pi/8}$. Then $\theta = -3\pi/4$. Then we have the two options, (4.11) and (4.13). We select (4.11). Then

$$[R_0] = -[R_{0+\pi}] = -\begin{bmatrix} 1/\sqrt{2} & -1/\sqrt{2} \\ 1/\sqrt{2} & 1/\sqrt{2} \end{bmatrix}$$

$$= -\begin{bmatrix} 1 & 1-\sqrt{2} \\ 0 & 1 \end{bmatrix}\begin{bmatrix} 1 & 0 \\ 1/\sqrt{2} & 1 \end{bmatrix}\begin{bmatrix} 1 & 1-\sqrt{2} \\ 0 & 1 \end{bmatrix} \qquad (4.14)$$

Substitute (4.14) in (4.1) to obtain a lifting structure for a multiplication of a complex number and the twiddle factor W_8^3. Figure 4.4 shows lifting/lattice structure of eight-point integer FFT using split-radix structure, where the two twiddle factors W_8^1 and $W_8^3(= -W_8^7)$ are implemented using lifting scheme. Inverse integer FFT is as usual the conjugate of the integer FFT whose block diagram is shown in Fig. 4.4.

4.3.1 Fixed-Point Arithmetic Implementation

One of the factors that primarily affects the cost of the DSP implementation is the resolution of the internal nodes (the size of the registers at each stage). In practice, it is impossible to retain infinite resolution of the signal samples and the transform coefficients. Since the floating-point operation is very expensive, these numbers are often quantized to a fixed number of bits. Two's-complement arithmetic for fixed-point representation of numbers is a system in which negative numbers are represented by the two's complement of the absolute value; this system represents signed integers on hardware, DSP and computers (Table 4.2).

Each addition can increase the number of bits by one, whereas each multiplication can increase the number of bits by $2n$ bits for the multiplication of two n-bit numbers. The nodes in latter stages require more bits than those in earlier stages to store the output after each arithmetic operation without overflows. As a result, the number of bits required to store the results grows cumulatively as the number of stages increases. In general, the number of bits at each internal node is fixed to a certain number of bits. The most significant bit (MSB) of the result after each operation will be kept up to a certain number of bits for each internal node, and the tail will be truncated. However, this conventional fixed-point arithmetic affects the

Table 4.2 Four-bit two's-complement integer. Four bits can represent values in the range of -8 to 7

Two's complement		Decimal
0	111	7
0	110	6
0	001	1
0	000	0
1	111	-1
1	001	-7
1	000	-8

The sign of a number (given in the first column) is encoded in the most significant bit (MSB)

invertibility of the transform because the DFT coefficients are quantized. Lifting scheme is a way to quantize the DFT coefficients that preserves the invertibility property [I-6].

The integer FFT and fixed-point FFT are compared in noise reduction application (Table 4.3). At low power, i.e., the coefficients are quantized to low resolution, the IntFFT yields significantly better results than the FxpFFT, and they yield similar results at high power [I-6].

While for two and higher dimensions, the row-column method, the vector-radix FFT and the polynomial transform FFT algorithms are commonly used fast algorithms for computing multidimensional discrete Fourier transform (M-D DFT). The application of the integer approach to the polynomial transform FFT for the $(N \times N)$ two-dimensional integer FFT is described in [I-34] using radix-2. The proposed method can be readily generalized to the split vector-radix and row-column algorithms.

The radix-2^2 algorithm is characterized by the property that it has the same complex multiplication computational complexity as the radix-4 FFT algorithm, but still retains the same butterfly (BF) structures as the radix-2 FFT algorithm (Table 4.4). The multiplicative operations are in a more regular arrangement as the non-trivial multiplications appear after every two BF stages. This spatial regularity provides a great advantage in hardware implementation if pipeline behavior is taken into consideration.

In the widely used OFDM systems [O2], the inverse DFT and DFT pair are used to modulate and demodulate the data constellation on the sub-carriers. The input to the IDFT at the transmitter side is a set of digitally modulated signals. Assuming the 64-QAM scheme is adopted, then the input levels are ±1, ±3, ±5, and ±7, which

Table 4.3 Computational complexities (the numbers of real multiplies and real adds) of the split-radix FFT and its integer versions (FxpFFT and IntFFT) when the coefficients are quantized/ rounded off to $N_c = 10$ bits at each stage. [I-6] © 2002 IEEE

N	Split-radix FFT		FxpFFT		IntFFT	
	Multiplies	Adds	Adds	Shifts	Adds	Shifts
16	20	148	262	144	202	84
32	68	388	746	448	559	261
64	196	964	1,910	1,184	1,420	694
128	516	2,308	4,674	2,968	3,448	1,742
256	1,284	5,380	10,990	7,064	8,086	4,160
512	3,076	12,292	25,346	16,472	18,594	9,720
1024	7,172	27,652	57,398	37,600	41,997	22,199

Table 4.4 Number of nontrivial complex multiplies. A set of complex multiply is three real multiplies. [I-33] © 2006 IEEE

N	Radix-2	Radix-2^2	Split-radix
16	10	8	8
64	98	76	72
256	642	492	456
1,024	3,586	2,732	2,504

can be represented by a six-bit vector. The output of the IDFT consists of the time-domain samples to be transmitted over the real channel. Accordingly, at the receiver side, the DFT is performed.

The input sequence uses 12 bits for both real and imaginary parts. The internal word length and the word length for the lifting coefficients and twiddle factors are set to 12 bits.

Based on the IntFFT, a VLSI feasible radix-2^2 FFT architecture is proposed and verified by Chang and Nguyen [I-33]. The most important feature of the IntFFT is that it guarantees the invertibility as well as provides accuracy comparable with the conventional FxpFFT. The required number of real multipliers is also reduced because the lifting scheme (LS) uses one fewer multiplier than general complex multipliers. Compared to FxpFFT designs, the system simulations prove that IntFFT-based architecture can also be adopted by OFDM systems [O2] and yield comparative bit error rate (BER) performance, even if the noisy channel is present.

4.4 Integer Discrete Fourier Transform

Integer Fourier transform approximates the DFT for the fixed-point multiplications [I-5]. The fixed-point multiplications can be implemented by the addition and binary shifting operations. For example

$$7 \times a = a \ll 2 + a \ll 1 + a$$

where a is an integer and $\ll$ is a binary left-shift operator.

Two types of integer transforms are presented in this section. Forward and inverse transform matrices can be the same and different. They are referred to as *near-complete* and *complete* integer DFTs.

4.4.1 Near-Complete *Integer DFT*

Let $[F]$ be the DFT matrix, and let $[F_i]$ be an integer DFT. Then for integer DFT to be orthogonal and, hence, be reversible, it is required that

$$[F_i]^*[F_i]^T = [F_i][F_i]^H = \text{diag}(r_0, r_1, \ldots, r_7) = [C] \qquad (4.15)$$

where $[F_i]^H$ denotes the transpose of $[F_i]^*$ and $r_l = 2^m$, where m is an integer. Thus it follows that

$$[C]^{-1}[F_i][F_i]^H = [I] \qquad (4.16)$$

To approximate the DFT, integer DFT $[F_i]$ keeps all the signs of entries of $[F]$ as follows.

$$[F_i] = \begin{bmatrix} 1 & 1 & 1 & 1 & 1 & 1 & 1 & 1 \\ a_1 & a_2 - ja_2 & -ja_1 & -a_2 - ja_2 & -a_1 & -a_2 + ja_2 & ja_1 & a_2 + ja_2 \\ 1 & -j & -1 & j & 1 & -j & -1 & j \\ b_1 & -b_2 - jb_2 & jb_1 & b_2 - jb_2 & -b_1 & b_2 + jb_2 & -jb_1 & -b_2 + jb_2 \\ 1 & -1 & 1 & -1 & 1 & -1 & 1 & -1 \\ b_1 & -b_2 + jb_2 & -jb_1 & b_2 + jb_2 & -b_1 & b_2 - jb_2 & jb_1 & -b_2 - jb_2 \\ 1 & j & -1 & -j & 1 & j & -1 & -j \\ a_1 & a_2 + ja_2 & ja_1 & -a_2 + ja_2 & -a_1 & -a_2 - ja_2 & -ja_1 & a_2 - ja_2 \end{bmatrix}$$
$$\tag{4.17}$$

In order for (4.15) to be satisfied, the complex inner products of the following row pairs of $[F_i]$ should be zero. When Row 2 represents the second row

$$\langle \text{Row 2, Row 6} \rangle = 0 \qquad \langle \text{Row 4, Row 8} \rangle = 0 \tag{4.18}$$

Here a complex inner product is defined by

$$\langle \underline{z}, \underline{w} \rangle = \underline{w}^H \underline{z}$$

for complex vectors $\underline{z}$ and $\underline{w}$. The vector $\underline{w}^H$ is the transpose of $\underline{w}^*$. From (4.18)

$$a_1 b_1 = 2 a_2 b_2 \Rightarrow \quad a_1 \geq a_2 \quad b_1 \geq b_2 \tag{4.19}$$

From (4.15)

$$r_0 = r_2 = r_4 = r_6 = N$$

$$r_1 = r_7 = (N/2)\, a_1^2 + N\, a_2^2$$

$$r_3 = r_5 = (N/2)\, b_1^2 + N\, b_2^2$$

Some possible choices of the parameters of eight-point integer DFT are listed in Table 4.5.

Table 4.5 Some sets of parameter values of eight-point integer DFT. [I-5] © 2000 IEEE

a_1	2	3	4	5	8	10	17	99	500
a_2	1	2	3	3	5	7	12	70	353
b_1	1	4	3	6	5	7	24	140	706
b_2	1	3	2	5	4	5	17	99	500

$[F_i]$ Only row vectors are orthogonal.

$\Downarrow$

Thus $[F_i][F_i]^H = $ diagonal matrix; $[F_i][F_i]^H \neq $ diagonal matrix

$\Downarrow$

$[\tilde{F}_i][\tilde{F}_i]^H = $ diagonal matrix; $[\tilde{F}_i]^H[\tilde{F}_i] = $ diagonal matrix
where $[\tilde{F}_i]$ is $[F_i]$ normalized by the first column as defined in (4.34).

$\Downarrow$

$$[\tilde{F}_i] = ([C]^{1/2})^{-1}[F_i] \tag{4.20}$$

where $[C]$ is defined in (4.15). Then $[\tilde{F}_i]^{-1} = [\tilde{F}_i]^H$, i.e., $[\tilde{F}_i]$ is unitary.

4.4.2 Complete *Integer DFT*

Let $[F]$ be the DFT matrix, and let $[F_i]^T$ the transpose of (4.17) and $[IF]^*$ be the forward and inverse integer DFTs.

$$[IF] = \begin{bmatrix} 1 & 1 & 1 & 1 & 1 & 1 & 1 & 1 \\ a_3 & a_4 - ja_4 & -ja_3 & -a_4 - ja_4 & -a_3 & -a_4 + ja_4 & ja_3 & a_4 + ja_4 \\ 1 & -j & -1 & j & 1 & -j & -1 & j \\ b_3 & -b_4 - jb_4 & jb_3 & b_4 - jb_4 & -b_3 & b_4 + jb_4 & -jb_3 & -b_4 + jb_4 \\ 1 & -1 & 1 & -1 & 1 & -1 & 1 & -1 \\ b_3 & -b_4 + jb_4 & -jb_3 & b_4 + jb_4 & -b_3 & b_4 - jb_4 & jb_3 & -b_4 - jb_4 \\ 1 & j & -1 & -j & 1 & j & -1 & -j \\ a_3 & a_4 + ja_4 & ja_3 & -a_4 + ja_4 & -a_3 & -a_4 - ja_4 & -ja_3 & a_4 - ja_4 \end{bmatrix} \tag{4.21}$$

Then for integer DFT to be orthogonal and, hence, be reversible, it is required that

$$[IF]^*[F_i]^T = \mathrm{diag}(r_0, r_1, \ldots, r_7) = [D] = \text{diagonal matrix} \tag{4.22}$$

where $r_l = 2^m$ and m is an integer. Since $[D]$ is a diagonal matrix, it follows that

$$[D]^{-1}[IF]^*[F_i]^T = [I] \tag{4.23}$$

From the constraint of (4.22), the complex inner products of the following pairs should be zero.

$$\langle \text{Row 2 of } [F_i], \text{Row 6 of } [IF]\rangle = \langle \text{Row 8 of } [F_i], \text{Row 4 of } [IF]\rangle = 0$$
$$\langle \text{Row 4 of } [F_i], \text{Row 8 of } [IF]\rangle = \langle \text{Row 6 of } [F_i], \text{Row 2 of } [IF]\rangle = 0$$
$$a_1 b_3 = 2a_2 b_4 \qquad a_3 b_1 = 2a_4 b_2 \tag{4.24}$$

Since the inner product of the corresponding rows of $[F_i]$ and $[IF]$ should be the power of two from the constraint of (4.22),

$$a_1 a_3 + 2a_2 a_4 = 2^k \qquad b_1 b_3 + 2b_2 b_4 = 2^h \qquad (4.25)$$

$$a_1 \geq a_2 \quad b_1 \geq b_2 \quad a_3 \geq a_4 \quad b_3 \geq b_4 \qquad (4.26)$$

From (4.24), we set

$$b_3 = 2a_2 \qquad b_4 = a_1 \qquad a_3 = 2b_2 \qquad a_4 = b_1 \qquad (4.27)$$

Then (4.25) becomes

$$2(a_1 b_2 + a_2 b_1) = 2^k \qquad 2(b_1 a_2 + c_2 a_1) = 2^h \qquad (4.28)$$

1. Choose a_1 and a_2 such that they are integers and

$$2a_2 \geq a_1 \geq a_2$$

2. Choose b_1 and b_2 such that they are integers and

$$2b_2 \geq b_1 \geq b_2 \qquad a_1 b_2 + a_2 b_1 = 2^n$$

where n is an integer.

3. Set a_3, a_4, b_3, b_4 as

$$b_3 = 2^{h+1} a_2 \qquad b_4 = 2^h a_1 \qquad a_3 = 2^{h+1} b_2 \qquad a_4 = 2^h b_1$$

where h is an integer.

Substitute (4.17) and (4.21) into (4.22).

$$r_0 = r_2 = r_4 = r_6 = N$$

$$r_1 = r_7 = (N/2)a_1 a_3 + N a_2 a_4$$

$$r_3 = r_5 = (N/2)b_1 b_3 + N b_2 b_4$$

Some possible choices of the parameters of eight-point integer DFT are listed in Table 4.6. The eight-point integer DFT retains some properties of the regular DFT.

Table 4.6 Some sets of parameter values of eight-point integer DFT. [I-5] © 2000 IEEE							
a_1	2	7	3	4	4	5	10
a_2	1	5	2	3	3	4	7
b_1	2	13	17	12	44	17	18
b_2	1	9	10	7	31	12	13
a_3	1	18	34	7	31	24	13
a_4	1	13	10	6	22	17	9
b_3	1	10	4	3	3	8	7
b_4	1	7	3	2	2	5	5

4.4.3 Energy Conservation

Only the rows of $[F_i]$ are orthogonal and the columns are not.

$$[F_i][IF]^H = [D] \tag{4.29}$$

where $[D]$ defined in (4.22) is diagonal and its entries are integers.
Let $\underline{X} = [F_i]^T \underline{x}$ and $\underline{Y} = ([D]^{-1}[IF])^T \underline{y}$. Then the energy conservation property is as follows.

$$\underline{x}^T \underline{y}^* = \underline{X}^T \underline{Y}^* \tag{4.30}$$

Proof.

$$\underline{X}^T \underline{Y}^* = \left([F_i]^T \underline{x} \right)^T \left([IF]^T [D]^{-1} \underline{y} \right)^* = \underline{x}^T [F_i][IF]^H [D]^{-1} \underline{y}^* = \underline{x}^T \underline{y}^* \tag{4.31}$$

4.4.4 Circular Shift

Let

$$X^i(k) = \text{intDFT} [x(n)] \qquad Y^i(k) = \text{intDFT} [x(n+h)] \tag{4.32}$$

where $x(n+h)$ is defined in (2.17). Then

$$Y^i(k) \approx X^i(k) W_N^{-hk} \quad \text{when both } k \text{ and } h \text{ are odd} \tag{4.33a}$$

$$Y^i(k) = X^i(k) W_N^{-hk} \quad \text{otherwise} \tag{4.33b}$$

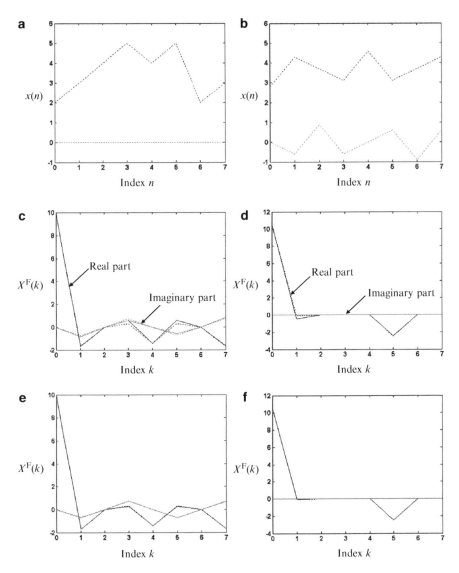

Fig. 4.5 The regular and integer DFTs of input signals are represented by *dashed* and *solid lines*. **a**, **b** Input signals, $\underline{x}_1$, $\underline{x}_2$. **c**, **d** *Near-complete* integer DFTs of $\underline{x}_1$, $\underline{x}_2$. **e**, **f** *Complete* integer DFTs of $\underline{x}_1$, $\underline{x}_2$. [I-5] © 2000 IEEE

Example 4.2   Figure 4.5 shows the near-complete and complete integer DFTs of two random input vectors $\underline{x}_1$ and $\underline{x}_2$.

$$\underline{x}_1 = (2, 3, 4, 5, 4, 5, 2, 3)^T$$
$$\underline{x}_2 = (2.8, \quad 4.3 - j0.6, \quad 3.7 + j0.9, \quad 3.1 - j0.6,$$
$$4.6, \quad 3.1 + j0.6, \quad 3.70 - j0.9, \quad 4.3 + j0.6)^T$$

A parameter set is chosen for the near-complete integer DFT:

$$\{a_1 = 2, \ a_2 = 1, \ b_1 = 1, \ b_2 = 1\}$$

A parameter set is chosen for the complete integer DFT:

$$\{a_1 = 7, \ a_2 = 5, \ b_1 = 13, \ b_2 = 9, \ a_3 = 18, \ a_4 = 13, \ b_3 = 10, \ b_4 = 7\}$$

Entries $F_i(k, n)$ of $[F_i]$ are normalized by the first column $F_i(k, 0)$ as

$$\tilde{F}_i(k, n) = F_i(k, n)/F_i(k, 0) \qquad k, n = 0, 1, ..., N - 1 \qquad (4.34)$$

Integer DFT vector is computed for both the near-complete and complete integer DFTs as follows.

$$\underline{X}_1^T = [\tilde{F}_i]^T \underline{x}_1^T \qquad (4.35)$$

$[F_i]^T$ can be normalized differently using $[IF][F_i]^H = [D]$ of (4.29) to get the normalized integer DFT $[\tilde{F}_i]^T$ as

$$[\tilde{F}_i] = \left([D]^{1/2} \right)^{-1} [F_i] \qquad (4.36)$$

where diagonal matrix $[D]$ is defined in (4.22). Similarly

$$[I\tilde{F}] = \left([D]^{1/2} \right)^{-1} [IF] \qquad (4.37)$$

Then $[\tilde{F}_i]^{-1} = [I\tilde{F}]^H$, i.e., $[\tilde{F}_i]$ and $[I\tilde{F}]$ are biorthogonal.

4.5   Summary

This chapter has developed the integer FFT (IntFFT) based on the lifting scheme. Its advantages are enumerated. A specific algorithm (eight-point IntFFT) using split-radix structure is developed. Extension of the 1-D DFT to the multi-D DFT (specifically 2-D DFT) is the focus of the next chapter. Besides the definitions and properties, filtering of 2-D signals such as images and variance distribution in the DFT domain are some relevant topics.

4.6 Problems

4.1 If $\det[A] = 1$ and $b \neq 0$,

$$[A] = \begin{bmatrix} a & b \\ c & d \end{bmatrix} = \begin{bmatrix} 1 & 0 \\ (d-1)/b & 1 \end{bmatrix} \begin{bmatrix} 1 & b \\ 0 & 1 \end{bmatrix} \begin{bmatrix} 1 & 0 \\ (a-1)/b & 1 \end{bmatrix} \qquad \text{(P4.1)}$$

Assume $c \neq 0$. Derive (4.5) from (P4.1).

4.2 Develop a flow-graph for implementing eight-point inverse integer FFT using split-radix structure (see Fig. 4.4).

4.3 Repeat Problem 4.2 for $N = 16$ for forward and inverse integer FFTs.

4.4 List five other parameter sets for the integer DFT than those in Table 4.5. What equation do you need?

4.7 Projects

4.1 Repeat the simulation described in Example 4.2 about integer DFTs and obtain the results shown in Fig. 4.5.

Chapter 5
Two-Dimensional Discrete Fourier Transform

5.1 Definitions

Two-dimensional DFT has applications in image/video processing. The extension from 1-D to 2-D for the DFT is straightforward. The 2-D DFT and its inverse can be defined as

$$X^F(k_1, k_2) = \sum_{n_1=0}^{N_1-1} \sum_{n_2=0}^{N_2-1} x(n_1, n_2) W_{N_1}^{n_1 k_1} W_{N_2}^{n_2 k_2} \qquad \begin{array}{l} \text{2D } (N_1 \times N_2) \text{ DFT} \\ \text{(Rectangular Array)} \end{array}$$

$$k_1 = 0, 1, \ldots, N_1 - 1 \quad \text{and} \quad k_2 = 0, 1, \ldots, N_2 - 1$$

(5.1a)

$$x(n_1, n_2) = \frac{1}{N_1 N_2} \sum_{k_1=0}^{N_1-1} \sum_{k_2=0}^{N_2-1} X^F(k_1, k_2) W_{N_1}^{-n_1 k_1} W_{N_2}^{-n_2 k_2} \quad \text{2D } (N_1 \times N_2) \text{ IDFT}$$

$$n_1 = 0, 1, \ldots, N_1 - 1 \quad \text{and} \quad n_2 = 0, 1, \ldots, N_2 - 1$$

(5.1b)

where $W_{N_1} = \exp(-j2\pi/N_1)$ and $W_{N_2} = \exp(-j2\pi/N_2)$.

$x(n_1, n_2)$ is the uniformly sampled sequence in the 2-D discrete spatial domain (note that the sampling intervals along horizontal and vertical coordinates in the spatial domain need not be the same), and $X^F(k_1, k_2)$ is the DFT coefficient in the 2-D discrete frequency domain. The normalization factor $N_1 N_2$ can be, as in the 1-D case, distributed equally between the forward and inverse DFTs or moved entirely to the forward DFT. A unitary 2-D DFT and its inverse can be defined as

$$X^F(k_1, k_2) = \frac{1}{\sqrt{N_1 N_2}} \sum_{n_1=0}^{N_1-1} \sum_{n_2=0}^{N_2-1} x(n_1, n_2) W_{N_1}^{n_1 k_1} W_{N_2}^{n_2 k_2}$$

$$k_1 = 0, 1, \ldots, N_1 - 1 \quad \text{and} \quad k_2 = 0, 1, \ldots, N_2 - 1$$

(5.2a)

K.R. Rao et al., *Fast Fourier Transform: Algorithms and Applications*,
Signals and Communication Technology,
DOI 10.1007/978-1-4020-6629-0_5, © Springer Science+Business Media B.V. 2010

$$x(n_1, n_2) = \frac{1}{\sqrt{N_1 N_2}} \sum_{k_1=0}^{N_1-1} \sum_{k_2=0}^{N_2-1} X^{\mathrm{F}}(k_1, k_2) W_{N_1}^{-n_1 k_1} W_{N_2}^{-n_2 k_2} \tag{5.2b}$$

$$n_1 = 0, 1, \ldots, N_1 - 1 \quad \text{and} \quad n_2 = 0, 1, \ldots, N_2 - 1$$

While N_1 and N_2 can be of any dimension, for purposes of simplicity we will assume $N_1 = N_2 = N$. All the concepts, theorems, properties, algorithms, etc., that are to be developed for $N_1 = N_2$ are equally valid for $N_1 \neq N_2$. Hence the 2-D DFT pair described in (5.1) can be simplified as

$$X^{\mathrm{F}}(k_1, k_2) = \sum_{n_1=0}^{N-1} \sum_{n_2=0}^{N-1} x(n_1, n_2) W_N^{(n_1 k_1 + n_2 k_2)} \quad k_1, k_2 = 0, 1, \ldots, N - 1 \tag{5.3a}$$

$$x(n_1, n_2) = \frac{1}{N^2} \sum_{k_1=0}^{N-1} \sum_{k_2=0}^{N-1} X^{\mathrm{F}}(k_1, k_2) W_N^{-(n_1 k_1 + n_2 k_2)} \quad n_1, n_2 = 0, 1, \ldots, N - 1 \tag{5.3b}$$

Denote this pair as $x(n_1, n_2) \Leftrightarrow X^{\mathrm{F}}(k_1, k_2)$.

The separable property of 2-D DFT can be illustrated by rewriting (5.3a) as

$$X^{\mathrm{F}}(k_1, k_2) = \sum_{n_2=0}^{N-1} \left(\sum_{n_1=0}^{N-1} x(n_1, n_2) W_N^{n_1 k_1} \right) W_N^{n_2 k_2} \tag{5.4a}$$

$$k_1, k_2 = 0, 1, \ldots, N - 1$$

and

$$x(n_1, n_2) = \frac{1}{N} \sum_{k_2=0}^{N-1} \left(\frac{1}{N} \sum_{k_1=0}^{N-1} X^{\mathrm{F}}(k_1, k_2) W_N^{-n_1 k_1} \right) W_N^{-n_2 k_2} \tag{5.4b}$$

$$n_1, n_2 = 0, 1, \ldots, N - 1$$

Equation (5.4a) can be recognized as 1-D DFT of the 2-D sequence $x(n_1, n_2)$ along columns followed by 1-D DFT of the semitransformed sequence along rows. Similarly, (5.4b) is 1-D IDFT of $X^{\mathrm{F}}(k_1, k_2)$ along columns followed by 1-D IDFT along rows. The column-row operations of the 1-D DFT and 1-D IDFT can be interchanged, by rearranging (5.4) as

$$X^{\mathrm{F}}(k_1, k_2) = \sum_{n_1=0}^{N-1} \left(\sum_{n_2=0}^{N-1} x(n_1, n_2) W_N^{n_2 k_2} \right) W_N^{n_1 k_1} \tag{5.5a}$$

$$k_1, k_2 = 0, 1, \ldots, N - 1$$

$$x(n_1, n_2) = \frac{1}{N} \sum_{k_1=0}^{N-1} \left(\frac{1}{N} \sum_{k_2=0}^{N-1} X^F(k_1, k_2) W_N^{-n_2 k_2} \right) W_N^{-n_1 k_1}$$

$$n_1, n_2 = 0, 1, \ldots, N - 1 \tag{5.5b}$$

This is shown in Fig. 5.1.

An $(N \times N)$ 2-D DFT can be therefore implemented by $2N$ 1-D DFTs each of length N. For efficient implementation of 2-D DFT, the FFT algorithms developed in Chapter 3 for the 1-D case can be utilized. Assuming N is an integer power of two, as radix-2 1-D FFT (Chapter 3) requires $N \log_2 N$ complex adds and $\frac{N}{2} \log_2 N$ complex multiplies, a radix-2 $(N \times N)$ 2-D FFT requires $2N^2 \log_2 N$ complex adds and $N^2 \log_2 N$ complex multiplies. It is obvious that all these properties are equally valid for the 2-D IDFT.

The 2-D DFT/IDFT described in (5.3) can be represented in matrix form as (separable property):

$$[X^F(k_1, k_2)] = [F][x(n_1, n_2)][F]$$

$$(N \times N) \qquad (N \times N) \tag{5.6a}$$

and

$$[x(n_1, n_2)] = \frac{1}{N^2} [F]^* [X^F(k_1, k_2)] [F]^*$$

$$(N \times N) \qquad (N \times N) \tag{5.6b}$$

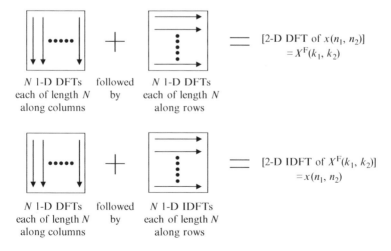

N 1-D DFTs   followed N 1-D DFTs
each of length N by   each of length N
along columns   along rows

= [2-D DFT of $x(n_1, n_2)$]
= $X^F(k_1, k_2)$

N 1-D DFTs   followed N 1-D IDFTs
each of length N by   each of length N
along columns   along rows

= [2-D IDFT of $X^F(k_1, k_2)$]
= $x(n_1, n_2)$

Fig. 5.1 Separable properties of 2-D DFT and 2-D IDFT

where

$$[x(n_1, n_2)] = \begin{bmatrix} x(0,0) & x(0,1) & \cdots & x(0,N-1) \\ x(1,0) & x(1,1) & \cdots & x(1,N-1) \\ \vdots & \vdots & \ddots & \vdots \\ x(N-1,0) & x(N-1,1) & \cdots & x(N-1,N-1) \end{bmatrix}$$ (5.7a)

$(N \times N)$

and

$$[X^F(k_1, k_2)] = \begin{bmatrix} X^F(0,0) & X^F(0,1) & \cdots & X^F(0,N-1) \\ X^F(1,0) & X^F(1,1) & \cdots & X^F(1,N-1) \\ \vdots & \vdots & \ddots & \vdots \\ X^F(N-1,0) & X^F(N-1,1) & \cdots & X^F(N-1,N-1) \end{bmatrix}$$ (5.7b)

$(N \times N)$

Substitute (5.6a) in (5.6b) to get,

$$\frac{1}{N^2} [F]^* [F] [x(n_1, n_2)] [F][F]^* = [x(n_1, n_2)]$$

Note that $[F]^*[F] = N[I_N]$. $\left([F] = [F]^T\right)$

It can be shown easily that (5.6) can be equivalently represented as

$$[\underline{X}^F(k_1, k_2)]_{\text{LO}} = ([F] \otimes [F]) [\underline{x}(n_1, n_2)]_{\text{LO}}$$ (5.8a)

$(N^2 \times 1)$ $\qquad\qquad\qquad (N^2 \times 1)$

and

$$[\underline{x}(n_1, n_2)]_{\text{LO}} = \frac{1}{N^2} ([F]^* \otimes [F]^*) [\underline{X}^F(k_1, k_2)]_{\text{LO}}$$ (5.8b)

$(N^2 \times 1)$ $\qquad\qquad\qquad (N^2 \times 1)$

where
$[\underline{x}(n_1, n_2)]_{\text{LO}} = [x(0,0), x(0,1), \ldots, x(0,N-1), x(1,0), x(1,1), \ldots, x(1,N-1),$
$x(N-1,1\ldots, x(N-1,0),), \ldots, x(N-1, N-1)]^T$ is a $(N^2 \times 1)$ column vector.
This is obtained by reordering each row of (5.7a) as a column of N elements. This
rearrangement is called *lexicographic ordering* (LO). By lexicographic ordering of
$[X^F(k_1, k_2)]$, $[\underline{X}^F(k_1, k_2)]_{\text{LO}}$ is obtained.

The symbol $\otimes$ denotes Kronecker product (also called matrix product or direct
product) defined as (see Appendix E)

$$[A] \otimes [B] = \begin{bmatrix} a_{11} & a_{12} & \cdots & a_{1n} \\ a_{21} & a_{22} & \cdots & a_{2n} \\ \vdots & \vdots & \ddots & \vdots \\ a_{m1} & a_{m2} & \cdots & a_{mn} \end{bmatrix} \otimes \begin{bmatrix} b_{11} & b_{12} & \cdots & b_{1q} \\ b_{21} & b_{22} & \cdots & b_{2q} \\ \vdots & \vdots & \ddots & \vdots \\ b_{p1} & b_{p2} & \cdots & b_{pq} \end{bmatrix}$$

$$(m \times n) \qquad\qquad (p \times q)$$

$$= \begin{bmatrix} a_{11}[B] & a_{12}[B] & \cdots & a_{1n}[B] \\ a_{21}[B] & a_{22}[B] & \cdots & a_{2n}[B] \\ \vdots & \vdots & \ddots & \vdots \\ a_{m1}[B] & a_{m2}[B] & \cdots & a_{mn}[B] \end{bmatrix} = [C]$$

$$(mp \times nq) \qquad\qquad (mp \times nq)$$

(5.9)

Note that in general $[A] \otimes [B] \neq [B] \otimes [A]$.

5.2 Properties

The various properties that have been developed for the 1-D DFT are equally valid for 2-D DFT.

5.2.1 Periodicity

Both $x(n_1, n_2)$ and $X^F(k_1, k_2)$ are periodic along both dimensions with period N i.e.,

$$x(n_1 + N, n_2) = x(n_1, n_2 + N) = x(n_1 + N, n_2 + N) = x(n_1, n_2) \qquad (5.10a)$$

$$X^F(k_1 + N, k_2) = X^F(k_1, k_2 + N) = X^F(k_1 + N, k_2 + N) = X^F(k_1, k_2) \qquad (5.10b)$$

5.2.2 Conjugate Symmetry

When $x(n_1, n_2)$ is real:

$$X^F\left(\tfrac{N}{2} \pm k_1, \tfrac{N}{2} \pm k_2\right) = \left[X^F\left(\tfrac{N}{2} \mp k_1, \tfrac{N}{2} \mp k_2\right)\right]^*, \quad k_1, k_2 = 0, 1, \ldots, \tfrac{N}{2} - 1 \quad (5.11a)$$

and

$$X^{\mathrm{F}}(k_1, k_2) = [X^{\mathrm{F}}(N - k_1, N - k_2)]^*, \quad k_1, k_2 = 0, 1, \ldots, N - 1 \qquad (5.11\mathrm{b})$$

This implies of the N^2 DFT coefficients only the DFT coefficients in the cross hatched region are unique (Fig. 5.2a). Specific conjugate pairs of DFT coefficients for real data are shown in Fig. 5.2b for $M = N = 8$.

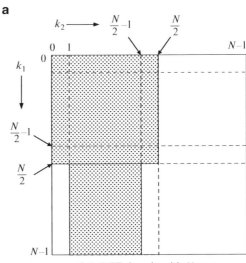

2-D DFT domain with N even

Fig. 5.2 a When $x(n_1, n_2)$ is real, the DFT coefficients in the cross hatched region are unique. The numbers of those coefficients are $(N^2/2) + 2$ for $N =$ even, and $(N^2 + 1)/2$ for $N =$ odd. **b** Conjugate symmetry of DFT coefficients for real image data, where M and N are even numbers $(M = N = 8)$

5.2.3 Circular Shift in Time/Spatial Domain (Periodic Shift)

$$x(n_1, n_2) \quad \Leftrightarrow \quad X^F(k_1, k_2)$$

$$x(n_1 - m_1, n_2 - m_2) \quad \Leftrightarrow \quad X^F(k_1, k_2)W_N^{k_1 m_1 + k_2 m_2} \tag{5.12}$$

where $x(n_1 - m_1, n_2 - m_2)$ is circular shift of $x(n_1, n_2)$ by m_1 samples along n_1 and m_2 samples along n_2. Since $\left| W_N^{k_1 m_1 + k_2 m_2} \right| = 1$, the amplitude and power spectra of $x(n_1, n_2)$ are invariant to its circular shift.

5.2.4 Circular Shift in Frequency Domain (Periodic Shift)

$$x(n_1, n_2)W_N^{-(n_1 u_1 + n_2 u_2)} \quad \Leftrightarrow \quad X^F(k_1 - u_1, k_2 - u_2) \tag{5.13a}$$

where $X^F(k_1 - u_1, k_2 - u_2)$ is circular shift of $X^F(k_1, k_2)$ by u_1 samples along k_1 and u_2 samples along k_2. A special case of this circular shift is of interest when $u_1 = u_2 = \frac{N}{2}$.
Then

$$x(n_1, n_2)(-1)^{n_1 + n_2} \Leftrightarrow X^F\left(k_1 - \tfrac{N}{2}, k_2 - \tfrac{N}{2}\right) \quad \text{for } N = \text{an even integer} \tag{5.13b}$$

as $W_N^{N/2} = -1$ and $W_N^{-(n_1 + n_2)N/2} = \exp\left[\frac{\pm j 2\pi}{N}(n_1 + n_2)\frac{N}{2}\right] = e^{j\pi(n_1 + n_2)} = (-1)^{n_1 + n_2}$.
The dc coefficient $X^F(0, 0)$ top left corner in Fig. 5.3a is now shifted to the center of the 2-D frequency plane (Fig. 5.3b). (For the same reason, $x\left(n_1 - \frac{N}{2}, n_2 - \frac{N}{2}\right) \Leftrightarrow (-1)^{k_1 + k_2}X^F(k_1, k_2)$ for N even.)

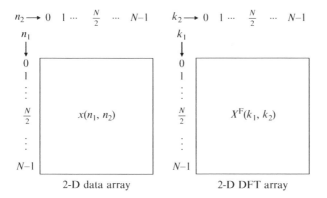

2-D data array   2-D DFT array

$$x(n_1, n_2) \quad \Leftrightarrow \quad X^F(k_1, k_2)$$

a

$f_{s1} = 1/T_1 = $ Sampling rate (vertically)
$f_{s2} = 1/T_2 = $ Sampling rate (horizontally)

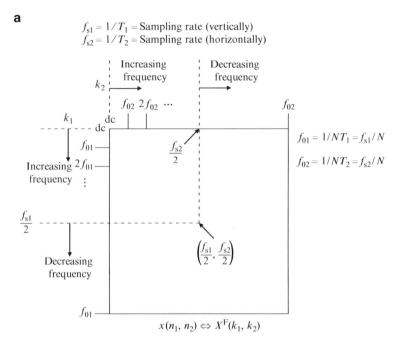

$f_{01} = 1/NT_1 = f_{s1}/N$

$f_{02} = 1/NT_2 = f_{s2}/N$

$x(n_1, n_2) \Leftrightarrow X^F(k_1, k_2)$

b

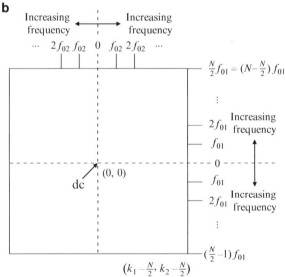

Fig. 5.3 The 2-D DFT of **a** $x(n_1, n_2)$ and **b** $x(n_1, n_2)(-1)^{n_1 + n_2}$

Assuming the sampling intervals along n_1 and n_2 are T_1 and T_2, ($f_{s1} = 1/T_1$ and $f_{s2} = 1/T_2$ are the corresponding sampling rates, # of samples/meter) the frequency resolutions along k_1 and k_2 are $f_{01} = \frac{1}{NT_1}$ and $f_{02} = \frac{1}{NT_2}$ respectively.

5.2.5 Skew Property

$$x(n_1 - mn_2, n_2) \quad \Leftrightarrow \quad X^F(k_1, k_1 + mk_2) \tag{5.14}$$

A skew of m in one dimension of an image is equivalent to skew of the spectrum of that image by $(-m)$ in the other dimension [IP26]. For example let m be 1.

$$[x] = \begin{pmatrix} 4 & 5 & 6 & 0 \\ 1 & 2 & 3 & 0 \\ 0 & 0 & 0 & 0 \\ 0 & 0 & 0 & 0 \end{pmatrix} \Leftrightarrow [X^F] = \begin{pmatrix} 21 & -4-j7 & 7 & -4+j7 \\ 15-j6 & -4-j3 & 5-j2 & -j7 \\ 9 & -j3 & 3 & j3 \\ 15+j6 & -j7 & 5+j2 & -4+j3 \end{pmatrix}$$

$$[y] = \begin{pmatrix} 4 & 0 & 0 & 0 \\ 1 & 5 & 0 & 0 \\ 0 & 2 & 6 & 0 \\ 0 & 0 & 3 & 0 \end{pmatrix} \Leftrightarrow [Y^F] = \begin{pmatrix} 21 & -4-j7 & 7 & -4+j7 \\ -4-j3 & 5-j2 & j7 & 15-j6 \\ 3 & j3 & 9 & -j3 \\ -4+j3 & 15+j6 & -j7 & 5+j2 \end{pmatrix}$$

Note that zeros are padded in the spatial domain, and DFT coefficients in each row of $[Y^F]$ are circularly shifted. The proof of this property is shown in Appendix F.2.

5.2.6 Rotation Property

Rotating the image by an angle θ in the spatial domain causes its 2D-DFT to be rotated by the same angle in the frequency domain [E5].

$$x(n_1 \cos\theta - n_2 \sin\theta, n_1 \sin\theta + n_2 \cos\theta) \quad \Leftrightarrow \quad X^F(k_1 \cos\theta - k_2 \sin\theta, k_1 \sin\theta + k_2 \cos\theta) \tag{5.15}$$

where an $N \times N$ square grid on which the image $x(n_1, n_2)$ is rotated by the angle θ in the counterclockwise direction.

Note that the grid is rotated so the new grid points may not be defined. The value of the image at the nearest valid grid point can be estimated by interpolation (see Section 8.4).

5.2.7 Parseval's Theorem

This is the energy conservation property of any unitary transform i.e., energy is preserved under orthogonal transformation. This states that

$$\sum_{n_1=0}^{N-1}\sum_{n_2=0}^{N-1}|x(n_1,n_2)|^2 = \sum_{n_1=0}^{N-1}\sum_{n_2=0}^{N-1}[x(n_1,n_2)]x^*(n_1,n_2)$$

$$= \sum_{n_1=0}^{N-1}\sum_{n_2=0}^{N-1}\left[\frac{1}{N^2}\sum_{k_1=0}^{N-1}\sum_{k_2=0}^{N-1}X^F(k_1,k_2)W_N^{-(n_1k_1+n_2k_2)}\right]x^*(n_1,n_2)$$

$$= \frac{1}{N^2}\sum_{k_1=0}^{N-1}\sum_{k_2=0}^{N-1}X^F(k_1,k_2)\left[\sum_{n_1=0}^{N-1}\sum_{n_2=0}^{N-1}x(n_1,n_2)W_N^{(n_1k_1+n_2k_2)}\right]^*$$

$$= \frac{1}{N^2}\sum_{k_1=0}^{N-1}\sum_{k_2=0}^{N-1}X^F(k_1,k_2)\left[X^F(k_1,k_2)\right]^*$$

$$= \frac{1}{N^2}\sum_{k_1=0}^{N-1}\sum_{k_2=0}^{N-1}|X^F(k_1,k_2)|^2 \qquad (5.16)$$

5.2.8 Convolution Theorem

Circular convolution of two *periodic* sequences in time/spatial domain is equivalent to multiplication in the 2-D DFT domain. Let $x(n_1,n_2)$ and $y(n_1,n_2)$ be two real periodic sequences with period N along n_1 and n_2. Their circular convolution is given by

$$z_{con}(m_1,m_2) = \frac{1}{N^2}\sum_{n_1=0}^{N-1}\sum_{n_2=0}^{N-1}x(n_1,n_2)y(m_1-n_1,m_2-n_2)$$

$$= x(n_1,n_2) * y(n_1,n_2) \qquad (5.17a)$$

$$m_1,m_2 = 0,1,\ldots,N-1$$

In the 2-D DFT domain, this is equivalent to

$$Z^F_{con}(k_1,k_2) = \frac{1}{N^2}\left[X^F(k_1,k_2)Y^F(k_1,k_2)\right] \qquad (5.17b)$$

$$k_1,k_2 = 0,1,\ldots,N-1$$

where

$$x(n_1,n_2) \Leftrightarrow X^F(k_1,k_2)$$
$$y(n_1,n_2) \Leftrightarrow Y^F(k_1,k_2)$$
$$z_{con}(m_1,m_2) \Leftrightarrow Z^F_{con}(k_1,k_2) \qquad (5.17c)$$
$$z_{con}(m_1,m_2) = \frac{1}{N^4}\sum_{k_1=0}^{N-1}\sum_{k_2=0}^{N-1}X^F(k_1,k_2)Y^F(k_1,k_2)W_N^{-(m_1k_1+m_2k_2)}$$

Example 5.1   Circular convolution of two periodic arrays in the spatial domain (5.17a).

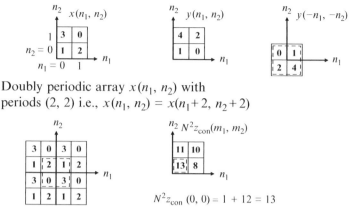

Doubly periodic array $x(n_1, n_2)$ with
periods (2, 2) i.e., $x(n_1, n_2) = x(n_1 + 2, n_2 + 2)$

$$N^2 z_{con}(0, 0) = 1 + 12 = 13$$

The output $z_{con}(m_1, m_2)$ can also be written as a column-ordered vector equation (see Example 2.7 in [B6]).

Example 5.2   Multiplication in the 2-D DFT domain (5.17b) and its inverse transform in MATLAB code are as follows. The input arrays are given in Example 5.1. Note that the matrix representation is a 90° clockwise rotation of the 2-D Cartesian coordinate representation.

```
x = [1 3; 2 0];
y = [1 4; 0 2];
X = fft2(x); Y = fft2(y); z = ifft2(X.*Y);
% z = [13 11;
%       8 10]
```

To obtain a noncircular (aperiodic) convolution of two sequences by the DFT/ FFT approach, the two sequences have to be extended by adding zeros as in the case of noncircular convolution via the DFT/FFT for the 1-D signal (Fig. 2.9). This technique can be illustrated as follows:

As both the DFT and IDFT are periodic, the DFT/FFT approach yields periodic convolution. However by extending the original sequences by adding sufficient number of zeros, a non-periodic convolution can be obtained by the DFT/FFT approach (Fig. 5.4).

Note that $X_e^F(k_1, k_2)$ and $Y_e^F(k_1, k_2)$ denote the 2-D DFTs of the extended sequences $x_{ext}(n_1, n_2)$ and $y_{ext}(n_1, n_2)$ respectively.

5.2.9   Correlation Theorem

Similar to the convolution-multiplication theorem (convolution in time/spatial domain is equivalent to multiplication in the DFT domain or vice versa), an

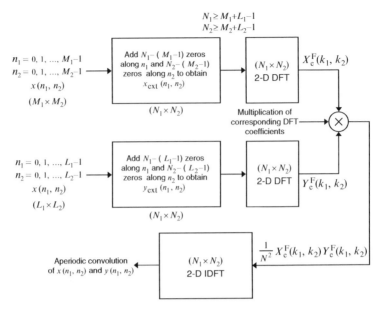

Fig. 5.4 2-D aperiodic (noncircular) convolution using the 2-D DFT. Implementation of DFT and IDFT is by the fast algorithms

analogous relationship exists for the correlation. Analogous to (5.17a), the circular correlation is given by

$$z_{\text{cor}}(m_1, m_2) = \frac{1}{N^2} \sum_{n_1=0}^{N-1} \sum_{n_2=0}^{N-1} x(n_1, n_2)\, y(m_1 + n_1, m_2 + n_2)$$

$$m_1, m_2 = 0, 1, \ldots, N-1$$

(5.18a)

In the 2-D DFT domain this is equivalent to

$$Z_{\text{cor}}^{\text{F}}(k_1, k_2) = \frac{1}{N^2} \left[X^{\text{F}}(k_1, k_2) \right]^* Y^{\text{F}}(k_1, k_2)$$

(5.18b)

where

$$z_{\text{cor}}(m_1, m_2) \quad \Leftrightarrow \quad Z_{\text{cor}}^{\text{F}}(k_1, k_2)$$

To obtain a noncircular (**aperiodic**) correlation of two sequences by the DFT/FFT approach, the two sequences have to be padded by adding zeros (similar to the convolution case). By taking complex conjugation of $X^{\text{F}}(k_1, k_2)$ prior to its multiplication with $Y^{\text{F}}(k_1, k_2)$, Fig. 5.4 can be used to obtain the noncircular correlation.

5.2.10 Spatial Domain Differentiation

$$\frac{\partial^m x(n_1, n_2)}{\partial n_1^m} \Leftrightarrow (jk_1)^m X^F(k_1, k_2) \tag{5.19a}$$

5.2.11 Frequency Domain Differentiation

$$(-jn_1)^m x(n_1, n_2) \Leftrightarrow \frac{\partial^m X^F(k_1, k_2)}{\partial k_1^m} \tag{5.19b}$$

5.2.12 Laplacian

$$\nabla^2 x(n_1, n_2) \quad \Leftrightarrow \quad -\left(k_1^2 + k_2^2\right) X^F(k_1, k_2) \tag{5.20}$$

5.2.13 Rectangle

This is an impulse response model for motion blur in an imaging system (see Figs. 5.5, 5.6a, b). The blurred image $y(n_1, n_2)$ and impulse response $\text{rect}(a, b)$ are expressed as

$$y(n_1, n_2) = \sum_{m_1=a-1}^{0} \sum_{m_2=b-1}^{0} x(n_1 - m_1, n_2 - m_2)$$

$$\text{rect}(a, b) \Leftrightarrow ab \, \text{sinc}\left(\frac{k_1 a}{M}\right) \text{sinc}\left(\frac{k_2 b}{N}\right) e^{-j\pi(k_1 a/M + k_2 b/N)} \tag{5.21}$$

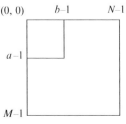

Fig. 5.5 Rectangular function, rect (a, b)

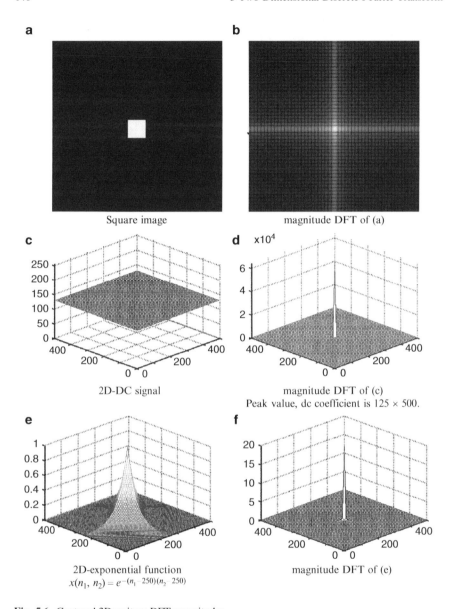

Fig. 5.6 Centered 2D-unitary DFT magnitudes

5.3 Two-Dimensional Filtering

By appropriately weighing the 2-D DFT coefficients, a 2-D signal, $x(n_1, n_2)$ can be correspondingly filtered in the frequency domain (Fig. 5.7).

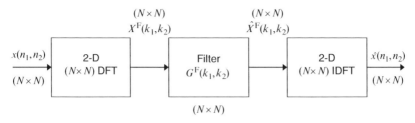

Fig. 5.7 2-D filtering in the 2-D DFT (frequency) domain

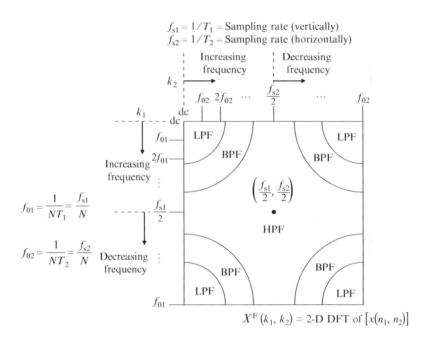

Fig. 5.8 The 2-D DFT coefficients in the corresponding zones are retained with the rest set to zero. LPF: low pass filter, BPF: band pass filter, HPF: high pass filter (zonal filter)

In Fig. 5.7, $\widehat{X}^F(k_1, k_2) = X^F(k_1, k_2)\, G^F(k_1, k_2)$, where $G^F(k_1, k_2)$ is the weighing function (2-D filter). The lowpass (LPF), bandpass (BPF) and highpass (HPF) filters are shown in Figs. 5.8 and 5.9. 2D-DFTs of images are shown in Figs. 5.10 and 5.11.

Fig. 5.9 Zonal filtering
for 2-D DFT of
$[x(n_1, n_2)(-1)^{n_1+n_2}]$.
As in Fig. 5.8, the 2-D
DFT coefficients in the
corresponding zones
are retained with the rest
set to zero

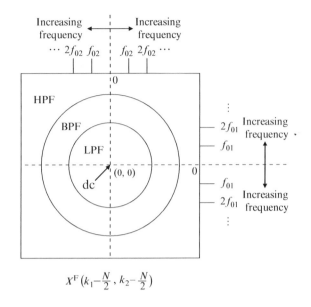

$$X^F\left(k_1 - \frac{N}{2}, k_2 - \frac{N}{2}\right)$$

5.3.1 Inverse Gaussian Filter (IGF)

This filter (Fig. 5.12) is defined as

$$G^F(k_1, k_2) = \begin{cases} e^{(k_1^2+k_2^2)/2\sigma^2} & k_1, k_2 = 0, 1, \ldots, \frac{N}{2} \\ G^F(N - k_1, N - k_2) & \text{otherwise} \end{cases} \qquad (5.22)$$

Radial cross sections of the inverse Gaussian filter are shown in Fig. 5.13c. The DFT domain operation is an element-by-element multiplication.

$$\widehat{X}^F(k_1, k_2) = G^F(k_1, k_2) X^F(k_1, k_2) \qquad k_1, k_2 = 0, 1, \ldots, N - 1 \qquad (5.23)$$

This filter weighs high frequencies heavily and restores images (Fig. 5.14) blurred by atmospheric turbulence or other phenomenon modeled by Gaussian distribution.

5.3.2 Root Filter

The 2-D DFT coefficients $X^F(k_1, k_2)$ can be represented by (5.24) i.e., magnitude and phase. Root filter is described in Fig. 5.15.

$$X^F(k_1, k_2) = |X^F(k_1, k_2)| \angle \theta^F(k_1, k_2) \qquad (5.24)$$

where $\angle \theta^F(k_1, k_2) = e^{j\theta^F(k_1,k_2)}$.

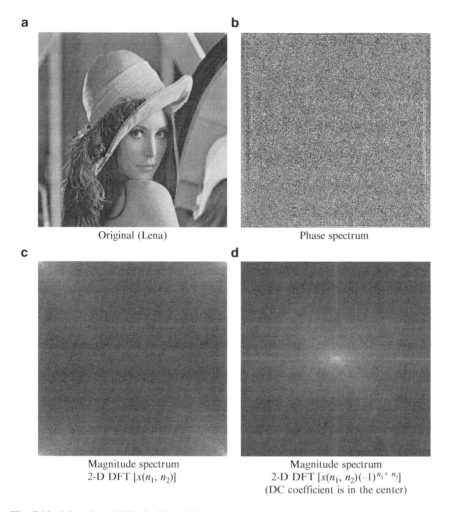

a

Original (Lena)

b

Phase spectrum

c

Magnitude spectrum
2-D DFT $[x(n_1, n_2)]$

d

Magnitude spectrum
2-D DFT $[x(n_1, n_2)(-1)^{n_1+ n_2}]$
(DC coefficient is in the center)

Fig. 5.10 2-D unitary DFT of a 256 × 256 image, 8 bpp (0–255)

Root filtering enhances (heavily weighs) high frequency coefficients (low magnitudes) compared to low frequency coefficients as in general high frequency coefficients have lower magnitudes compared to low frequency coefficients (Table 5.1, Fig. 5.16).

5.3.3 Homomorphic Filtering

If logarithm is applied to the magnitude spectrum of the DFT as

$$S^F(k_1, k_2) = (\ln |X^F(k_1, k_2)|)e^{j0(k_1, k_2)} \qquad |X^F(k_1, k_2)| \geq 0 \qquad (5.25a)$$

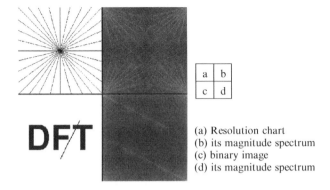

<table>
<tr><td>a</td><td>b</td></tr>
<tr><td>c</td><td>d</td></tr>
</table>

(a) Resolution chart
(b) its magnitude spectrum
(c) binary image
(d) its magnitude spectrum

Fig. 5.11 2-D unitary DFT of images

Fig. 5.12 Inverse Gaussian filter

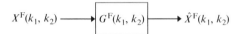

$X^F(k_1, k_2) \longrightarrow \boxed{G^F(k_1, k_2)} \longrightarrow \hat{X}^F(k_1, k_2)$

then the inverse transform of $S^F(k_1, k_2)$, denoted by $s(n_1, n_2)$, is called the *cepstrum* of the image (Fig. 5.17a). Examples of cepstra are shown in Fig. 5.18. The homomorphic transform reduces the dynamic range of the image in the transform domain.

Homomorphic filtering is a useful method to restore images in the presence of multiplicative noise. Figure 5.19 describes the process. The input image is modeled as the product of a noise-free image $r(n_1, n_2)$ and illumination interference array $l(n_1, n_2)$.

$$x(n_1, n_2) = r(n_1, n_2)\, l(n_1, n_2) \tag{5.25b}$$

Apply the logarithm to (5.25b) to obtain the additive noise observation model

$$\ln\{x(n_1, n_2)\} = \ln\{r(n_1, n_2)\} + \ln\{l(n_1, n_2)\} \tag{5.25c}$$

The 2-D DFT domain zonal mask is now applied to reduce the logarithm of the interference component. Zonal masking is followed by exponentiation. An example of the enhancement process is shown in Fig. 5.20. In this example, the illumination field $l(n_1, n_2)$ increases from left to right from a value of 0.05 to 2. As a smooth cutoff filter, a Butterworth high-pass filter specified in (P8.16) is used instead of the zonal high-pass filter. In addition, the scheme shown in Fig. 4.61 of [IP19] is used to

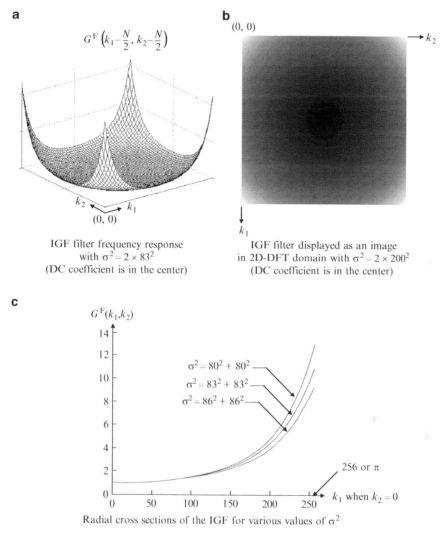

a

$$G^{F}\left(k_1 - \frac{N}{2}, k_2 - \frac{N}{2}\right)$$

k_2 k_1
(0, 0)

IGF filter frequency response
with $\sigma^2 = 2 \times 83^2$
(DC coefficient is in the center)

b

(0, 0) $\longrightarrow k_2$

k_1

IGF filter displayed as an image
in 2D-DFT domain with $\sigma^2 = 2 \times 200^2$
(DC coefficient is in the center)

c

$G^{F}(k_1, k_2)$

$\sigma^2 = 80^2 + 80^2$
$\sigma^2 = 83^2 + 83^2$
$\sigma^2 = 86^2 + 86^2$

256 or π

k_1 when $k_2 = 0$

Radial cross sections of the IGF for various values of σ^2

Fig. 5.13 Inverse Gaussian filter (IGF) with $N_1 = N_2 = N = 512$

pass low frequencies instead of stopping. Thus this filter is similar to a high-frequency emphasis filter.

The illumination component of an image varies slowly whereas the reflectance component varies abruptly at the junctions of objects. Thus the low frequencies of the 2-D DFT of the logarithm of an image are associated with illumination, and the high frequencies are associated with reflectance (see [IP18, B42] for similar examples).

Original

Enhanced with
$\sigma^2 = k_1^2 + k_2^2, k_1 = k_2 = 83$

Fig. 5.14 Inverse Gaussian filtering on the "Lena" image of size (512 × 512).

Fig. 5.15 Root filter. Here α is between 0 and 1

Table 5.1 Root filtering enhances high frequency coefficients

	$\alpha = 1/2$	
	Before filtering	After filtering
|Low frequency coefficient|	100	10
|High frequency coefficient|	10	3.162

5.3.4 Range Compression

The dynamic range of a 2D-DFT of an image is so large that only a few coefficients are visible. The dynamic range can be compressed via the logarithmic transformation

$$V^{\mathrm{F}}(k_1, k_2) = c \log_{10}\left(1 + \left|X^{\mathrm{F}}(k_1, k_2)\right|\right) \qquad (5.26)$$

where c is a scaling constant (Fig. 5.21). In practice a positive constant is added to the magnitude spectrum to prevent the logarithm from going to negative infinity.

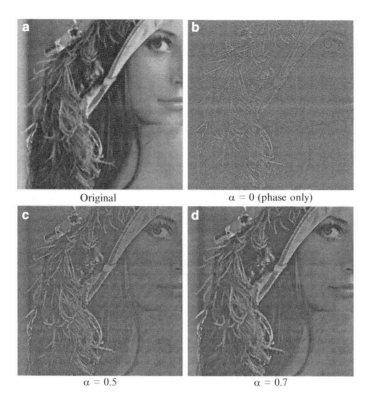

Fig. 5.16 Root filtering

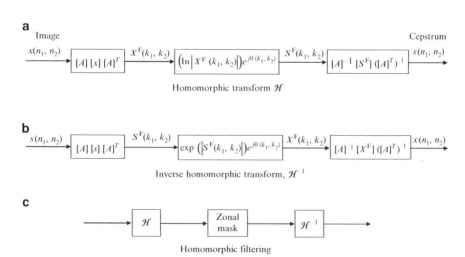

Fig. 5.17 Cepstrum and homomorphic filtering. Note that homomorphic filtering can be in any transform domain such as DCT and Hadamard transform (Fig. 5.18)

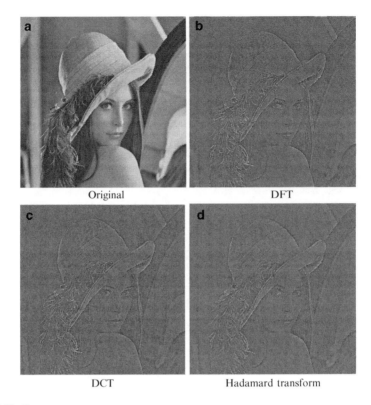

Original   DFT

DCT   Hadamard transform

Fig. 5.18 Cepstra

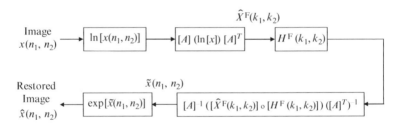

Fig. 5.19 Homomorphic filtering for image enhancement. ○ represents the element-by-element product of two matrices

5.3.5 *Gaussian Lowpass Filter*

A Gaussian lowpass filter for an $N \times N$ image has the frequency response

$$H^{\mathrm{F}}(k_1, k_2) = \begin{cases} e^{-\left(k_1^2 + k_2^2\right)/2\sigma^2}, & k_1, k_2 = 0, 1, \ldots, \frac{N}{2} \\ H^{\mathrm{F}}(N - k_1, N - k_2), & \text{otherwise} \end{cases} \quad (5.27)$$

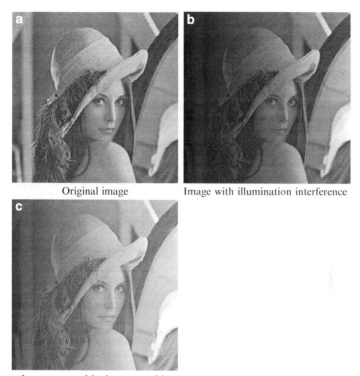

Original image   Image with illumination interference

Image restored by homomorphic
filtering

Fig. 5.20 Homomorphic filtering of an image with illumination interference

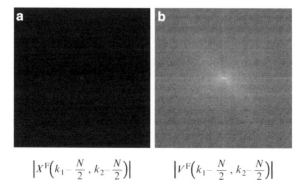

$$\left|X^F\!\left(k_1-\frac{N}{2}\,,k_2-\frac{N}{2}\right)\right| \qquad \left|V^F\!\left(k_1-\frac{N}{2}\,,k_2-\frac{N}{2}\right)\right|$$

Fig. 5.21 Range compression of the DFT of Lena image (see Fig. 5.10). All the magnitude spectrum images in this book are range compressed

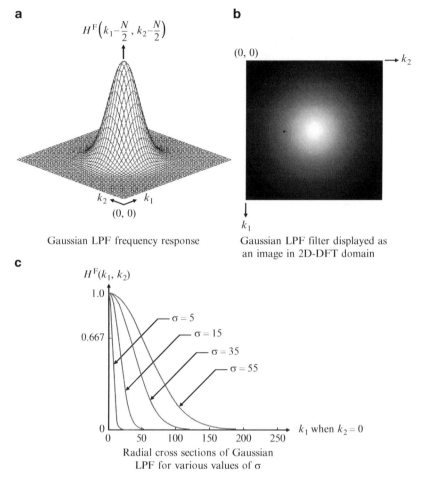

a

$$H^{\mathrm{F}}\left(k_1 - \frac{N}{2}, k_2 - \frac{N}{2}\right)$$

k_2 k_1

$(0, 0)$

Gaussian LPF frequency response

b

$(0, 0)$ k_2

k_1

Gaussian LPF filter displayed as
an image in 2D-DFT domain

c

$H^{\mathrm{F}}(k_1, k_2)$

1.0

0.667

$\sigma = 5$

$\sigma = 15$

$\sigma = 35$

$\sigma = 55$

0

0   50   100   150   200   250 k_1 when $k_2 = 0$

Radial cross sections of Gaussian
LPF for various values of σ

Fig. 5.22 Gaussian low pass filter with $N = 500$

where $\sqrt{k_1^2 + k_2^2}$ is the distance from point (k_1, k_2) to the center of the filter (see Fig. 5.8), and σ is the cutoff parameter. When $\sqrt{k_1^2 + k_2^2} = \sigma$, the filter falls to 0.667 of its maximum value of one (see Fig. 5.22c). Note that Fig. 5.22a and b correspond to Fig. 5.9.

5.4   Inverse and Wiener Filtering

Inverse filter restores a blurred image perfectly from the output of a noiseless linear system. However, in the presence of additive white noise, it does not work well. The ratio of spectrum $N^{\mathrm{F}}/H^{\mathrm{F}}$ affects on the image restoration [B43].

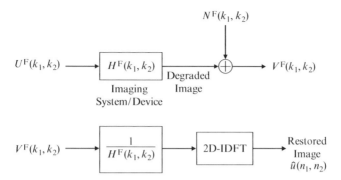

Fig. 5.23 Inverse filtering

Suppose we have the following system as shown in Fig. 5.23 in the frequency domain.

$$V^{\mathrm{F}}(k_1, k_2) = U^{\mathrm{F}}(k_1, k_2)H^{\mathrm{F}}(k_1, k_2) + N^{\mathrm{F}}(k_1, k_2) \qquad (5.28)$$

where $U^{\mathrm{F}}(k_1, k_2)$ is the 2D-DFT of an input image $u(n_1, n_2)$, $N^{\mathrm{F}}(k_1, k_2)$ is an additive noise, and $H^{\mathrm{F}}(k_1, k_2)$ is a degradation function. The 2D-IDFT of $V^{\mathrm{F}}(k_1, k_2)$ is the blurred image and is the input to the inverse filter, $\frac{1}{H^{\mathrm{F}}(k_1,k_2)}$. When $|H^{\mathrm{F}}(k_1, k_2)| < \epsilon$, $\frac{N^{\mathrm{F}}(k_1,k_2)}{H^{\mathrm{F}}(k_1,k_2)}$ becomes very large (see Fig. 5.24c). Hence use the pseudo inverse filter defined as

$$H^-(k_1, k_2) = \begin{cases} \dfrac{1}{H^{\mathrm{F}}(k_1, k_2)}, & |H^{\mathrm{F}}| \neq 0 \\ 0, & |H^{\mathrm{F}}| = 0 \end{cases} \qquad (5.29)$$

$H^-(k_1, k_2)$ is set to zero whenever $|H^{\mathrm{F}}|$ is less than a chosen quantity ϵ.

Inverse filtering is the process of recovering the input of a system from its output. Inverse filter is obtained by dividing the degraded image with the original image in the 2D-transform domain. If the degradation has zero or very low values, then $N^{\mathrm{F}}/H^{\mathrm{F}}$ dominates.

5.4.1 The Wiener Filter

The inverse and pseudo-inverse filters remain *sensitive to noise*. Noise can be amplified. Wiener filter can overcome this limitation. Wiener filtering is a method of restoring images in the presence of blur as well as noise. This restoration filter is a linear space-invariant filter which make use of the power spectrum of both the image and the noise to prevent excessive noise amplification as seen in (5.32b).

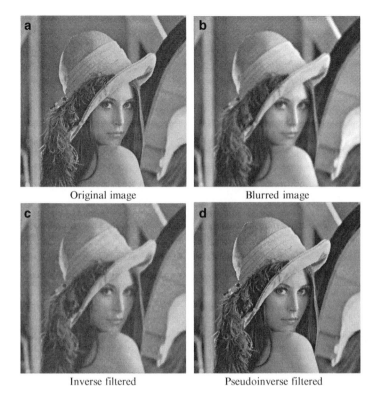

Original image   Blurred image

Inverse filtered   Pseudoinverse filtered

Fig. 5.24 Inverse and pseudoinverse filtered images

Given $E[u(n_1, n_2)] = 0$ and $E[v(n_1, n_2)] = 0$, obtain $\hat{u}(n_1, n_2)$ an estimate of original image $u(n_1, n_2)$ from observed/degraded image $v(n_1, n_2)$ such that the mean square error (MSE)

$$\sigma_e^2 = E\left([u(n_1, n_2) - \hat{u}(n_1, n_2)]^2\right) \tag{5.30}$$

is minimized (Fig. 5.25a). Here $E(\cdot)$ denotes the expectation over an ensemble of images. That is, we obtain the best linear estimate of original image $u(n_1, n_2)$ which is of the form

$$\hat{u}(n_1, n_2) = \sum_{m_1, \, m_2 = -\infty}^{\infty} g(n_1 - m_1, n_2 - m_2)\, v(m_1, m_2) \tag{5.31}$$

where the Wiener filter $g(n_1, n_2; m_1, m_2)$ is determined such that the MSE of (5.30) is minimized.

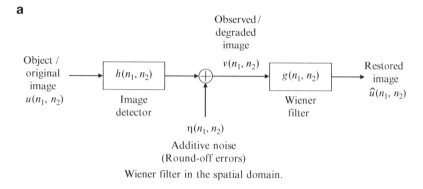

a

Wiener filter in the spatial domain.

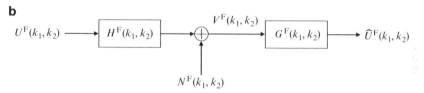

b

Wiener filter in the 2D-DFT domain.

Fig. 5.25 Wiener filtering. For implementation, see Fig. 8.13 in [B6]

Let $U^F, V^F, \hat{U}^F$ and N^F be the 2-D DFTs of $u, v, \hat{u}$ and additive noise η respectively. Let S_{uu} and $S_{\eta\eta}$ be the power spectral densities (PSD) of u and η respectively. The power spectral density S_{uu} of the input image $u(n_1, n_2)$ is defined as

$$S_{uu}(k_1, k_2) = \frac{1}{N^2}\left|U^F(k_1, k_2)\right|^2$$

and S_{uu} is the 2D-DFT of the autocorrelation function of $u(n_1, n_2)$.

For a linear shift invariant (LSI) system with frequency response $H^F(k_1, k_2)$, the Fourier Wiener filter is defined as

$$G^F(k_1, k_2) = \frac{[H^F(k_1, k_2)]^* S_{uu}(k_1, k_2)}{|H^F(k_1, k_2)|^2 S_{uu}(k_1, k_2) + S_{\eta\eta}(k_1, k_2)} \quad (5.32a)$$

$$= \frac{[H^F(k_1, k_2)]^*}{|H^F(k_1, k_2)| + \frac{S_{\eta\eta}(k_1, k_2)}{S_{uu}(k_1, k_2)}} \quad (5.32b)$$

This filter needs to know the power spectra of the object/original image and the noise and the frequency response (or the 2-D DFT of the point spread function

[PSF]) of the imaging system. For a typical image restoration problem, those are known. If the noise variance is not known, it can be estimated from a flat region of the observed image. In addition, it is possible to estimate S_{uu} in a number of different ways. The most common of these is to use the power spectrum S_{vv} of the observed image $v(n_1, n_2)$, as an estimate of S_{uu} [LA16]. The deblurred image is computed as

$$\hat{u}(n_1, n_2) = \text{2D IDFT}\left[G^{\text{F}}(k_1, k_2)V^{\text{F}}(k_1, k_2)\right]$$

The mean square error can be written as (total energy is preserved under unitary transform)

$$\sigma_e^2 = z_{\text{cor}}(0, 0) = \text{2D IDFT of [PSD of the error] at } ((n_1, n_2) = (0, 0))$$

$$= \frac{1}{N^2} \sum_{k_1,\ k_2=0}^{N-1} S_e(k_1, k_2) \tag{5.33a}$$

where S_e, the power spectral density of the error, is

$$S_e(k_1, k_2) = \left|1 - G^{\text{F}}H^{\text{F}}\right|^2 S_{uu} + \left|G^{\text{F}}\right|^2 S_{\eta\eta} \tag{5.33b}$$

By using (5.32) this can be simplified as

$$S_e = \frac{S_{uu}S_{\eta\eta}}{\left|H^{\text{F}}\right|^2 S_{uu} + S_{\eta\eta}} \tag{5.33c}$$

Wiener filtering of noisy blurred images is shown in Fig. 5.26.

5.4.2 Geometric Mean Filter (GMF)

This is the geometric mean of the pseudoinverse and Wiener filters, i.e.,

$$G_s(k_1, k_2) = (H^{-}(k_1, k_2))^s \left(\frac{[H^{\text{F}}(k_1, k_2)]^* S_{uu}(k_1, k_2)}{|H^{\text{F}}(k_1, k_2)|^2 S_{uu}(k_1, k_2) + S_{\eta\eta}(k_1, k_2)}\right)^{1-s}$$

$$0 \le s \le 1 \tag{5.34}$$

Blurred with small noise   Restored image (a)

Blurred with increased noise   Restored image (c)

Fig. 5.26 Wiener filtering of noisy blurred images

For $s = \frac{1}{2}$, the GMF is described as

$$
\begin{aligned}
G_{1/2}&(k_1, k_2) \\
&= \left| H^{\mathrm{F}}(k_1, k_2) H^{-}(k_1, k_2) \right|^{1/2} \exp(-j\,\theta_H) \left(\frac{S_{uu}(k_1, k_2)}{\left| H^{\mathrm{F}}(k_1, k_2) \right|^2 S_{uu}(k_1, k_2) + S_{\eta\eta}(k_1, k_2)} \right)^{1/2}
\end{aligned}
\tag{5.35}
$$

where $\theta_H(k_1, k_2)$ is the phase spectrum of $H^{\mathrm{F}}(k_1, k_2)$. Equation (5.35) is the true meaning of the geometric mean (see Appendix A.1) (Fig. 5.27). For $s = 0$, the GMF becomes the Wiener filter; for $s = 1$, the GMF becomes the inverse filter.

Noisy blurred image	Pseudoinverse filtered
Wiener filtered	Geometric mean filtered with $s = \dfrac{1}{2}$

Fig. 5.27 Geometric mean filtering

5.5 Three-Dimensional DFT

Similar to 2-D DFT (see (5.1)), a 3-D DFT pair can be defined as follows [DS10]:

5.5.1 3-D DFT

$$X^{\mathrm{F}}(k_1, k_2, k_3) = \sum_{n_1=0}^{N_1-1} \sum_{n_2=0}^{N_2-1} \sum_{n_3=0}^{N_3-1} x(n_1, n_2, n_3) W_{N_1}^{n_1 k_1} W_{N_2}^{n_2 k_2} W_{N_3}^{n_3 k_3}$$

$$k_i = 0, 1, \ldots, N_i - 1, \quad i = 1, 2, 3$$

(5.36a)

where $W_{N_i} = \exp(-j2\pi/N_i), i = 1, 2, 3.$

5.5.2   3-D IDFT

$$x(n_1, n_2, n_3) = \frac{1}{N_1 N_2 N_3} \sum_{k_1=0}^{N_1-1} \sum_{k_2=0}^{N_2-1} \sum_{k_3=0}^{N_3-1} X^F(k_1, k_2, k_3) W_{N_1}^{-n_1 k_1} W_{N_2}^{-n_2 k_2} W_{N_3}^{-n_3 k_3}$$

(5.36b)

$$n_i = 0, 1, \ldots, N_i - 1, \qquad i = 1, 2, 3$$

5.5.3   3D Coordinates

Horizontal, vertical and temporal (time).

Assume for simplicity $N_1 = N_2 = N_3 = N$. Then the 3-D DFT pair can be simplified as follows.

5.5.4   3-D DFT

$$X^F(k_1, k_2, k_3) = \sum_{n_1, n_2, n_3=0}^{N-1} x(n_1, n_2, n_3) W_N^{\sum_{i=1}^{3} n_i k_i}$$

(5.37a)

$$k_i = 0, 1, \ldots, N - 1, \quad i = 1, 2, 3$$

where the notation $\sum_{n_1, n_2, n_3=0}^{N-1}$ implies $\sum_{n_1=0}^{N-1} \sum_{n_2=0}^{N-1} \sum_{n_3=0}^{N-1}$.

5.5.5   3-D IDFT

$$x(n_1, n_2, n_3) = \frac{1}{N^3} \sum_{k_1, k_2, k_3=0}^{N-1} X^F(k_1, k_2, k_3) W_N^{-\sum_{i=1}^{3} n_i k_i}$$

(5.37b)

$$n_i = 0, 1, \ldots, N - 1, \qquad i = 1, 2, 3$$

The 3-D DFT pair can be symbolically represented as

$$x(n_1, n_2, n_3) \quad \Leftrightarrow \quad X^F(k_1, k_2, k_3)$$

(5.38)

All the properties, concepts, theorems, etc., developed for 1-D and 2-D DFTs can be extended to 3-D DFT in a simple and straight forward fashion. 3-D DFT has been applied to watermarking [E4, E8].

The 2-D and 3-D DFTs can be extended to the more general case. For example, an L-D DFT and L-D IDFT can be expressed as

$$X^{F}(k_1, k_2, \ldots, k_L) = \sum_{n_1=0}^{N_1-1} \sum_{n_2=0}^{N_2-1} \cdots \sum_{n_L=0}^{N_L-1} x(n_1, n_2, \ldots, n_L) W_{N_1}^{n_1 k_1} W_{N_2}^{n_2 k_2} \cdots W_{N_L}^{n_L k_L}$$

$$k_i = 0, 1, \ldots, N_i - 1, \quad i = 1, 2, \ldots, L$$

$$(5.39a)$$

$$x(n_1, n_2, \ldots, n_L) = \frac{1}{N_1 N_2 \cdots N_L} \sum_{k_1=0}^{N_1-1} \sum_{k_2=0}^{N_2-1} \cdots \sum_{k_L=0}^{N_L-1} X^{F}(k_1, k_2, \ldots, k_L) W_{N_1}^{-n_1 k_1} W_{N_2}^{-n_2 k_2} \cdots W_{N_L}^{-n_L k_L}$$

$$n_i = 0, 1, \ldots, N_i - 1, \quad i = 1, 2, \ldots, L$$

$$(5.39b)$$

As before, when $N_1 = N_2 = \cdots = N_L = N$, the L-D DFT pair can be simplified as

$$X^{F}(k_1, k_2, \ldots, k_L) = \sum_{n_1, n_2, \ldots, n_L=0}^{N-1} x(n_1, n_2, \ldots, n_L) W_N^{\sum_{i-1}^{L} n_i k_i}$$

$$k_i = 0, 1, \ldots, N - 1, \quad i = 1, 2, \ldots, L$$

$$(5.40a)$$

$$x(n_1, n_2, \ldots, n_L) = \frac{1}{N^L} \sum_{k_1, k_2, \ldots, k_L=0}^{N-1} X^{F}(k_1, k_2, \ldots, k_L) W_N^{-\sum_{i-1}^{L} n_i k_i}$$

$$n_i = 0, 1, \ldots, N - 1, \quad i = 1, 2, \ldots, L$$

$$(5.40b)$$

All the properties, concepts, theorems, etc., that are valid for the 1-D, 2-D, and 3-D DFTs, needless to say, are equally applicable to the L-D DFT.

5.6   Variance Distribution in the 1-D DFT Domain

$\underline{x} = [x(0), x(1), \ldots, x(N-1)]^{T}$ is a real random vector where $x(0), x(1), \ldots, x(N-1)$ are the N random variables. Assume $\underline{x}$ is real. Covariance matrix of $\underline{x}$ is $[\Sigma]$.

$$[\Sigma] = E\left[(\underline{x} - \bar{\underline{x}})(\underline{x} - \bar{\underline{x}})^{T}\right]$$

$$(5.41a)$$

where $\underline{\bar{x}} = E(\underline{x}) = $ mean of $\underline{x}$.

$$[\Sigma] = E\left[\begin{pmatrix} x_0 - \bar{x}_0 \\ x_1 - \bar{x}_1 \\ \vdots \\ x_{N-1} - \bar{x}_{N-1} \end{pmatrix} \left(x_0 - \bar{x}_0, \ x_1 - \bar{x}_1, \ \ldots, \ x_{N-1} - \bar{x}_{N-1} \right) \right] \qquad (5.41b)$$
$(N \times N)$

The covariance matrix in the data domain is

$$[\Sigma] = \begin{bmatrix} \sigma_{00}^2 & \sigma_{01}^2 & \sigma_{02}^2 & \cdots & \sigma_{0,N-1}^2 \\ \sigma_{10}^2 & \sigma_{11}^2 & \sigma_{12}^2 & \cdots & \sigma_{1,N-1}^2 \\ \sigma_{20}^2 & \sigma_{21}^2 & \sigma_{22}^2 & \cdots & \sigma_{2,N-1}^2 \\ \vdots & \vdots & \vdots & \ddots & \vdots \\ \sigma_{N-1,0}^2 & \sigma_{N-1,1}^2 & \sigma_{N-1,2}^2 & \cdots & \sigma_{N-1,N-1}^2 \end{bmatrix} \qquad (5.41c)$$

In $[\Sigma]$, the diagonal elements are variances and off diagonal elements are covariances.

$$E\left[(x_j - \bar{x}_j)(x_k - \bar{x}_k) \right] = \sigma_{jk}^2 \qquad (j \neq k)$$
$$= \text{covariance between } x_j \text{ and } x_k$$
$$E\left[(x_j - \bar{x}_j)(x_j - \bar{x}_j) \right] = \sigma_{jj}^2 = \text{variance of } x_j$$

The covariance matrix in the DFT domain is

$$[\Sigma]^{\mathrm{F}} = E\left[\left(\underline{X}^{\mathrm{F}} - \underline{\bar{X}}^{\mathrm{F}} \right) \left(\underline{X}^{\mathrm{F}} - \underline{\bar{X}}^{\mathrm{F}} \right)^{*^T} \right]$$
$(N \times N)$
$$= E\left\{ [F](\underline{x} - \underline{\bar{x}})([F](\underline{x} - \underline{\bar{x}}))^{*^T} \right\}$$
$$= [F]E\left[(\underline{x} - \underline{\bar{x}})(\underline{x} - \underline{\bar{x}})^T \right][F]^* \qquad (5.42a)$$
$$= [F] \quad [\Sigma] \quad [F]^*$$
$\quad (N \times N) \ (N \times N) \ (N \times N)$

$$[\Sigma]^{\mathrm{F}} = \begin{bmatrix} \tilde{\sigma}_{00}^2 & \tilde{\sigma}_{01}^2 & \tilde{\sigma}_{02}^2 & \cdots & \tilde{\sigma}_{0,N-1}^2 \\ \tilde{\sigma}_{10}^2 & \tilde{\sigma}_{11}^2 & \tilde{\sigma}_{12}^2 & \cdots & \tilde{\sigma}_{1,N-1}^2 \\ \tilde{\sigma}_{20}^2 & \tilde{\sigma}_{21}^2 & \tilde{\sigma}_{22}^2 & \cdots & \tilde{\sigma}_{2,N-1}^2 \\ \vdots & \vdots & \vdots & \ddots & \vdots \\ \tilde{\sigma}_{N-1,0}^2 & \tilde{\sigma}_{N-1,1}^2 & \tilde{\sigma}_{N-1,2}^2 & \cdots & \tilde{\sigma}_{N-1,N-1}^2 \end{bmatrix} \qquad (5.42b)$$
$(N \times N)$

In $[\Sigma]^{\mathrm{F}}$, the diagonal elements are variances, and the off diagonal elements are covariances in the DFT domain.

5.7 Sum of Variances Under Unitary Transformation Is Invariant

$$\underline{X} = [A]\,\underline{x}, \quad [A]^{-1} = ([A]^*)^T \quad \text{and} \quad \underline{x} = ([A]^*)^T\,\underline{X}$$
$$\scriptstyle (N\times 1)(N\times N)(N\times 1)$$

where $\underline{x} = (x_0, x_1, x_2, \ldots, x_{N-1})^T$ is a random vector and $[A]$ is an unitary transform.

$$E\left[(\underline{x}^*)^T\underline{x}\right] = E\left[\underline{x}^T\underline{x}\right] = E\left[\sum_{i=0}^{N-1} x_i^2\right] = \sum_{i=0}^{N-1} E[x_i^2] = \sum_{i=0}^{N-1} \sigma_{ii}^2$$

$$= \text{Sum of variances in data domain}$$

$$E\left[(\underline{x}^*)^T\underline{x}\right] = E\left[(\underline{X}^*)^T ([A]^*)^T [A]\,\underline{X}\right] = E\left[(\underline{X}^*)^T\underline{X}\right]$$

$$= E\left[\sum_{i=0}^{N-1} |X_i|^2\right] = \sum_{i=0}^{N-1} E\left[|X_i|^2\right] = \sum_{i=0}^{N-1} \tilde{\sigma}_{ii}^2$$

$$= \text{Sum of variances in unitary transform domain}$$

where $\underline{X} = (X_0, X_1, X_2, \ldots, X_{N-1})^T$ is the transform coefficient vector.

5.8 Variance Distribution in the 2-D DFT Domain

For a 2-D $(N \times N)$ data array, $[x]$ can be described as

$$[x] = \begin{bmatrix} x_{00} & x_{01} & x_{02} & \cdots & x_{0,N-1} \\ x_{10} & x_{11} & x_{12} & \cdots & x_{1,N-1} \\ x_{20} & x_{21} & x_{22} & \cdots & x_{2,N-1} \\ \vdots & \vdots & \vdots & \ddots & \vdots \\ x_{N-1,0} & x_{N-1,1} & x_{N-1,2} & \cdots & x_{N-1,N-1} \end{bmatrix} \tag{5.43}$$

This has N^4 covariances of which N^2 are variances. Evaluation of the N^2 variances can be simplified by assuming independent row and column statistics.

Assume row and column statistics of the 2-D data are independent of each other. This assumption simplifies the computation of the N^2 variances both in the data and transform domains. Let the variances of elements of any row be (each row has the same statistics) $(\sigma_{00R}^2, \sigma_{11R}^2, \sigma_{22R}^2, \ldots, \sigma_{N-1,N-1,R}^2)$ in the data domain. Similarly each column has the same statistics (not necessarily the same statistics of any row). Let the variances of elements of any column be $(\sigma_{00C}^2, \sigma_{11C}^2, \sigma_{22C}^2, \ldots, \sigma_{N-1,N-1,C}^2)$ in the data domain. Then the variances of $[x]$ are

$$
\left[
\begin{pmatrix}
\sigma_{00R}^2 \\
\sigma_{11R}^2 \\
\sigma_{22R}^2 \\
\vdots \\
\sigma_{N-1,N-1,R}^2
\end{pmatrix}
\begin{pmatrix}
\sigma_{00C}^2, \sigma_{11C}^2, \sigma_{22C}^2, \ldots, \sigma_{N-1,N-1,C}^2
\end{pmatrix}
\right]
$$

$$(N \times 1) \qquad\qquad\qquad (1 \times N)$$

$$
=
\begin{bmatrix}
\left(\sigma_{00R}^2 \sigma_{00C}^2\right) & \left(\sigma_{00R}^2 \sigma_{11C}^2\right) & \cdots & \left(\sigma_{00R}^2 \sigma_{N-1,N-1,C}^2\right) \\
\left(\sigma_{11R}^2 \sigma_{00C}^2\right) & \left(\sigma_{11R}^2 \sigma_{11C}^2\right) & \cdots & \left(\sigma_{11R}^2 \sigma_{N-1,N-1,C}^2\right) \\
\vdots & \vdots & \ddots & \vdots \\
\left(\sigma_{N-1,N-1,R}^2 \sigma_{00C}^2\right) & \left(\sigma_{N-1,N-1,R}^2 \sigma_{11C}^2\right) & \cdots & \left(\sigma_{N-1,N-1,R}^2 \sigma_{N-1,N-1,C}^2\right)
\end{bmatrix}
$$

$$(N \times N)$$

$$(5.44)$$

Let the 2-D DFT of $[x]$ be $[X^F]$. (Assume as before $[x]$ has independent row and column statistics.)

$$\left[X^F(k_1, k_2)\right] = [F][x(n_1, n_2)][F]$$

$$
[X^F(k_1, k_2)] =
\begin{bmatrix}
X_{00}^F & X_{01}^F & X_{02}^F & \cdots & X_{0,N-1}^F \\
X_{10}^F & X_{11}^F & X_{12}^F & \cdots & X_{1,N-1}^F \\
X_{20}^F & X_{21}^F & X_{22}^F & \cdots & X_{2,N-1}^F \\
\vdots & \vdots & \vdots & \ddots & \vdots \\
X_{N-1,0}^F & X_{N-1,1}^F & X_{N-1,2}^F & \cdots & X_{N-1,N-1}^F
\end{bmatrix}
$$

$$(N \times N)$$

$$(5.45)$$

Let the variances of any row of $[X^F]$ be $(\tilde{\sigma}_{00R}^2, \tilde{\sigma}_{11R}^2, \tilde{\sigma}_{22R}^2, \ldots, \tilde{\sigma}_{N-1,N-1,R}^2)$. Similarly let the variances of any column of $[X^F]$ be $(\tilde{\sigma}_{00C}^2, \tilde{\sigma}_{11C}^2, \tilde{\sigma}_{22C}^2, \ldots, \tilde{\sigma}_{N-1,N-1,C}^2)$. Then the variances of $[X^F]$ are

$$\left[\begin{pmatrix} \tilde{\sigma}^2_{00R} \\ \tilde{\sigma}^2_{11R} \\ \tilde{\sigma}^2_{22R} \\ \vdots \\ \tilde{\sigma}^2_{N-1,N-1,R} \end{pmatrix} \begin{pmatrix} \tilde{\sigma}^2_{00C}, \tilde{\sigma}^2_{11C}, \tilde{\sigma}^2_{22C}, \dots, \tilde{\sigma}^2_{N-1,N-1,C} \end{pmatrix}\right]$$

$$(N \times 1) \qquad\qquad\qquad (1 \times N)$$

$$= \begin{bmatrix} \left(\tilde{\sigma}^2_{00R} \tilde{\sigma}^2_{00C} \right) & \left(\tilde{\sigma}^2_{00R} \tilde{\sigma}^2_{11C} \right) & \cdots & \left(\tilde{\sigma}^2_{00R} \tilde{\sigma}^2_{N-1,N-1,C} \right) \\ \left(\tilde{\sigma}^2_{11R} \tilde{\sigma}^2_{00C} \right) & \left(\tilde{\sigma}^2_{11R} \tilde{\sigma}^2_{11C} \right) & \cdots & \left(\tilde{\sigma}^2_{11R} \tilde{\sigma}^2_{N-1,N-1,C} \right) \\ \vdots & \vdots & \ddots & \vdots \\ \left(\tilde{\sigma}^2_{N-1,N-1,R} \tilde{\sigma}^2_{00C} \right) & \left(\tilde{\sigma}^2_{N-1,N-1,R} \tilde{\sigma}^2_{11C} \right) & \cdots & \left(\tilde{\sigma}^2_{N-1,N-1,R} \tilde{\sigma}^2_{N-1,N-1,C} \right) \end{bmatrix}$$

$$(N \times N)$$

$$(5.46)$$

Efficient data compression can be achieved by adopting a bit allocation based on the variances of the transform coefficients. This concept is described here in detail for the DFT. It is, of course, valid for any orthogonal transform.

5.9 Quantization of Transform Coefficients can be Based on Their Variances

If the average number of bits per transform coefficient used by the 1-D transform coding system is B and the number of bits used by the kth coefficient is B_k, then

$$B = \frac{1}{N} \sum_{k=0}^{N-1} B_k = \text{Given average bit rate} \quad \Leftarrow \textbf{Constraint} \qquad (5.47)$$

where N is the number of transform coefficients. The reconstruction error variance $\tilde{\sigma}^2_{B_k}$ for the kth coefficient is related to the kth quantizer input variance $\tilde{\sigma}^2_k$ by [D37, p. 28, B6, p. 103]

$$\tilde{\sigma}^2_{B_k} = \alpha_k 2^{-2B_k} \tilde{\sigma}^2_k \qquad (5.48)$$

where α_k is a factor that depends on the input distribution and the quantizer. The total reconstruction error variance $\tilde{\sigma}_B^2 = \sum_{k=0}^{N-1} \tilde{\sigma}_{B_k}^2$ is given by

$$\tilde{\sigma}_B^2 = \sum_{k=0}^{N-1} \alpha_k 2^{-2B_k} \tilde{\sigma}_k^2 \qquad \Leftarrow \textbf{To be minimized} \qquad (5.49)$$

The bit allocation problem is to find B_k that minimize the distortion (5.49) subject to the constraint (5.47). Assume α_k in (5.49) is a constant α for all k. Then

$$\tilde{\sigma}_B^2 = \alpha \sum_{k=0}^{N-1} 2^{-2B_k} \tilde{\sigma}_k^2 \qquad (5.50)$$

The constraint (5.47) can be rewritten as

$$\left[B - \frac{1}{N} \sum_{k=0}^{N-1} B_k \right] = 0 \qquad (5.51)$$

Then we can set up the optimization problem in terms of Lagrange multiplier λ [IP32, IP33, IP34]. Hence the function J to be minimized with the *constraint embedded* is

$$J = \alpha \sum_{k=0}^{N-1} 2^{-2B_k} \tilde{\sigma}_k^2 - \lambda \left[B - \frac{1}{N} \sum_{k=0}^{N-1} B_k \right] \qquad (5.52)$$

where $B_k, k = 0, 1, \ldots, N - 1$ are the variables. Hence set the derivative of J with respect to B_l equal to zero and solve for B_l. Here index k is changed to l to avoid any confusion in following equations.

$$\frac{\partial}{\partial B_l}(J) = 0 \qquad l - 0, 1, \ldots, N - 1 \qquad (5.53)$$

$$\alpha \frac{\partial}{\partial B_l}\left(\tilde{\sigma}_l^2 2^{-2B_l} \right) + \frac{\lambda}{N} = 0 \qquad (5.54)$$

$$\alpha \tilde{\sigma}_l^2 \frac{\partial}{\partial B_l}\left(\frac{1}{2} \right)^{2B_l} + \frac{\lambda}{N} = 0 \qquad (5.55)$$

Note that

$$\frac{\partial}{\partial B_l}\left(2^{-2} \right)^{B_l} = \left(2^{-2} \right)^{B_l} \ln\left(2^{-2} \right) \qquad (\ln = \log_e) \qquad (5.56)$$

from the derivative of the exponential function (see Appendix F for a proof)

$$\frac{d}{du}(a^u) = a^u \ln a \tag{5.57}$$

where $a = 2^{-2}$ for (5.56).

$$\alpha\tilde{\sigma}_l^2 \left(\frac{1}{2}\right)^{2B_l} \left[\ln\left(\frac{1}{2}\right)\right] 2 + \frac{\lambda}{N} = 0 \tag{5.58}$$

or

$$(2\alpha\tilde{\sigma}_l^2)(\ln 2)2^{-2B_l} = \frac{\lambda}{N} \tag{5.59}$$

Apply the base-2 logarithm function to both sides in (5.59).

$$\log_2\left[(2\alpha\tilde{\sigma}_l^2)(\ln 2)\right] - 2\,B_l = \log_2\left(\frac{\lambda}{N}\right) \tag{5.60}$$

$$B_l = \frac{1}{2}\log_2\left(2\alpha\tilde{\sigma}_l^2 \ln 2\right) - \frac{1}{2}\log_2\left(\frac{\lambda}{N}\right) \tag{5.61}$$

Since the average bit rate $B = \frac{1}{N}\sum_{k=0}^{N-1} B_k$

$$B = \frac{1}{2N}\sum_{k=0}^{N-1}\log_2\left(2\alpha\tilde{\sigma}_k^2 \ln 2\right) - \frac{1}{2}\log_2\left(\frac{\lambda}{N}\right) \tag{5.62}$$

Hence

$$\log_2\left(\frac{\lambda}{N}\right) = \left[\frac{1}{N}\sum_{k=0}^{N-1}\log_2\left(2\alpha\tilde{\sigma}_k^2 \ln 2\right) - 2B\right] \tag{5.63}$$

$$\frac{\lambda}{N} = 2^{-2B}\prod_{k=0}^{N-1}\left(2\alpha\tilde{\sigma}_k^2 \ln 2\right)^{1/N} \tag{5.64}$$

Substitute (5.64) in (5.61).

$$B_l = \frac{1}{2}\log_2\left(2\alpha\tilde{\sigma}_l^2 \ln 2\right) - \frac{1}{2}\log_2\left[2^{-2B}\prod_{k=0}^{N-1}\left(2\alpha\tilde{\sigma}_k^2 \ln 2\right)^{1/N}\right] \tag{5.65}$$

$$B_l = \frac{1}{2}\log_2\left(2\alpha\tilde{\sigma}_l^2 \ln 2\right) - \frac{1}{2N}\sum_{k=0}^{N-1}\log_2\left(2\alpha\tilde{\sigma}_l^2 \ln 2\right) + B \tag{5.66}$$

$$B_l = \frac{1}{2}\log_2\left[\frac{\tilde{\sigma}_l^2}{\prod_{k=0}^{N-1}\left(\tilde{\sigma}_k^2\right)^{1/N}}\right] + B \tag{5.67}$$

As index k is a local variable, k can be changed to m. Then index l can be switched back to k.

$$\boxed{B_k = B + \frac{1}{2}\log_2\tilde{\sigma}_k^2 - \left[\frac{1}{2N}\sum_{m=0}^{N-1}\log_2\tilde{\sigma}_m^2\right]} \quad k = 0, 1, \dots, N-1 \tag{5.68}$$

The last term in (5.68) is a constant independent of k. This implies that the number of bits B_k assigned to the kth transform coefficient is proportional to the logarithm of its variance. Note that B_k is rounded to the nearest integer while maintaing (5.47).

Similar to (5.68) the bit allocation for the 2-D transform coefficients can be expressed by (5.69) (see Fig. 5.28).

$$B_{i,j} = \#\quad \text{of}\quad \text{bits}\quad \text{allocated}\quad \text{to}\quad \text{transform}\quad \text{coefficient}$$
$$v_{i,j}, i,j = 0, 1, \dots, N-1$$

$$B_{i,j} = B + \frac{1}{2}\log_2 \tilde{\sigma}_{i,j}^2 - \left[\frac{1}{2N^2}\sum_{k=0}^{N-1}\sum_{l=0}^{N-1}\log_2\tilde{\sigma}_{k,l}^2\right] \tag{5.69}$$

$$B = \frac{1}{N^2}\sum_{k=0}^{N-1}\sum_{l=0}^{N-1}B_{k,l} = \text{Average bit rate} \quad \Leftarrow \textbf{Constraint} \tag{5.70}$$

$$B_{i,j} \sim \log_2 \tilde{\sigma}_{i,j}^2$$

$B_{i,j}$ is proportional to the logarithm of the variance of transform coefficient $v_{i,j}$.

# of quantization levels $= 2^{B_{i,j}}$

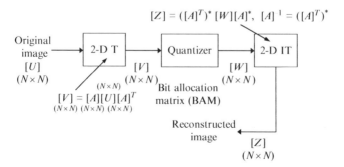

Fig. 5.28 Bit allocation is based on the variances of the two-dimensional transform coefficients. (Separable transform) Transform is implemented on a $(N \times N)$ block-by-block basis

Fig. 5.29 Bit assignment map in the 2D-DFT domain based on variance distribution of the transform coefficients (average bit rate is 1 bit per transform coefficient, or $B = 1$ in (5.70), and $\rho = 0.95$). Only the magnitudes of coefficients are coded. ρ is the adjacent correlation coefficient. The 2D ($N \times N$) data domain samples are governed by I order Markov process (see (5.75))

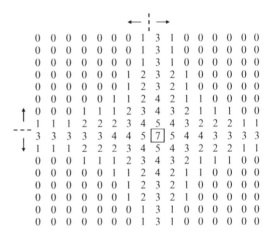

Fig. 5.30 Bit assignment map in the 2D-DFT. The dc coefficient *top left* corner in Fig. 5.7 is now shifted to the *center* of the 2-D frequency plane. Figure 5.28 corresponds to Fig. 5.7 and this figure (Fig. 5.30) corresponds to Fig. 5.9 (use fftshift)

If some transform coefficients have very low variances, then they can be replaced by zeros. An example of bit allocation for 2-D (16×16) DFT coefficients is shown in Figs. 5.29 and 5.30.

5.10 Maximum Variance Zonal Sampling (MVZS)

The covariance matrices in transform (orthogonal) domain $[\tilde{\Sigma}]$ and in the data domain $[\Sigma]$ are related by

$$[\tilde{\Sigma}] = [A]E\left[(\underline{x} - \bar{\underline{x}})(\underline{x} - \bar{\underline{x}})^T\right][A]^T = [A][\Sigma][A]^T \tag{5.71}$$

Assume $[A]$ is real and unitary (normalized) $[A]^T = [A]^{-1}$.

$$\underline{x} = (x_0, x_1, \ldots, x_{N-1})^T$$

$$[\tilde{\Sigma}] = \begin{bmatrix} \tilde{\sigma}^2_{00} & \tilde{\sigma}^2_{01} & \cdots & \tilde{\sigma}^2_{0,N-1} \\ \tilde{\sigma}^2_{10} & \tilde{\sigma}^2_{11} & \cdots & \tilde{\sigma}^2_{1,N-1} \\ \vdots & \vdots & \ddots & \vdots \\ \tilde{\sigma}^2_{N-1,0} & \tilde{\sigma}^2_{N-1,1} & \cdots & \tilde{\sigma}^2_{N-1,N-1} \end{bmatrix} \tag{5.72}$$

In any orthogonal transform domain, the variances of a 2-D data can be obtained based on independence of row and column statistics (similar to the DFT) case. For the 1-D data, the variance distribution in the transform domain for I order Markov process is shown in Table 5.2 and Fig. 5.33. Sum of the variances along any column is the same (energy invariance). The normalized basis restriction error is defined as

$$J_m = \frac{\sum_{k=m}^{N-1} \tilde{\sigma}^2_{kk}}{\sum_{k=0}^{N-1} \tilde{\sigma}^2_{kk}}, \qquad m = 0, 1, \ldots, N-1 \tag{5.73}$$

where $\tilde{\sigma}^2_{kk}$ are rearranged in nonincreasing order (see Fig. 5.34). Here the covariance function of a first order stationary Markov sequence $x(n)$ is given as

$$r(n) = \rho^{|n|} \qquad |\rho| < 1 \qquad \forall n \tag{5.74}$$

This is often used as the covariance model of a scan line of images. For an $(N \times 1)$ vector $\underline{x}$, its covariance matrix is $\{r(m-n)\}, m, n = 0, 1, \ldots, N-1$, that is

$$[R] = \begin{bmatrix} 1 & \rho & \rho^2 & \cdots & \rho^{N-1} \\ \rho & 1 & \rho & \ddots & \vdots \\ \rho^2 & \rho & 1 & \ddots & \rho^2 \\ \vdots & & \ddots & \ddots & \rho \\ \rho^{N-1} & \cdots & \rho^2 & \rho & 1 \end{bmatrix} \tag{5.75}$$

Thus $[R]$ is used instead of $[\Sigma]$ in (5.71) to get the variance distribution in the transform domain for the I order Markov process (see Table 5.2 and Fig. 5.33).

In MVZS, transform coefficients with large variances can be selected for quantization and coding with remainder (those with small variances) set to zero. The bitstream representing these coefficients is transmitted to the receiver and inverse operations i.e., decoding, inverse quantization, inverse transform etc., lead to reconstruction of the signal or image (Fig. 5.31).

Table 5.2 Variances $\tilde{\sigma}_{kk}^2$ of transform coefficients of a stationary Markov sequence with $\rho = 0.9$ and $N = 16$. For basic definitions of these transforms please refer to [B6]

Transform k	KLT	DCT- II	DST- I	Unitary DFT	Hadamard	Haar	Slant
0	9.927	9.835	9.218	9.835	9.835	9.835	9.835
1	2.949	2.933	2.642	1.834	0.078	2.536	2.854
2	1.128	1.211	1.468	0.519	0.206	0.864	0.105
3	0.568	0.582	0.709	0.250	0.105	0.864	0.063
4	0.341	0.348	0.531	0.155	0.706	0.276	0.347
5	0.229	0.231	0.314	0.113	0.103	0.276	0.146
6	0.167	0.169	0.263	0.091	0.307	0.276	0.104
7	0.129	0.130	0.174	0.081	0.104	0.276	0.063
8	0.104	0.105	0.153	0.078	2.536	0.100	1.196
9	0.088	0.088	0.110	0.081	0.098	0.100	0.464
10	0.076	0.076	0.099	0.091	0.286	0.100	0.105
11	0.068	0.068	0.078	0.113	0.105	0.100	0.063
12	0.062	0.062	0.071	0.155	1.021	0.100	0.342
13	0.057	0.057	0.061	0.250	0.102	0.100	0.146
14	0.055	0.055	0.057	0.519	0.303	0.100	0.104
15	0.053	0.053	0.054	1.834	0.104	0.100	0.063

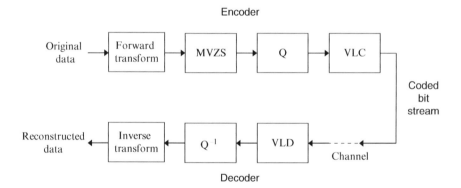

Fig. 5.31 Maximum variance zonal sampling transform coding

5.11 Geometrical Zonal Sampling (GZS)

Replace MVZS by GZS (Figs. 5.32, 5.35 through 5.39) in Fig. 5.31. See Appendix A.7.

5.12 Summary

Definitions and properties of the 2-D DFT analogous to those of the 1-D DFT are developed. Fast algorithms based on the row-column 1-D approach are outlined. Various filters for processing of 2-D signals such as images are defined and

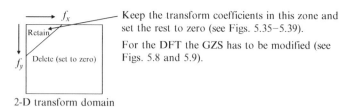

Keep the transform coefficients in this zone and set the rest to zero (see Figs. 5.35–5.39).

For the DFT the GZS has to be modified (see Figs. 5.8 and 5.9).

2-D transform domain

Fig. 5.32 Geometrical zonal sampling

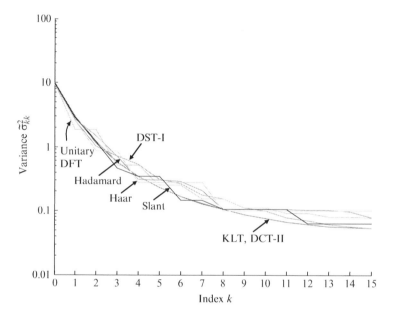

Fig. 5.33 Distribution of variances of the transform coefficients (in decreasing order) of a stationary Markov sequence with $\rho = 0.9$ and $N = 16$

illustrated with examples. DFT is compared with other orthogonal transforms based on some standard criteria such as variance distribution, basis restriction error, etc. Vector-radix 2-D FFT algorithms are the major thrust in the next chapter.

5.13 Problems

5.1 Appendix F.2 shows a proof by using the change of variables formula. Similarly prove (5.15).

5.2 Derive the expression for Wiener filter (5.32). See pages 276–279 in [B6].

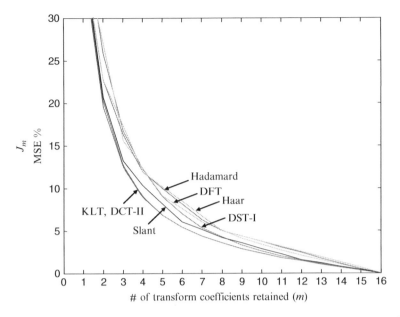

Fig. 5.34 Performance of different unitary transforms with respect to basis restriction errors (J_m) versus the number of basic (m) for a stationary Markov sequence with $\rho = 0.9$ and $N = 16$. See (5.73)

5.3 From Fig. 5.25b, 2D DFT of error $e(n_1, n_2)$ between the object (the original image) and the restored image can be expressed as

$$U^F - \hat{U}^F = U^F\left(1 - G^F H^F\right) - G^F N^F$$

Show that the power spectral density (PSD) of the error can be expressed as (5.33c). Use (5.32) for $G^F(k_1, k_2)$.

5.4 Starting from (5.47), derive (5.68). Show all the details.

5.5 Similar to the 1-D case (5.68), the bit allocation for the 2-D transform coefficients can be expressed by (5.69). Derive (5.69).

5.14 Projects

5.1 (1) Apply 2-D DFT to the image "Lena" (get the image on line from [IP30] and [IP31]).

(2) Apply zonal masks $G^F(k_1, k_2)$ for low-pass filtering (LPF), band-pass filtering (BPF), and high-pass filtering (HPF) to the result of the first step. The function is $G^F(k_1, k_2)$ zero outside the region of support for the particular filter (Fig. P5.1).

(3) Apply 2-D IDFT to all results of the second step.

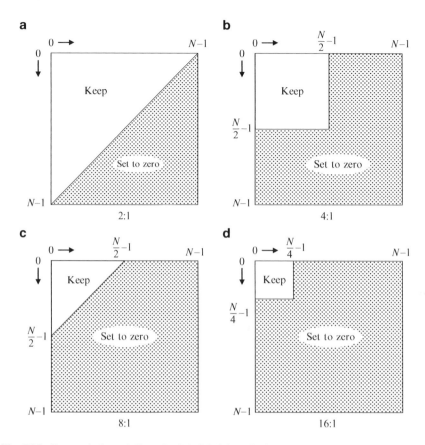

Fig. 5.35 Geometrical zonal filters for 2:1, 4:1, 8:1, and 16:1 sample reduction. White areas are passbands, whereas dark areas are stopbands. For the DFT these zones have to be modified (Fig. 5.36)

(4) Calculate MSE between the reconstructed and original images.
(5) Calculate the peak-to-peak signal-to-noise ratios (PSNR) of the reconstructed images (8 bpp).

The block diagram of this project is shown in Fig. P5.2.

For $(N \times N)$ images $x(n_1, n_2)$ and $\hat{x}(n_1, n_2)$, the mean square error is defined as

$$\sigma_a^2 = \frac{1}{N^2} \sum_{n_1=0}^{N-1} \sum_{n_2=0}^{N-1} \left(E[x(n_1, n_2) - \hat{x}(n_1, n_2)]^2 \right) \qquad (P5.1)$$

When ensembles for $x(n_1, n_2)$ and $\hat{x}(n_1, n_2)$ are not available, the average least square error is used as an estimate of (P5.1). This is for an image of size $(N \times N)$.

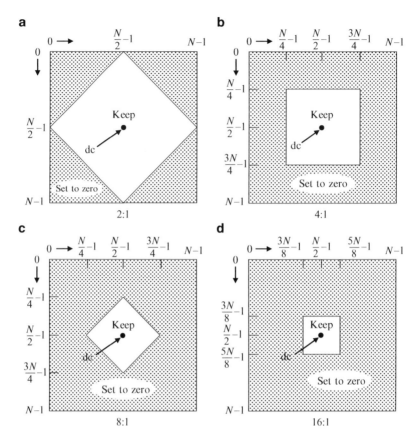

Fig. 5.36 2-D DFT domain geometrical zonal filters for 2:1, 4:1, 8:1, and 16:1 sample reduction. White areas are passbands, whereas dark areas are stopbands (see Figs. 5.8 and 5.9)

$$\sigma_{ls}^2 = \frac{1}{N^2} \sum_{n_1=0}^{N-1} \sum_{n_2=0}^{N-1} E|x(n_1, n_2) - \hat{x}(n_1, n_2)|^2 \qquad \text{(P5.2)}$$

The mean square error is expressed in terms of a peak-to-peak signal-to-noise ratio (PSNR), which is defined in decibels (dB) as

$$\text{PSNR} = 10\log_{10} \frac{255^2}{\sigma_e^2} \qquad \sigma_e^2 = \sigma_a^2 \text{ or } \sigma_{ls}^2 \qquad \text{(P5.3)}$$

where 255 is the range of $x(n)$ for 8 bit PCM.

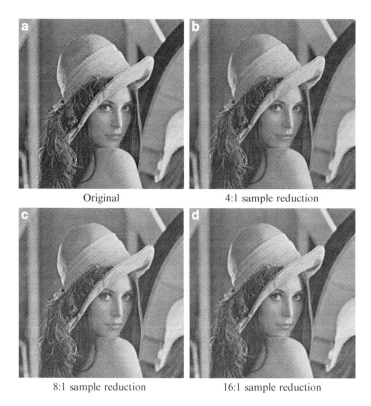

Original   4:1 sample reduction

8:1 sample reduction   16:1 sample reduction

Fig. 5.37 Basis restriction zonal filtered images in the DCT domain (geometrical zonal sampling – see Fig. 5.35)

5.2 Root filtering (see Fig. 5.15) is a contrast enhancement approach in the frequency domain. It takes the α-root of the magnitude part of the transform coefficient $X^F(k_1, k_2)$ while retaining the phase component, or

$$\text{For 2D-DFT} : R^F(k_1, k_2) = \left| X^F(k_1, k_2) \right|^\alpha e^{j\theta^F(k_1, k_2)} \quad 0 \le \alpha \le 1 \qquad \text{(P5.4)}$$

where the transform coefficient $X^F(k_1, k_2)$ can be written as

$$\text{For 2D-DFT} : \quad X^F(k_1, k_2) = |X^F(k_1, k_2)| \angle \theta^F(k_1, k_2) \qquad (5.24)$$

Procedure

(1) Apply 2D-DFT to the input image of size 256 by 256 (Lena).
(2) Perform the root filtering $R^F(k_1, k_2)$ for $\alpha = 0.6$ and 0.8 on the result of the first step (see pp. 258–259 in [B6]).

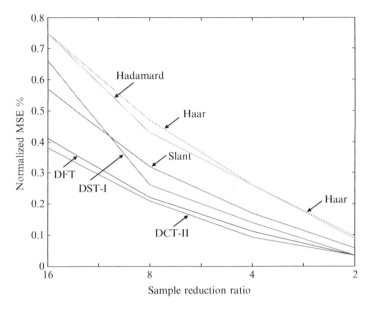

Fig. 5.38 Performance comparison of different transforms with respect to basis restriction zonal filtering for 512×512 Lena image. The normalized mean square error is defined in (A.8) (geometrical zonal sampling – see Figs. 5.35 and 5.36)

(3) Show the ratio of energy before and after root filtering and your results of steps 1 and 2 for both $\alpha = 0.6$ and 0.8 (Show the transformed images (see Fig. 5.21b) and reconstructed images). (Energy: see Examples 5.9 and 5.10, pp. 171–175 in [B6], Fig. P5.3.)

5.3 Inverse Gaussian filter weighs high frequencies heavily. It restores images blurred by atmospheric turbulence modeled by Gaussian distribution (see Fig. 5.14). The filter is defined by (5.22). Take Lena image and perform this filtering.

5.4 The dynamic range of a 2D-DFT of an image is so large that only a few transform coefficients are visible (see Fig. 5.21a). The dynamic range can be compressed via the logarithmic transformation (5.26). Take Lena image and perform the transformation.

5.5 **Inverse and Winner Filtering**

Inverse filter restores a blurred image perfectly from an output of a **noiseless** linear system. However, in the presence of additive white noise, it does not work well. In this project, how **the ratio of spectrum N^{F}/H^{F} affects the image restoration** is demonstrated.

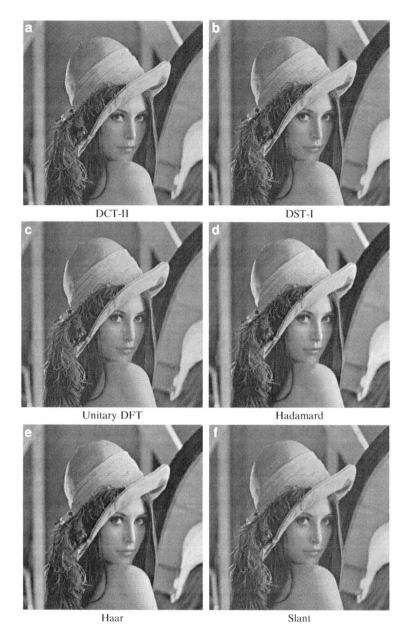

DCT-II   DST-I

Unitary DFT   Hadamard

Haar   Slant

Fig. 5.39 Basis restriction zonal filtering using different transforms with 4:1 sample reduction (geometrical zonal sampling – see Figs. 5.35 and 5.36)

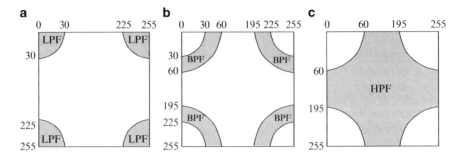

Fig. P5.1 Zonal masks in the 2D-DFT domain, $N_1 = N_2 = N = 256$ for **a** low-pass filtering, **b** band-pass filtering, and **c** high-pass filtering (see Fig. 5.7). Note that *dark areas* are passbands and *white areas* are stop bands **(a) (b) (c)**

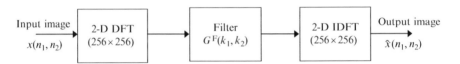

Fig. P5.2 2-D DFT domain filtering

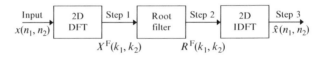

Fig. P5.3 Root filter in the 2D-DFT domain

Given: Suppose we have the following system in the frequency domain (Fig. P5.4).

(1) $V^F(k_1, k_2) = U^F(k_1, k_2)H^F(k_1, k_2) + N^F(k_1, k_2)$ (5.28)

where $U^F(k_1, k_2)$ is the 2D-DFT of an input image $u(n_1, n_2)$, $N^F(k_1, k_2)$ is a white noise with zero mean and unit variance in the 2D-DFT domain, and $H^F(k_1, k_2)$ is a degradation function which is given by

$$H^F(k_1, k_2) = \exp\left(-c\left[(k_1)^2 + (k_2)^2\right]^{5/6}\right)$$ (P5.5)

(2) 2D-IDFT of $V^F(k_1, k_2)$ is the blurred image. $\hat{u}(n_1, n_2)$ is the restored image and is described in Fig. P5.4.

Fig. P5.4 Inverse filtering

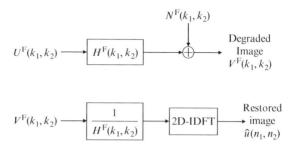

(3) $\frac{1}{H^F(k_1,k_2)}$ is the inverse filter. When $|H^F(k_1,k_2)| < \epsilon$, $\frac{N^F(k_1,k_2)}{H^F(k_1,k_2)}$ becomes very large (see Fig. 8.10c, page 277 in [B6]). Hence use the pseudo-inverse filter defined as

$$H^-(k_1,k_2) = \begin{cases} \dfrac{1}{H^F(k_1,k_2)}, & |H^F| \neq 0 \\ 0, & |H^F| = 0 \end{cases} \tag{5.29}$$

$H^-(k_1,k_2)$ is set to zero whenever $|H^F|$ is less than a chosen quantity ϵ.

Wiener Filter

For Wiener filter: $G^F(k_1,k_2) = \dfrac{[H^F(k_1,k_2)]^* S_{uu}(k_1,k_2)}{[H^F(k_1,k_2)]^2 S_{uu}(k_1,k_2) + S_{\eta\eta}(k_1,k_2)}$ (5.32)

where S_{uu} and $S_{\eta\eta}$ are the 2D-DFTs of the autocorrelation functions of $u(n_1,n_2)$ and $\eta(n_1,n_2)$ respectively.

Procedure

(1) Read an image (any size up to 512×512).

(2) Calculate the three different degradation functions $H^F(k_1,k_2)$ by setting $c = 0.0025$, 0.001 and 0.00025 in Eq. (P5.5).

(3) Calculate the Wiener filter $G^F(k_1, k_2)$ with $H^F(k_1, k_2)$ obtained in step 2, and show the three different blurred images.
(4) Implement the inverse filtering and Wiener filtering for each $H^F(k_1, k_2)$.
(5) Show the restored images with ratios N^F/H^F, where N^F and H^F are the 2D-DFTs of additive noise $\eta(n_1, n_2)$ and $h(n_1, n_2)$ respectively.

References

(1) Section 8.3, pp 275–284 in [B6]
(2) *Digital Image Processing* by 'Gonzalez and Woods' pp. 258–264, II Edition [IP18] and pp. 351–357, III Edition [IP19] (Prentice-Hall)

Notes

Inverse filtering is the process of recovering the input of a system from its output. Inverse filter is obtained by dividing the degraded image with the original image in the 2D-transform domain. If the degradation has zero or very low values, then N^F/H^F dominates.

The inverse and pseudo-inverse filters remain *sensitive to noise*. Wiener filter can overcome this limitation. Wiener filtering is a method of restoring images in the presence of blur and as well as noise.

5.6 Geometric Mean Filter
(1) For the degraded images given in Project 5.5 (Inverse and Wiener filtering), apply the geometric mean filter for $s = \frac{1}{2}$ and show the restored images.

For $s = \frac{1}{2}$, the GMF is described by (5.35).

(2) Repeat Part 1) for $s = \frac{1}{4}$.
(3) Repeat Part 1) for $s = \frac{3}{4}$.

Comment on the restored images.
(Hint: For $s = 0$, the GMF becomes the Wiener filter; for $s = 1$, the GMF becomes the inverse filter)

References

(1) Section 8.5, p. 291 in [B6]
(2) Gonzalez and Woods' Book: Section 5.10, p. 270, II Edition [IP18] and p. 361, III Edition [IP19] (Prentice-Hall)

5.7 (a) Circular time/spatial shift property is defined in (5.12). Apply this property to the Lena image of size 512×512. Show circular shift of the image by $m_1 = 100$ samples along n_1 and $m_2 = 50$ samples along n_2 and obtain Fig. P5.5a.
(b) Similarly, circular frequency shift property is defined in (5.13a). Show this property using the Lena image. Show circular shift of the magnitude spectrum of the image by $u_1 = \frac{N}{2} + 100$ samples along k_1 and $u_2 = \frac{N}{2} + 50$ samples along k_2 Fig. P5.5b, where $N = 512$.

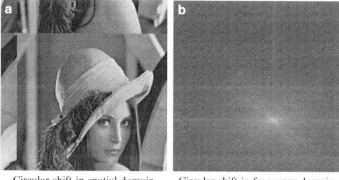

Circular shift in spatial domain   Circular shift in frequency domain

Fig. P5.5 Circular shift properties

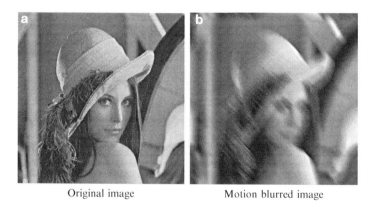

Original image   Motion blurred image

Fig. P5.6 Motion blurred 512×512 image with $a = b = 0.2$ and $C = 1$

5.8 Images blurred by motion and atmospheric turbulence are restored by inverse and Wiener filterings. These are encountered in the application of astronomical imaging.

(a) **Motion blur** represents the uniform local averaging of neighboring pixels, and it is a common result of camera panning or fast object motion. It is shown here for diagonal motion. It can be mathematically modeled as the degradation function

$$H^{\mathrm{F}}(k_1, k_2) = C \frac{\sin[\pi(k_1 a + k_2 b)]}{\pi(k_1 a + k_2 b)} e^{-j\pi(k_1 a + k_2 b)} \qquad (\text{P5.6a})$$

$$= C \operatorname{sinc}(k_1 a + k_2 b) e^{-j\pi(k_1 a + k_2 b)}, \quad \operatorname{sinc}(0) = 1 \qquad (\text{P5.6b})$$

where $a = b = 0.1$ and $C = 1$.

Show that original image and image blurred by the degradation function are same as shown in Fig. P5.6.

Use sinc function in MATLAB. We will recover the original image from this blurred image in part c).

(b) A degradation function for **atmospheric turbulence blur** occurring in the imaging of astronomical objects has the form

$$H^{F}(k_1, k_2) = \exp\left(-c\left[\left(k_1 - \frac{M}{2}\right)^2 + \left(k_2 - \frac{N}{2}\right)^2\right]^{5/6}\right) \qquad \text{(P5.7)}$$

$$M = N = 480$$

(1) This blur is common in land remote sensing and aerial imaging. Simulate the effects of noise.

(2) Inverse filter is very sensitive to noise. Thus inverse filter is applied only in the low pass (LP) zone with radii of 40, 70 and 85 as shown in Fig. P5.7. For this simulation *add a white noise to the blurred image* as described in Fig. P5.4.

var = 0.001;
noise = imnoise(zeros(size(Image)), 'gaussian', 0, var);

A Butterworh low pass function of order 10 is used for the inverse filter with a radius of 40 to eliminate *ringing* artifacts in a filtered image [Fig. P5.8(d)]. That of order 15 is used for the inverse filter with a radius of 100 for a sharp transition at the radius [Fig. P5.8(e)]. Figure P5.8(e) shows less ringing artifacts compared to Fig. P5.8(g).

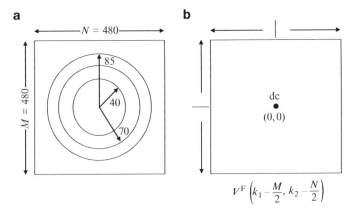

Fig. P5.7 A stabilized version of the inverse filter. **a** Inverse filter applied only in the low pass zone. **b** $V^{F}\left(k_1 - \frac{M}{2}, k_2 - \frac{N}{2}\right)$ 2D-DFT of a blurred image with origin in the center (see Fig. 5.23)

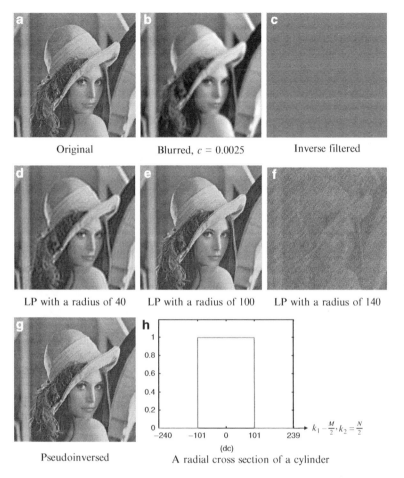

Fig. P5.8 Inverse filter is applied only in the low pass (LP) zone. LP filter with a radius of 100 shows the best results. This filter is an alternative to the pseudo inverse filter as a stabilized version of the inverse filter. The images are of 480 × 480 pixels

Pseudo inverse filter is set to zero outside of a cylinder whose radial cross section can be shown in Fig. P5.8(h) for the blurring function (P5.7). Thus a radius of around 100 can be a good choice for a sharp-transition Butterworth low pass filter.

(3) Compare inverse and Wiener filters.

See Fig. 5.28 on p. 355 in [IP19].
See Fig. 5.29 on p. 356 in [IP19] (Fig. P5.8).

For a linear shift invariant system with frequency response $H^{\mathrm{F}}(k_1, k_2)$, the Wiener filter is defined as

$$G^{\mathrm{F}}(k_1, k_2) = \frac{[H^{\mathrm{F}}(k_1, k_2)]^* S_{uu}(k_1, k_2)}{|H^{\mathrm{F}}(k_1, k_2)|^2 S_{uu}(k_1, k_2) + S_{\eta\eta}(k_1, k_2)}$$

$$= \frac{[H^{\mathrm{F}}(k_1, k_2)]^*}{|H^{\mathrm{F}}(k_1, k_2)|^2 + \frac{S_{\eta\eta}(k_1, k_2)}{S_{uu}(k_1, k_2)}} \qquad (\text{P5.8})$$

If we do not know the power spectral density of the original image, we may replace the term $\frac{S_{\eta\eta}(k_1, k_2)}{S_{uu}(k_1, k_2)}$ in (P5.8) by a constant K and experiment with various values of K.

$$G^{\mathrm{F}}(k_1, k_2) = \frac{[H^{\mathrm{F}}(k_1, k_2)]^*}{|H^{\mathrm{F}}(k_1, k_2)|^2 + K} \qquad (\text{P5.9})$$

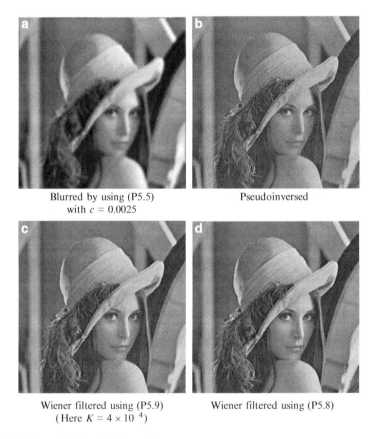

Blurred by using (P5.5) with $c = 0.0025$	Pseudoinversed
Wiener filtered using (P5.9) (Here $K = 4 \times 10^{-4}$)	Wiener filtered using (P5.8)

Fig. P5.9 Wiener filtering of a blurred image

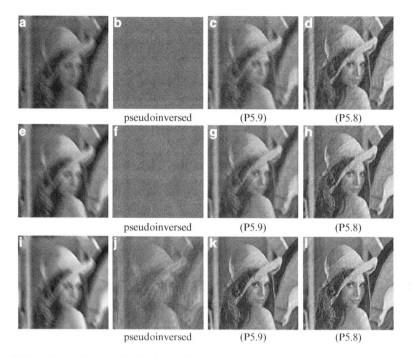

Fig. P5.10 Wiener filtering of noisy blurred images. **a** Motion blurred with noise of variance 386, **b** pseudoinversed, **c** Wiener filtered using (P5.9), **d** Wiener filtered using (P5.8), **e–h** noise of variance 17.7, **i–l** noise of variance 1.8×10^{-3}

Table P5.1 PSNR results

Noise variance	(P5.9)		(P5.8)
	K	PSNR (dB)	PSNR (dB)
386	0.085	21.78	23.51
17.7	0.035	23.17	25.89
1.8×10^{-3}	0.002	27.97	30.61

where K is a constant. This is rather an oversimplification, as the ratio $\frac{s_{\eta\eta}(k_1,k_2)}{s_{uu}(k_1,k_2)}$ is a function of (k_1, k_2) and not a constant [LA20]. From Figs. P5.9c and P5.9d, we see that the Wiener filter without the power spectral density of the original image yields a result close to the Wiener filter with that of the original image (Fig. P5.9).

(c) *Restoring motion blurred images.* Perform the simulation shown in Fig. P5.10. Degradation function $H^{F}(k_1, k_2)$ is given by (P5.6). Compute the PSNRs of the restored images to compare two Wiener filtering methods defined in (P5.8) and (P5.9) respectively. Show in a tabular form as Table P5.1.

For less values of K compared to a specific constant (e.g. K in Table P5.1), the Wiener filter of (P5.9) acts as a high pass filter. Since the blurring process is usually a low pass filter, the filtered image gets less blur and more noisy. For greater values of K, the Wiener filter acts as a low pass filter.

5.9 Take a (256 × 256) or (512 × 512) 8-bit/pel image and implement homomorphic filtering described in Figs. 5.19 and 5.20.

Chapter 6
Vector-Radix 2D-FFT Algorithms

Similar to radix-2 FFT, a vector-radix 2-D FFT can be developed for multidimensional signals. As is the case for radix-2 FFT, the vector radix algorithms can be developed based on both DIT and DIF [SR2, DS1, B39]. Also DIT and DIF can be mixed in the same algorithm. In fact, vector-radix algorithms exist for any radix i.e., radix-2, radix-3 [A14], radix-4 etc. As an illustration, vector radix FFT for a 2-D signal based on both DIT and DIF will be developed. It is then straight forward to extend this technique to all the other vector radix algorithms. As with all the fast algorithms, the advantages of vector radix 2-D FFT are reduced computational complexity, reduced memory (storage) requirements and reduced errors due to finite bit size arithmetic. Vector radix algorithms are much more amenable to vector processors.

6.1 Vector Radix DIT-FFT

The 2-D DFT defined in (5.1) assumes that both $x(n_1, n_2)$ and $X^F(k_1, k_2)$ are periodic with periods N_1 and N_2 along the two dimensions i.e.,

$$x(n_1, n_2) = x(n_1 + N_1, n_2) = x(n_1, n_2 + N_2) = x(n_1 + N_1, n_2 + N_2) \quad (6.1a)$$

$$X^F(k_1, k_2) = X^F(k_1 + N_1, k_2) = X^F(k_1, k_2 + N_2) = X^F(k_1 + N_1, k_2 + N_2) \quad (6.1b)$$

K.R. Rao et al., *Fast Fourier Transform: Algorithms and Applications*,   185
Signals and Communication Technology,
DOI 10.1007/978-1-4020-6629-0_6, © Springer Science+Business Media B.V. 2010

The vector radix-2 2-D DIT-FFT can be symbolically represented as follows:

where e = even and o = odd.

Break this down at each stage until (2×2) 2-D DFTs are obtained. Go backwards to get the $(N_1 \times N_2)$ 2-D DFT. (N_1 and N_2 are integer powers of 2.)

Assume $N_1 = N_2 = N$ for simplicity.

$$X^{\mathrm{F}}(k_1, k_2) = \sum_{n_1=0}^{N-1} \sum_{n_2=0}^{N-1} x(n_1, n_2) W_N^{(n_1 k_1 + n_2 k_2)}$$

$$k_1, k_2 = 0, 1, \ldots, N - 1$$

(5.3a)

This can be decomposed as

$$X^{\mathrm{F}}(k_1, k_2) = \sum_{m_1=0}^{\frac{N}{2}-1} \sum_{m_2=0}^{\frac{N}{2}-1} x(2m_1, 2m_2) W_N^{[2m_1 k_1 + 2m_2 k_2]} \qquad (\mathrm{e}, \mathrm{e})$$

$$+ \sum_{m_1=0}^{\frac{N}{2}-1} \sum_{m_2=0}^{\frac{N}{2}-1} x(2m_1, 2m_2 + 1) W_N^{[2m_1 k_1 + (2m_2+1) k_2]} \qquad (\mathrm{e}, \mathrm{o})$$

$$+ \sum_{m_1=0}^{\frac{N}{2}-1} \sum_{m_2=0}^{\frac{N}{2}-1} x(2m_1 + 1, 2m_2) W_N^{[(2m_1+1) k_1 + 2m_2 k_2]} \qquad (\mathrm{o}, \mathrm{e})$$

$$+ \sum_{m_1=0}^{\frac{N}{2}-1} \sum_{m_2=0}^{\frac{N}{2}-1} x(2m_1 + 1, 2m_2 + 1) W_N^{[(2m_1+1) k_1 + (2m_2+1) k_2]} \qquad (\mathrm{o}, \mathrm{o})$$

(6.2a)

$$X^{\mathrm{F}}(k_1, k_2) = \left[\sum_{m_1=0}^{N/2-1} \sum_{m_2=0}^{N/2-1} x(2m_1, 2m_2) W_{N/2}^{[m_1 k_1 + m_2 k_2]} \right]$$

$$+ W_N^{k_2} \left[\sum_{m_1=0}^{N/2-1} \sum_{m_2=0}^{N/2-1} x(2m_1, 2m_2 + 1) W_{N/2}^{[m_1 k_1 + m_2 k_2]} \right]$$

$$+ W_N^{k_1} \left[\sum_{m_1=0}^{N/2-1} \sum_{m_2=0}^{N/2-1} x(2m_1+1, 2m_2) W_{N/2}^{[m_1k_1+m_2k_2]} \right]$$

$$+ W_N^{k_1+k_1} \left[\sum_{m_1=0}^{N/2-1} \sum_{m_2=0}^{N/2-1} x(2m_1+1, 2m_2+1) W_{N/2}^{[m_1k_1+m_2k_2]} \right] \qquad (6.2b)$$

$$X^F(k_1, k_2) = S_{00}(k_1, k_2) + S_{01}(k_1, k_2)W_N^{k_2} + S_{10}(k_1, k_2)W_N^{k_1} + S_{11}(k_1, k_2)W_N^{k_1+k_2}$$
$$(6.2c)$$

$S_{00}(k_1,k_2), S_{01}(k_1,k_2), S_{10}(k_1,k_2), S_{11}(k_1,k_2)$ are periodic with period $N/2$ along both k_1 and k_2 i.e.,

$$S_{ij}(k_1, k_2) = S_{ij}\left(k_1 + \frac{N}{2}, k_2\right) = S_{ij}\left(k_1, k_2 + \frac{N}{2}\right) = S_{ij}\left(k_1 + \frac{N}{2}, k_2 + \frac{N}{2}\right) \quad i,j = 0,1$$
$$(6.3)$$

$$S_{00}(k_1, k_2) = \sum_{m_1=0}^{N/2-1} \sum_{m_2=0}^{N/2-1} x(2m_1, 2m_2) W_{N/2}^{[m_1k_1+m_2k_2]} \qquad (6.4a)$$

$$S_{01}(k_1, k_2) = \sum_{m_1=0}^{N/2-1} \sum_{m_2=0}^{N/2-1} x(2m_1, 2m_2+1) W_{N/2}^{[m_1k_1+m_2k_2]} \qquad (6.4b)$$

$$S_{10}(k_1, k_2) = \sum_{m_1=0}^{N/2-1} \sum_{m_2=0}^{N/2-1} x(2m_1+1, 2m_2) W_{N/2}^{[m_1k_1+m_2k_2]} \qquad (6.4c)$$

$$S_{11}(k_1, k_2) = \sum_{m_1=0}^{N/2-1} \sum_{m_2=0}^{N/2-1} x(2m_1+1, 2m_2+1) W_{N/2}^{[m_1k_1+m_2k_2]}$$

$$k_1, k_2 = 0, 1, \dots, \frac{N}{2} - 1 \qquad (6.4d)$$

The 2-D $(N \times N)$ DFT of $\{x(n_1,n_2)\}$ given by (5.3a) can be expressed as

$$S_{00}(k_1, k_2) + S_{01}(k_1, k_2)W_N^{k_2} + S_{10}(k_1, k_2)W_N^{k_1} + S_{11}(k_1, k_2)W_N^{k_1+k_2}$$

S_{00}, S_{01}, S_{10} and S_{11} are $\left(\frac{N}{2} \times \frac{N}{2}\right)$ 2-D DFTs. This can be illustrated for $N_1 = N_2 = 4$.

$$x_{00}, x_{01}, x_{02}, x_{03}, x_{10}, x_{11}, x_{12}, x_{13}, x_{20}, x_{21}, x_{22}, x_{23}, x_{30}, x_{31}, x_{32}, x_{33}$$

S_{00}	S_{01}	S_{10}	S_{11}
$x_{00}, x_{02}, x_{20}, x_{22}$	$x_{01}, x_{03}, x_{21}, x_{23}$	$x_{10}, x_{12}, x_{30}, x_{32}$	$x_{11}, x_{13}, x_{31}, x_{33}$
(e, e)	(e, o)	(o, e)	(o, o)

Using (6.2) and (6.3), the vector radix 2-D DFT can be expressed as follows:

$$
\begin{bmatrix}
X^{\mathrm{F}}(k_1, k_2) \\
X^{\mathrm{F}}(k_1, k_2 + N/2) \\
X^{\mathrm{F}}(k_1 + N/2, k_2) \\
X^{\mathrm{F}}(k_1 + N/2, k_2 + N/2)
\end{bmatrix}
=
\begin{bmatrix}
1 & 1 & 1 & 1 \\
1 & -1 & 1 & -1 \\
1 & 1 & -1 & -1 \\
1 & -1 & -1 & 1
\end{bmatrix}
\begin{bmatrix}
S_{00}(k_1, k_2) \\
W_N^{k_2} S_{01}(k_1, k_2) \\
W_N^{k_1} S_{10}(k_1, k_2) \\
W_N^{k_1+k_2} S_{11}(k_1, k_2)
\end{bmatrix}
$$

$$k_1, k_2 = 0, 1, \ldots, \frac{N}{2} - 1 \tag{6.5}$$

$$W_N^{2m_1 k_1} = W_{N/2}^{m_1 k_1}, \quad W_N^{2m_2 k_2} = W_{N/2}^{m_2 k_2}$$

$$W_N^{2m} = \exp\left(\frac{-j2\pi}{N} 2m\right) = \exp\left(\frac{-j2\pi}{N/2} m\right) = W_{N/2}^m$$

The matrix relationship (6.5) can be described in flow graph format (this requires three multiplies and eight adds) (Fig. 6.1).

This is vector radix (2×2) 2-D DIT-DFT. For $N_1 = N_2 = 2^n = N$, the decimation procedure is repeated $\log_2 N$ times. Each stage of decimation has $N^2/4$ butterflies. Each butterfly needs three complex multiplies and eight complex adds. Hence $(N \times N)$ point radix (2×2) DIT-DFT requires $3\left(\frac{N^2}{4}\log_2 N\right)$ complex multiplies and $8\left(\frac{N^2}{4}\log_2 N\right)$ complex adds. Brute force requires N^4 complex multiplies and adds. When $N_1 = R_1^n, N_2 = R_2^n, (R_1, R_2)$ vector radix DIT-FFT can be developed. This is similar to the (2×2) vector radix DIT-FFT, and is called mixed vector radix FFT. Various possibilities, such as vector radix DIF-FFT, mixed vector radix DIF-FFT, and vector radix DIF/DIT FFT can be developed.

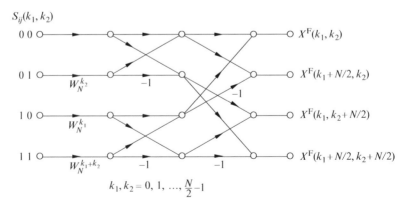

$$k_1, k_2 = 0, 1, \ldots, \frac{N}{2} - 1$$

Fig. 6.1 Flowgraph representation of (6.5)

6.2   Vector Radix DIF-FFT

Similar to vector radix DIT-FFT, a vector radix DIF-FFT can be developed. Assume as before $N_1 = N_2 = 2^n = N$. The 2-D DFT of $x(n_1, n_2), n_1, n_2 = 0, 1, \ldots, N-1$ can be expressed as

$$X^{\mathrm{F}}(k_1, k_2) = \sum_{n_1=0}^{N-1} \sum_{n_2=0}^{N-1} x(n_1, n_2) W_N^{n_1 k_1} W_N^{n_2 k_2}$$

$$= \left(\sum_{n_1=0}^{(N/2)-1} + \sum_{n_1=N/2}^{N-1} \right) \left(\sum_{n_2=0}^{(N/2)-1} + \sum_{n_2=N/2}^{N-1} \right) \left[x(n_1, n_2) W_N^{n_1 k_1} W_N^{n_2 k_2} \right]$$

$$= \left(\sum_{n_1=0}^{(N/2)-1} \sum_{n_2=0}^{(N/2)-1} + \sum_{n_1=0}^{(N/2)-1} \sum_{n_2=N/2}^{N-1} + \sum_{n_1=N/2}^{N-1} \sum_{n_2=0}^{(N/2)-1} + \sum_{n_1=N/2}^{N-1} \sum_{n_2=N/2}^{N-1} \right)$$
$$\times \left[x(n_1, n_2) W_N^{n_1 k_1} W_N^{n_2 k_2} \right] \qquad k_1, k_2 = 0, 1, \ldots, \tfrac{N}{2} - 1$$

(6.6)

$= I + II + III + IV.$ (Here, I, II, III, and IV represent the corresponding four summations.)

In II, III, and IV the summation variables are replaced respectively as follows:

$n_2 = m_2 + N/2$ and $n_1 = m_1 + N/2$, and $n_1 = m_1 + N/2$ and $n_2 = m_2 + N/2$.

Then (6.6) can be simplified as

$$X^{\mathrm{F}}(k_1, k_2) = \sum_{n_1=0}^{(N/2)-1} \sum_{n_2=0}^{(N/2)-1} x(n_1, n_2) W_N^{n_1 k_1 + n_2 k_2}$$

$$+ \sum_{n_1=0}^{(N/2)-1} \sum_{n_2=0}^{(N/2)-1} x\left(n_1, n_2 + \tfrac{N}{2}\right) W_N^{n_1 k_1 + n_2 k_2} (-1)^{k_2}$$

$$+ \sum_{n_1=0}^{(N/2)-1} \sum_{n_2=0}^{(N/2)-1} x\left(n_1 + \tfrac{N}{2}, n_2\right) W_N^{n_1 k_1 + n_2 k_2} (-1)^{k_1}$$

$$+ \sum_{n_1=0}^{(N/2)-1} \sum_{n_2=0}^{(N/2)-1} x\left(n_1 + \tfrac{N}{2}, n_2 + \tfrac{N}{2}\right) W_N^{n_1 k_1 + n_2 k_2} (-1)^{k_1 + k_2}$$

$$k_1, k_2 = 0, 1, \ldots, N - 1 \qquad\qquad (6.7)$$

The 2-D DFT can be considered for four cases as follows:

k_1	k_2
(a) Even integer $(2r_1)$	Even integer $(2r_2)$
(b) Odd integer $(2r_1 + 1)$	Even integer $(2r_2)$
(c) Even integer $(2r_1)$	Odd integer $(2r_2 + 1)$
(d) Odd integer $(2r_1 + 1)$	Odd integer $(2r_2 + 1)$

$$r_1, r_2 = 0, 1, \ldots, \tfrac{N}{2} - 1$$

For these four cases (6.7) can be expressed as

$$
X^{\mathrm{F}}(2r_1, 2r_2) = \sum_{n_1=0}^{(N/2)-1} \sum_{n_2=0}^{(N/2)-1} \left[x(n_1, n_2) + x\left(n_1, n_2 + \tfrac{N}{2}\right) \right.
$$
$$
\left. + x\left(n_1 + \tfrac{N}{2}, n_2\right) + x\left(n_1 + \tfrac{N}{2}, n_2 + \tfrac{N}{2}\right) \right] W_{N/2}^{n_1 r_1 + n_2 r_2} \tag{6.8a}
$$

$$
X^{\mathrm{F}}(2r_1 + 1, 2r_2) = \sum_{n_1=0}^{(N/2)-1} \sum_{n_2=0}^{(N/2)-1} W_N^{n_1} \left[x(n_1, n_2) + x\left(n_1, n_2 + \tfrac{N}{2}\right) \right.
$$
$$
\left. - x\left(n_1 + \tfrac{N}{2}, n_2\right) - x\left(n_1 + \tfrac{N}{2}, n_2 + \tfrac{N}{2}\right) \right] W_{N/2}^{n_1 r_1 + n_2 r_2} \tag{6.8b}
$$

$$
X^{\mathrm{F}}(2r_1, 2r_2 + 1) = \sum_{n_1=0}^{(N/2)-1} \sum_{n_2=0}^{(N/2)-1} W_N^{n_2} \left[x(n_1, n_2) - x\left(n_1, n_2 + \tfrac{N}{2}\right) \right.
$$
$$
\left. + x\left(n_1 + \tfrac{N}{2}, n_2\right) - x\left(n_1 + \tfrac{N}{2}, n_2 + \tfrac{N}{2}\right) \right] W_{N/2}^{n_1 r_1 + n_2 r_2} \tag{6.8c}
$$

$$
X^{\mathrm{F}}(2r_1 + 1, 2r_2 + 1) = \sum_{n_1=0}^{(N/2)-1} \sum_{n_2=0}^{(N/2)-1} W_N^{n_1 + n_2} \left[x(n_1, n_2) - x\left(n_1, n_2 + \tfrac{N}{2}\right) \right.
$$
$$
\left. - x\left(n_1 + \tfrac{N}{2}, n_2\right) + x\left(n_1 + \tfrac{N}{2}, n_2 + \tfrac{N}{2}\right) \right] W_{N/2}^{n_1 r_1 + n_2 r_2} \tag{6.8d}
$$

$$r_1, r_2 = 0, 1, \ldots, \tfrac{N}{2} - 1$$

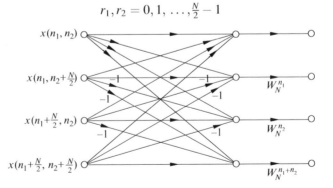

Fig. 6.2 Flowgraph representation of (6.9)

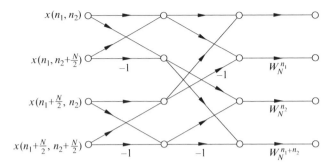

Fig. 6.3 Simplified flowgraph representation of (6.9)

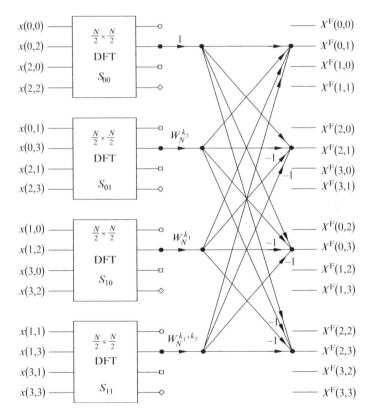

Fig. 6.4 First stage of decimation of a radix-(2×2) FFT. Only one of four butterflies is shown to minimize confusion. $N_1 = N_2 = 4$. (4×4) 2-D DFT

where

$X^F(2r_1, 2r_2)$ is $\left(\frac{N}{2} \times \frac{N}{2}\right)$ 2-D DFT of
$\left[x(n_1, n_2) + x\left(n_1, n_2 + \frac{N}{2}\right) + x\left(n_1 + \frac{N}{2}, n_2\right) + x\left(n_1 + \frac{N}{2}, n_2 + \frac{N}{2}\right)\right]$

$X^F(2r_1 + 1, 2r_2)$ is $\left(\frac{N}{2} \times \frac{N}{2}\right)$ 2-D DFT of
$W_N^{n_1}\left[x(n_1, n_2) + x\left(n_1, n_2 + \frac{N}{2}\right) - x\left(n_1 + \frac{N}{2}, n_2\right) - x\left(n_1 + \frac{N}{2}, n_2 + \frac{N}{2}\right)\right]$

$X^F(2r_1, 2r_2 + 1)$ is $\left(\frac{N}{2} \times \frac{N}{2}\right)$ 2-D DFT of
$W_N^{n_2}\left[x(n_1, n_2) - x\left(n_1, n_2 + \frac{N}{2}\right) + x\left(n_1 + \frac{N}{2}, n_2\right) - x\left(n_1 + \frac{N}{2}, n_2 + \frac{N}{2}\right)\right]$

$X^F(2r_1 + 1, 2r_2 + 1)$ is $\left(\frac{N}{2} \times \frac{N}{2}\right)$ 2-D DFT of
$W_N^{n_1 + n_2}\left[x(n_1, n_2) - x\left(n_1, n_2 + \frac{N}{2}\right) - x\left(n_1 + \frac{N}{2}, n_2\right) + x\left(n_1 + \frac{N}{2}, n_2 + \frac{N}{2}\right)\right]$
$$r_1, r_2 = 0, 1, \ldots, \frac{N}{2} - 1 \tag{6.9}$$

This set of equations (6.9) can be represented in a flowgraph format (Fig. 6.2). This flowgraph can be further simplified as given in Fig. 6.3. This requires eight adds only instead of the 12 adds besides the three multiplies $\left(W_N^{n_1}, W_N^{n_2}, W_N^{n_1 + n_2}\right)$.

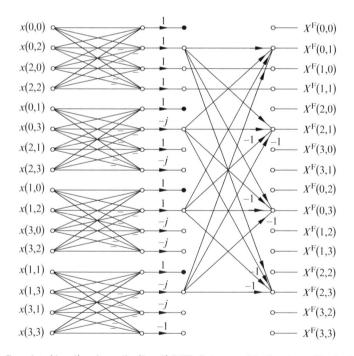

Fig. 6.5 Complete (4×4)-point radix-(2×2) DFT. Only one of the four butterflies in the second stage is shown

An $(N \times N)$ 2-D DFT i.e., $X^{\mathrm{F}}(k_1, k_2), k_1, k_2 = 0, 1, \ldots, N - 1$ is split into four $\left(\frac{N}{2} \times \frac{N}{2}\right)$ 2-D DFTs, i.e.,

$$\begin{bmatrix} X^{\mathrm{F}}(2r_1, 2r_2) \\ X^{\mathrm{F}}(2r_1 + 1, 2r_2) \\ X^{\mathrm{F}}(2r_1, 2r_2 + 1) \\ X^{\mathrm{F}}(2r_1 + 1, 2r_2 + 1) \end{bmatrix}, \qquad r_1, r_2 = 0, 1, \ldots, \frac{N}{2} - 1$$

Split each of these into four $\left(\frac{N}{4} \times \frac{N}{4}\right)$ 2-D DFTs until (2×2) 2-D DFTs are obtained (Figs. 6.4 and 6.5). This is vector radix 2-D DIF FFT. As in the DIT case, several possibilities exist for the DIF.

6.3 Summary

While radix-x ($x = 2, 3, 4, \ldots$) based algorithms are described in Chapters 3 and 5, vector-radix 2-D FFT algorithms are the main focus of this chapter. These are suitable for implementation by vector processors. The following chapter defines and develops nonuniform DFT (NDFT) including its properties.

Chapter 7
Nonuniform DFT

7.1 Introduction

For signals consisting of a number of frequency components, the Fourier transform (FT) effectively reveals their frequency contents and is generally able to represent the signals with an acceptable resolution divided by equal bandwidth in the frequency domain. The discrete Fourier transform (DFT) is an important tool for digital signal processing, in which the N-point DFT of a length-N sequence is given by the frequency samples at N-uniformly spaced points [W23]. It has been widely applied in solving both time-domain and frequency-domain problems, signal analysis/synthesis, detection/estimation, and data compression [B6].

Over the decades since the development of the DFT processing algorithms, there has been increasing demand for nonuniform DFT (NDFT) [N26], where data are sampled at nonuniform points in either the time domain or spectral domain, or both. Samples are located at N arbitrary but distinct points. It thus can be treated as the generalized form of DFT, which is a special case of the NDFT.

Although the need for the NDFT has been recognized for a long time, development of fast transform algorithms has not been concrete, since the unitary property is not inherently guaranteed. Since mid-1990s, some fast computation algorithms have been developed by using the approximation algorithm [N27, N35, N30].

Application fields of the NDFT include efficient spectral analysis, filter design, antenna pattern synthesis, decoding of dual-tone multi-frequency (DTMF) signals [N16], synthetic aperture radar (SAR), ground penetrating radar (GPR), antenna design [N43], magnetic resonance imaging (MRI), and X-ray computerized tomography (CT) [N37]. Some of these deal with images with uniform samples to be nonuniformly transformed into the frequency domain, while others deal with images with nonuniform samples to be uniformly transformed.

In this chapter, we describe the nonuniform DFT in which samples may be nonuniformly selected in the time or frequency domain. The main focus will be on nonuniform sampling in the frequency domain, since many applications are derived to analyze frequency components at arbitrary points. We define one-dimensional

K.R. Rao et al., *Fast Fourier Transform: Algorithms and Applications*,
Signals and Communication Technology,
DOI 10.1007/978-1-4020-6629-0_7, © Springer Science+Business Media B.V. 2010

and two-dimensional NDFT in the following sections, explaining those properties compared to uniform DFT discussed in other chapters. Fast algorithms and inverse NDFT which are the main obstacles for developing the NDFT are investigated in this chapter. Finally we discuss some applications and implementations in MATLAB.

7.2   One-Dimensional NDFT

7.2.1   DFT of Uniformly Sampled Sequences

Let us take into consideration the definition of Fourier transform (FT) in the continuous domain first: Under certain conditions about the function $x(t)$, the Fourier transform of this function exists and can be defined as

$$X(\omega) = \int_{-\infty}^{\infty} x(t)e^{-j\omega t}dt \qquad (7.1)$$

where $\omega = 2\pi f$ and f is a temporal frequency in hertz. The original signal is recovered by the inverse Fourier transform (IFT), given by

$$x(t) = \frac{1}{2\pi} \int_{-\infty}^{\infty} X(\omega)\, e^{j\omega t}d\omega \qquad (7.2)$$

From the definition of FT, we consider now the discrete Fourier transform. In this case we have a finite number N of samples of the signal $x(t)$ taken at regular intervals of duration T_s (which can be considered as a sampling interval). Note that the signal can also be a function in the spatial domain. In practical cases the signal $x(t)$ is not of an infinite duration, but its total duration is $T = NT_s$ and we have a set $\{x_n\}$ of samples of the signal $x(t)$ taken at regular intervals. We can define $x_n = x(t_n)$, where $t_n = nT_s$ for $n = 0, 1, \ldots, N - 1$ is the sampling coordinate.

In the case of the discrete Fourier transform (DFT), not only we want the signal to be discrete and not continuous, but we also want the Fourier transform, which is a function of the temporal frequency, to be defined only at regular points in the frequency domain. Thus the function $X(\omega)$ is not defined for every value of ω but only for certain values ω_m. We want the samples $X(\omega_m)$ to be regularly spaced as well, so that all the samples ω_m are multiples of a dominant frequency $\frac{1}{T}$, that is to say $\omega_m = \frac{2\pi}{T}m$ for $m = 0, 1, \ldots, N - 1$. Let us note now that T is equal to the finite duration of the signal $x(t)$ from which we want to define its discrete Fourier transform. Note also that we assume the number of samples in frequency to be equal to the number of samples in the temporal domain, that is N. This is not a necessary condition, but it simplifies the notation. The direct extension of (7.1) to the discrete domain is:

$$X(\omega_m) = \sum_{n=0}^{N-1} x(t_n)e^{-j\omega_m t_n} \qquad (7.3)$$

Considering that ω_m is defined at only discrete values $\frac{2\pi}{T}m$ and t_n is also defined at only discrete values nT_s, it is possible to rewrite (7.3) as

$$X(\omega_m) = \sum_{n=0}^{N-1} x(t_n) e^{-j\left(m\frac{2\pi}{T}\right)(nT_s)} = \sum_{n=0}^{N-1} x(t_n) e^{-j\left(m\frac{2\pi}{NT_s}\right)(nT_s)} \tag{7.4}$$

It is now possible to simplify and express the dependence on ω_m only in terms of m and the dependence on t_n only in terms of n. In this way the final definition of the DFT is:

$$X^{\mathrm{F}}(m) = \sum_{n=0}^{N-1} x(n) \, e^{-j\frac{2\pi}{N}nm} = \sum_{m=0}^{N-1} x(n) \, W_N^{nm} \qquad m = 0, 1, \ldots, N-1 \tag{7.5}$$

where $W_N = e^{-j\frac{2\pi}{N}}$. The inverse discrete Fourier transform (IDFT) is given as

$$x(n) = \frac{1}{N} \sum_{m=0}^{N-1} X^{\mathrm{F}}(m) \, e^{j\frac{2\pi}{N}nm} = \frac{1}{N} \sum_{m=0}^{N-1} X^{\mathrm{F}}(m) \, W_N^{-nm} \qquad n = 0, 1, \ldots, N-1 \tag{7.6}$$

7.2.2 Definition of the NDFT

Now we want to generalize the definition and the computation of the Fourier transform from the regular sampling to the irregular sampling domain. In the general case, the definition of the nonuniform discrete Fourier transform (NDFT) is the same as the one given by (7.3), taking into consideration that the samples can be taken at irregular intervals in time (t_n) and/or in frequency (ω_m).

The NDFT is an extension of the uniform DFT, that is, the latter is a special case of the former. Here, the uniform DFT is that the Fourier transform of equally spaced data $x(n)$ is evaluated on an equally spaced grid. Assuming the temporal sampling locations $\{t_n, n = 0, 1, \ldots, N-1\} \in [0, N)$ and the frequency sampling locations $\{\omega_m, m = 0, 1, \ldots, N-1\} \in [0, N)$, we have four different types of the generalized DFT as given below [N19]. The complex sequence $x = \{x(t_0), \ldots, x(t_{N-1})\}$ is convolved with corresponding exponential function to derive its spectrum $X = \{X(\omega_0), \ldots, X(\omega_{N-1})\}$. The indices t_n and ω_m are replaced by n and m, respectively, if the Fourier transform of equally spaced data $x(n)$ is evaluated on equally spaced grid points.

1. NDFT-1: *Nonuniform temporal* sampling points and *uniform frequency* sampling points. Images obtained by MRI belong to this case [N37]. The NDFT-1 is defined as

$$X(m) = \frac{1}{\sqrt{N}} \sum_{n=0}^{N-1} x(t_n) \exp\left(\frac{-j2\pi m t_n}{N}\right) \qquad m, n = 0, 1, \ldots, N-1 \tag{7.7}$$

2. NDFT-2: *Uniform temporal* sampling points and *nonuniform frequency* sampling points. Images obtained by a regular camera belong to this case. The NDFT-2 is defined as

$$X(\omega_m) = \frac{1}{\sqrt{N}} \sum_{n=0}^{N-1} x(n) \exp\left(\frac{-j2\pi n\omega_m}{N}\right) \quad m, n = 0, 1, \ldots, N-1 \qquad (7.8)$$

3. NDFT-3: *Nonuniform temporal* sampling points and *nonuniform frequency* sampling points. Images obtained by irregular sampling device and transformed into irregular frequency domain belong to this case. The NDFT-3 is defined as

$$X(\omega_m) = \frac{1}{\sqrt{N}} \sum_{n=0}^{N-1} x(t_n) \exp\left(\frac{-j2\pi \omega_m t_n}{N}\right) \quad m, n = 0, 1, \ldots, N-1 \qquad (7.9)$$

4. NDFT-4: *Uniform temporal* sampling points and *uniform frequency* sampling points. The conventional DFT belongs to this case. It is defined in (7.5).

However, in practice, we want to take into consideration a more restricted case. One is the case where the Fourier transform of non-equally spaced data $x(t_n), n = 0, 1, \ldots, N-1$ is evaluated on an equally spaced grid (NDFT-1). The extension from non-equally spaced sampling to equally spaced sampling, therefore, provides an efficient transform tool for the non-equally spaced time samples, such as medical image analysis [N37]. Arbitrary sampling points in the time domain should be defined by certain criteria which are very hard to adapt to signal property, unless having some prior knowledge.

Another possible consideration is that the Fourier transform of equally spaced data $x(n)$ is evaluated on non-equally spaced grid points $\omega_m, m = 0, 1, \ldots, N-1$ (NDFT-2). This scheme provides more flexibility to get the spectrum at arbitrary frequencies as in Fig. 7.1. In certain applications such as spectral analysis, it is preferable to sample the Z-transform $X(z)$ at N non-equally distributed points on the

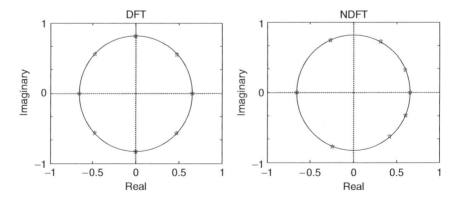

Fig. 7.1 Eight frequency points along the unit circle in the z-plane for DFT and NDFT

unit circle in the z-plane, depending on the frequency spectrum. The more spectra concentrated in the frequency domain, the more number of samples should be allocated. Coarse analysis does not cause severe errors for lower magnitude components.

Implementation of the nonuniform DFT can be done by either direct multiplication of the NDFT matrix with data matrix or fast multiplication which will be discussed later in this chapter. Direct multiplication is useful for understanding the concept of transform. The MATLAB functions for forward and inverse transform of DFT and NDFT are given for comparison as follows.

```
function [Xk] = dft1d(xn, N)
% Computes Uniform Discrete Fourier Transform by multiplication of
DFT matrix.
%- - - - - - - - - - - - - - - - - - - - - -
% xn = N-point 1D input sequence over 0 <= n <= N-1 (row vector)
% Xk = 1D DFT coefficient array over 0 <= k <= N-1
% N = Length of DFT
% Usage: Xk = dft1d(xn, N)
%
n = 0 : 1 : N-1; % Index for input data
k = 0 : 1 : N-1; % Index for DFT coefficients
Wn = exp(-j*2*pi/N); % Twiddle factor
nk = k'*n; % Creates an N x N matrix
DFTmtx = Wn .^ nk; % DFT matrix (N x N)
Xk = (DFTmtx * xn.'); % DFT coefficients (column vector)
% xn.' is the non-conjugate transpose of xn and equals to "transpose
(xn)".

function [xn] = idft1d(Xk, N)
% Computes Uniform Inverse Discrete Fourier Transform.
% - - - - - - - - - - - - - - - - - - - - -
% xn = N-point 1D input sequence over 0 <= n <= N-1 (column vector)
% Xk = 1D DFT coefficient array over 0 <= k <= N-1
% N = Length of DFT
% Usage: xn = idft1d(Xk, N)
%
n = 0 : 1 : N-1; % Index for input data
k = 0 : 1 : N-1; % Index for DFT coefficients
Wn = exp(j*2*pi/N); % Twiddle factor
nk = n'*k; % Creates an N x N matrix
IDFTmtx = Wn .^ nk; % IDFT matrix (N x N), equivalent to (DFTmtx)'
xn = (IDFTmtx * Xk).' / N; % Reconstructed sequence (row vector)
function [Xk] = ndft1d(xn, N, fs)

% Computes NonUniform Discrete Fourier Transform
%- - - - - - - - - - - - - - - - - - - - - - -
% xn = N-point 1D input sequence over 0 <= n <= N-1 (Uniform)
% Xk = 1D NDFT coefficient array (NonUniform)
% fs = nonuniform frequency vector (N-point)
% N = Length of DFT
% Usage: Xk = ndft1d(xn, N, fs)
%
```

(continued)

```
n = 0 : 1 : N-1; % Index for input data
Wn = exp(-j*2*pi/N); % Twiddle factor
nk = fs' * n; % Creates an N × N matrix
DFTmtx = Wn .^ nk; % DFT matrix (N × N)
Xk = (DFTmtx * xn.'); % DFT coefficients (column vector)

function [xn] = indft1d(Xk, N, fs)
% Computes 1D NonUniform Inverse Discrete Fourier Transform
% - - - - - - - - - - - - - - - - - - - - - - - - - - - -
% xn = N-point 1D input sequence over 0 <= n <= N-1
% fs = Nonuniform frequency vector (N-point)
% Xk = 1D DFT coefficient array over 0 <= k <= N-1
% N = Length of DFT
% Usage: xn = indft1d(Xk, N, fs)
%
n = 0: 1: N-1; % Index for input data

Wn = exp(-j*2*pi/N); % Twiddle factor
nk = fs' * n; % Creates an N × N matrix
DFTmtx = Wn .^ nk; % DFT matrix (N × N)
xn = (inv(DFTmtx) * Xk) .'; % Reconstructed sequence (row vector)
```

7.2.3 Properties of the NDFT

As defined above, the NDFT of the sequence $x(n)$ is denoted by $X(z_k)$ in the z-domain, while the DFT of the sequence is denoted by $X^F(k)$ on the unit circle (Fig. 2.3b). Thus some of the properties in the DFT are also satisfied in the NDFT operation.

1. *Linearity.* Given $x_1(n) \overset{\text{NDFT}}{\longleftrightarrow} X_1(z_k)$ and $x_2(n) \overset{\text{NDFT}}{\longleftrightarrow} X_2(z_k)$, then

$$[a_1 x_1(n) + a_2 x_2(n)] \overset{\text{NDFT}}{\longleftrightarrow} [a_1 X_1(z_k) + a_2 X_2(z_k)] \qquad (7.10)$$

where a_1 and a_2 are constants. Hence the NDFT is a linear operation, as is the DFT.

If the two signals are different in length, i.e., $x_1(n)$ has length N_1 and $x_2(n)$ has length N_2, then the linearly combined sequence will have a maximum length of $N = \max(N_1, N_2)$. A sequence with less number of samples is padded with a number of zeros prior to the linear operation to obtain the same length.

2. *Complex conjugation.* For N-point DFT and NDFT

$$x^*(n) \overset{\text{NDFT}}{\longleftrightarrow} X^*(z_k^*) \qquad (7.11)$$

where $x^*(n)$ is the conjugate of $x(n)$. The NDFT of $x^*(n)$ is given by

$$F_N[x^*(n)] = \sum_{n=0}^{N-1} x^*(n) z_k^{-n} = \left\{ \sum_{n=0}^{N-1} x(n) \left(z_k^*\right)^{-n} \right\}^* = X^*(z_k^*) \qquad (7.12)$$

where $F_N[x^*(n)]$ is NDFT of $x^*(n)$. Thus the conjugation of the original sequence leads to the conjugation of its NDFT, which is evaluated at the conjugated sample points in the z-domain.

3. *Real part of sequence.* The NDFT of real part of a sequence is the summation of the halved real and conjugated NDFT components, i.e.,

$$\text{Re}\{x(n)\} \overset{\text{NDFT}}{\longleftrightarrow} \frac{1}{2}\{X(z_k) + X^*(z_k^*)\} \qquad (7.13)$$

Since the real part of $x(n)$ is given by $x_r(n) = \frac{1}{2}\{x(n) + x^*(n)\}$, the NDFT can be obtained by applying the result of linearity and complex conjugation properties, as

$$F_N[x_r(n)] = \frac{1}{2}\{X(z_k) + X^*(z_k^*)\} \qquad (7.14)$$

4. *Imaginary part of sequence*

$$j\,\text{Im}\{x(n)\} \overset{\text{NDFT}}{\longleftrightarrow} \frac{1}{2}\{X(z_k) - X^*(z_k^*)\} \qquad (7.15)$$

Since the imaginary part of $x(n)$ can be expressed as $x_i(n) = \frac{1}{j2}\{x(n) - x^*(n)\}$, the NDFT can be obtained by applying the result of linearity and complex conjugation properties, so that

$$F_N[j\,x_i(n)] = \frac{1}{2}\{X(z_k) - X^*(z_k^*)\} \qquad (7.16)$$

5. *Shifting in time domain.* Given a positive integer q_0, consider the arbitrary spaced discrete-time signal $x(n - q_0)u(n - q_0)$, which is the q_0-step shifted version of $x(n)u(n)$. Then

$$x(n - q_0)u(n - q_0) \overset{\text{NDFT}}{\longleftrightarrow} \sum_{n=0}^{N-1+q_0} x(n - q_0)u(n - q_0)z_k^{-n} \qquad (7.17)$$

Since $u(n - q_0) = 0$ for $n < q_0$ and $u(n - q_0) = 1$ for $n \geqslant q_0$,

$$x(n - q_0)u(n - q_0) \overset{\text{NDFT}}{\longleftrightarrow} \sum_{n=q_0}^{N-1+q_0} x(n - q_0)z_k^{-n} \qquad (7.18)$$

Consider a change of index in the right-side summation in (7.18). Let $m = n - q_0$ so that $n = m + q_0$, then $m = 0$ when $n = q_0$ and $m = N - 1$ when $n = N - 1 + q_0$. Hence

$$\sum_{n=q_0}^{N-1+q_0} x(n - q_0)z_k^{-n} = \sum_{m=0}^{N-1} x(m)z_k^{-(m+q_0)} = z_k^{-q_0}\sum_{m=0}^{N-1} x(m)z_k^{-m} = z_k^{-q_0}X(z_k) \quad (7.19)$$

Thus combining this with (7.18) yields the transform pair:

$$x(n - q_0)u(n - q_0) \quad \overset{\text{NDFT}}{\longleftrightarrow} \quad z_k^{-q_0}X(z_k) \qquad (7.20)$$

This means that a shift in time domain corresponds to the multiplication of the NDFT $X(z_k)$ by a complex factor, $z_k^{-q_0}$.

6. *Multiplication by a_s^n*

$$a_s^n x(n) \quad \overset{\text{NDFT}}{\longleftrightarrow} \quad X\left(\frac{z_k}{a_s}\right) \qquad (7.21)$$

For any nonzero real or complex number a_s, the NDFT of the multiplied sequence is given by

$$F_N[a_s^n x(n)] = \sum_{n=0}^{N-1} a_s^n x(n)\, z_k^{-n} = \sum_{n=0}^{N-1} x(n)\, (z_k/a_s)^{-n}$$
$$= X(z_k/a_s) \qquad (7.22)$$

7. *Time reversal*

$$x(-n) \quad \overset{\text{NDFT}}{\longleftrightarrow} \quad X\left(\frac{1}{z_k}\right) \qquad (7.23)$$

The NDFT of $x(-n)$ is calculated by

$$F_N[x(-n)] = \sum_{n=-N+1}^{0} x(-n)z_k^{-n} \qquad (7.24)$$

Let $m = -n$. Then

$$F_N[x(-n)] = \sum_{m=0}^{N-1} x(m)z_k^m = X(1/z_k) \qquad (7.25)$$

Thus time reversal of the sequence results in the inversion of the sample locations in the z-plane.

7.2.4 Examples of the NDFT-2

The aim of this chapter is to derive nonuniform sampling of signal in the frequency domain transformed by DFT, which will be transformed by the NDFT. For example, a signal in Fig. 7.2 is assumed to be mix of two sinusoidal sequences, i.e., 0.2π and 0.7π. DFT of the signal can be obtained based on the regular sample intervals (Fig. 7.1). We assume the samples are infinite, i.e., the time duration is infinitely long, resulting in infinitely short response in the frequency domain.

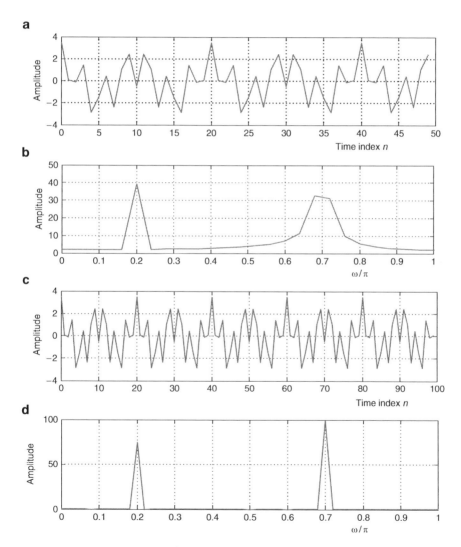

Fig. 7.2 DFT results. **a** Input sequence with 50 samples. **b** DFT output of (**a**). **c** Input sequence with 100 samples. **d** DFT output of (**c**)

However, the DFT of the signal with limited number of samples in Fig. 7.2a does not show proper frequency analysis, resulting in broad range of spectrum near 0.2π and 0.7π in Fig. 7.2b. Enhanced result is obtained with more number of samples in Fig. 7.2c but still extra frequency components exist around the two main frequencies. This means that the impulse response of frequency components contained in a signal is obtained with infinite number of time-domain samples, which are not feasible in the actual situation. In other words, the signal should not be time-domain band-limited to get exact properties. Thus we need to introduce a new concept of nonuniform sample allocation in the frequency domain by densely populating in the specified frequency range, while having less number of samples in the less important range. The NDFT yields optimal approximation of the main spectra.

The sequence of MATLAB commands is as follows.

```
%Program test_fft.m
k = 0:49;
w1 = 0.2*pi;w2 = 0.7*pi;
x1 = 1.5*cos(w1*k); x2 = 2*cos(w2*k);
x = x1+x2;
G = fft(x);
figure
subplot(2,1,1);
plot(k,x); grid; axis([0 49 -4 4]);
xlabel('Time Index n'); ylabel('Amplitude');
title('Input Sequence(50 pts)');
subplot(2,1,2);
plot(2*k/50,abs(G)); grid; axis([0 1 0 50])
xlabel('\rmomega/\rmpi'); ylabel('Amplitude');
title('Output Sequence');

k = 0:99;
w1 = 0.2*pi;w2 = 0.7*pi;
x1 = 1.5*cos(w1*k); x2 = 2*cos(w2*k);
x = x1+x2;
G = fft(x);
figure
subplot(2,1,1);
plot(k,x); grid;
xlabel('Time Index n'); ylabel('Amplitude');
title('Input Sequence(100 pts)');
subplot(2,1,2);
plot(k/50,abs(G)); grid; axis([0 1 0 100])
xlabel('\rmomega/\rmpi'); ylabel('Amplitude');
title('Output Sequence');
```

To compute the NDFT, we set nonuniform sample intervals based on the two main frequencies. This can be carried out by either manual adjustment or resampling by centroid as shown in Fig. 7.3 [N44]. Note that the centroid C of a finite set of points $x_1, x_2, \ldots, x_k$ in R^n is defined as

Fig. 7.3 Block diagram for nonuniform resampling by oversampling and centroid concept

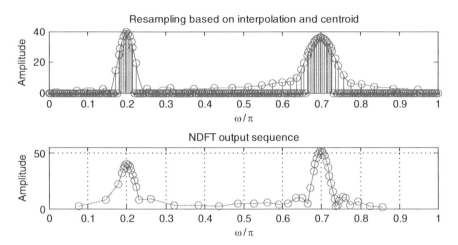

Fig. 7.4 Resampling results based on interpolation and centroid. Dense sampling is given for higher magnitude of spectrum. Lower figure shows NDFT output using the resampled points

$$C = \frac{x_1 + x_2 + \cdots + x_k}{k} \tag{7.26}$$

where R^n denotes the set of all $n \times 1$ matrices with real entries. In this example, the order of samples in the NDFT is nonlinearly allocated and concentrated on the main frequencies. This nonlinear resampling can be executed in the DFT domain and represented by:

1. Execute DFT with equispaced samples.
2. Interpolate the sample intervals with a certain ratio (more than two) and reshape the amplitude spectrum based on the curve fitting, such as the cubic splined curvature [B6, 296].
3. Resample the interpolated samples to obtain the same number of total samples as in the equispaced case based on the centroid concept.
4. Perform the NDFT with resampled nonequispaced frequencies.

Resampling in the frequency domain is executed based on interpolation with a ratio of one-to-ten and centroid to get the number of original samples. The result in Fig. 7.4 shows optimal energy distribution and resampling to limited number of frequency components.

The following commands perform the NDFT to detect two main frequencies based on FFT of input signal, oversampling by a factor of ten, resampling by centroid concept to get 50 samples which are the same as the input, forward NDFT:

```
%Program ndft_cent_run.m
 clear ; k = 0:49;
 w1 = 0.2*pi ; w2 = 0.7*pi;
 x1 = 1.5*cos(w1*k) ; x2 = 2*cos(w2*k);
 x = x1+x2;
 %
 G = fft(x) ; y = abs(G); % 50 samples
 %resample_fun(k,G1,0,24,20,20)
 xx=0:0.1:49; %491 samples
 yy = spline(k,y,xx); yy = abs(yy); %491 samples
 % Resampling (downsampling) to nsample
 tsum=sum(yy); inc=0; nsample=100; snum=length(xx);
 avstep = tsum/nsample;
 for i=1:nsample % nsample is # of cent_position
    tmp=0; step=0;
    for j=1+inc:snum
       tmp=tmp+yy(j);
       step=step+1;
       if tmp >= avstep
          tmp1=tmp-yy(j);
          if (avstep-tmp1) <= (tmp-avstep)
             cent_position(i)=j-1;
             inc=inc+step-1;
             break
          else
             cent_position(i)=j;
             inc=inc+step;
             break
          end
       end
    end
 end
 cent_position

 yy1=zeros(1,snum);
 for j=1:nsample-1
    yy1(cent_position(j))=yy(cent_position(j));
 end
 subplot(3,1,1);
 stem(2*xx/50,yy1);grid; axis([0 1 0 40])
 hold on;
 plot(2*xx/50,yy); axis([0 1 0 40])
 xlabel('\rmomega/\rmpi'); ylabel('Magnitude');
 title('Resampling based on interpolation and centroid')

 Zi=2*cent_position/snum;
 Zi=Zi(1:50);
 [H,Z] = ndft(x,pi*Zi,'uc');

 subplot(3,1,2);
 plot(Zi,abs(H), 'Marker','o'); grid; axis([0 1 0 55])
 xlabel('\rmomega/\rmpi'); ylabel('Magnitude');
 title('NDFT output sequence');
```

Another example to test the NDFT and INDFT using up/down linear interpolation shows a reasonable approximation (Fig. 7.5) to obtain the forward and inverse NDFT which will be discussed in Section 7.3. The MATLAB codes are as follows.

```
% Program to test NDFT/INDFT using up/down interpolation
%
clf; % clear current figure
% Generate the input sequence
k = 0:49;
w1 = 0.2*pi; w2 = 0.7*pi;
```

(*continued*)

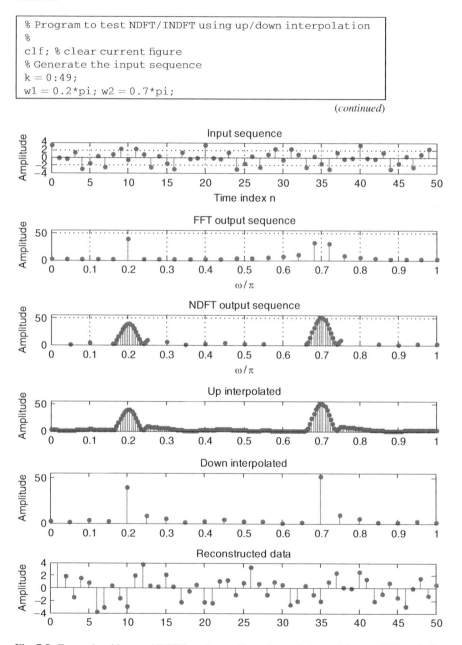

Fig. 7.5 Forward and inverse NDFT based on up/down interpolation and inverse FFT method

```
x1 = 1.5*cos(w1*k); x2 = 2*cos(w2*k);
x = x1+x2;
% Calculating the Discrete Fourier Transform of the
% sequence to find the frequencies involved in the signal.
G = fft(x);
% Calculating the NDFT around the frequencies of interest.
Zi1 = [0.16:.005:0.25]; Zi2 = [0.35:.05:0.55];
Zi3 = [0.66:.005:0.75]; Zi4 = [0.85:.05:1];
Zi = [0.05 0.1 Zi1 0.3 Zi2 Zi3 Zi4];
[H,Z] = ndft(x,pi*Zi,'uc');

% Plot the input and output sequences
subplot(4,1,1);
stem(k,x, 'Marker','.'); grid; axis([0 50 -4 4]);
xlabel('Time index n'); ylabel('Amplitude');
title('Input sequence');
subplot(4,1,2);
stem(2*k/50,abs(G), 'Marker','.');grid; axis([0 1 0 55])
xlabel('\rmomega/\rmpi'); ylabel('Magnitude');
title('FFT output sequence');
subplot(4,1,3);
stem(Zi,abs(H), 'Marker','.');grid; axis([0 1 0 55])
xlabel('\rmomega/\rmpi'); ylabel('Magnitude');
title('NDFT output sequence');
Zinc = [0:0.005:1]; Zdec = [0:0.05:1];
H_int = interp1(Zi, H, Zinc,'linear', 'extrap');
ix = ifft(G);
subplot(4,1,4);stem(Zinc,abs(H_int), 'Marker','.');
axis([0 1 0 55]);title('Up interpolated');
H_rec = interp1(Zinc, H_int, Zdec,'linear', 'extrap');
figure;subplot(411);stem(Zdec,abs(H_rec), 'Marker','.');
axis([0 1 0 55]);title('Down interpolated');
Fi = ifft(H_rec, 50);
subplot(412);stem([1:50],real(Fi)*2, 'Marker','.');
axis([0 50 -4 4]); title('Reconstructed data');
```

7.3   Fast Computation of NDFT

7.3.1   Forward NDFT

Computation of the NDFT can be directly performed by multiplication and summation. To compute each complex sample of the NDFT, we need N complex multiplies and $(N - 1)$ complex adds, i.e., $4N$ real multiplies and $(4N - 2)$ real adds. Therefore, we can say that the computation complexity is approximately proportional to N^2 for evaluating N samples. Note that this is the same as that for directly computing the DFT. While the FFT can be applied by taking into account the orthogonal basis functions of the DFT, the Cooley and Tukey algorithm [N3] cannot be used for the NDFT.

The DFT can be defined as a subset of the Z-transform, since samples in the frequency domain are located on the unit circle. Now we expand the sample locations to arbitrary points in the z-plane. Hence, the N-point NDFT is defined as the frequency samples of the Z-transform at arbitrary N points and expressed in the form

$$X_{ndft}(z_m) = \sum_{n=0}^{N-1} x_n z_m^{-n} \tag{7.27}$$

where z_m are the complex points of interest that can be irregular spaced.

Equation (7.27) can be rewritten in a matrix form as $\underline{X} = [D]\underline{x}$ where the matrix $[D]$ and the vectors $\underline{X}$ and $\underline{x}$ are given as

$$\underline{X} = \begin{bmatrix} X_{ndft}(z_0) \\ X_{ndft}(z_1) \\ \vdots \\ X_{ndft}(z_{N-1}) \end{bmatrix}, \quad \underline{x} = \begin{bmatrix} x[0] \\ x[1] \\ \vdots \\ x[N-1] \end{bmatrix},$$

$$[D] = \begin{bmatrix} 1 & z_0^{-1} & z_0^{-2} & \cdots & z_0^{-(N-1)} \\ 1 & z_1^{-1} & z_1^{-2} & \cdots & z_1^{-(N-1)} \\ 1 & \vdots & \vdots & \ddots & \vdots \\ 1 & z_{N-1}^{-1} & z_{N-1}^{-2} & \cdots & z_{N-1}^{-(N-1)} \end{bmatrix} \tag{7.28}$$

Note that computation of the matrix multiplication requires $O(N^2)$ operational complexity.

7.3.1.1 Horner's Nested Multiplication Method

A recursive algorithm called Horner's method [N7] can reduce memory size keeping only two multiplier coefficients. Equation (7.27) can be written as

$$X_{ndft}(z_m) = \sum_{n=0}^{N-1} x_n z_m^{-n} = z_m^{-(N-1)} \sum_{n=0}^{N-1} x_n z_m^{N-1-n} = z_m^{-(N-1)} A_m \tag{7.29}$$

where $A_m = \sum_{n=0}^{N-1} x_n z_m^{N-1-n} = x_0 z_m^{N-1} + x_1 z_m^{N-2} + \cdots + x_{N-2} z_m + x_{N-1}$.

It can also be expressed as a nested structure that can be calculated easily by a recursive algorithm:

$$A_m = (\cdots(x_0 z_m + x_1)z_m + \cdots)z_m + x_{N-1} \tag{7.30}$$

Example: $A_m = ((x_0 z_m + x_1)z + x_2)z + x_3$ for $N = 3$

The smallest nest output is $y_1 = y_0 z_m + x_1$ with the initial condition $y_0 = x_0$ and we can express (7.30) by the recursive difference equation as $y_n = y_{n-1} z_m + x_n$.

When this algorithm has accumulated N times, we finally get the result $A_m = y_{N-1}$. Thus, only two multiplier coefficients (z_m and $z_m^{-(N-1)}$) are needed to calculate the mth NDFT sample, reducing the memory (buffer) size for the input signal. But the complexity of this method requires a total of $4N$ real multiplies and $(4N - 2)$ real adds, which is the same as in the direct method.

Horner's algorithm can be implemented in MATLAB as follows.

```
function [Xndft, zm] = ndft(x, z, 'Horner') % z(m) = zm

% Arbitrary points in the z-plane: Using Horner's method
N = length(x);
for m = 1: N
    x_tmp = x(1);
    for n = 2:N
        x_tmp = x_tmp * z(m) + x(n);
    end
    A(m) = x_tmp; % A(m) = Am
end

X = z.^(1-N).*A; Xndft = X; zm = z; % Output

subplot(2,1,1);
stem(angle(zm), abs(X)); title('Magnitude plot of the NDFT');
xlabel('Frequency \omega'); ylabel('Magnitude');
axis([min(angle(Zm)) max(angle(zm)) 0 max(abs(X))]);
subplot(2,1,2);
stem(angle(zm), unwrap(angle(X))); title('Phase plot of the NDFT');
xlabel('Frequency \omega'); ylabel('Phase in radians');
axis([min(angle(zm)) max(angle(zm)) min(unwrap(angle(X)))
max(unwrap(angle(X)))]);
```

7.3.1.2 Goertzel's Algorithm

The NDFT of a sequence x_n is the Z-transform at N arbitrarily located points in the z-plane. To reduce computational complexity, the NDFT can be evaluated at points on the unit circle (i.e., $z_m = e^{j\omega_m}$) with center at the origin in the z-plane where ω_m is the angular frequency. The Goertzel's algorithm [N2, M13] using trigonometric series interpretation, requires only three coefficients, $\cos\omega_m$, $\sin\omega_m$, and $e^{-j(N-1)\omega_m}$ for each NDFT sample. Now (7.27) can be written as

$$X_{\text{ndft}}(z_m) = \sum_{n=0}^{N-1} x_n z_m^{-n} = z_m^{-(N-1)} \sum_{n=0}^{N-1} x_n z_m^{N-1-n}$$

$$= e^{-j(N-1)\omega_m} \sum_{n=0}^{N-1} x_n e^{j(N-1-n)\omega_m} = e^{-j(N-1)\omega_m} A_m \qquad (7.31)$$

where $A_m = \sum_{n=0}^{N-1} x_n e^{j(N-1-n)\omega_m}$, since $z_m^{-n} = z_m^{-(N-1)} z_m^{N-1-n}$.

We can further decompose A_m into cosine and sine components as

$$A_m = C_m + jS_m \tag{7.32}$$

where $\quad C_m = \sum_{n=0}^{N-1} x_n \cos[(N-1-n)\omega_m], \quad$ and $\quad S_m = \sum_{n=0}^{N-1} x_n \sin[(N-1-n)\omega_m].$
Since each component requires N coefficients, we need to calculate $2N$-point arithmetic. To reduce the complexity, the partial sums are defined as

$$C_m^{(i)} = \sum_{n=0}^{i} x_n \cos[(N-1-n)\omega_m] \quad i = 0, 1, \ldots, N-1 \tag{7.33a}$$

$$S_m^{(i)} = \sum_{n=0}^{i} x_n \sin[(N-1-n)\omega_m] \quad i = 0, 1, \ldots, N-1 \tag{7.33b}$$

so that $C_m^{(N-1)} = C_m, S_m^{(N-1)} = S_m$, if $i = N-1$. By analyzing (7.33), we observe that

$$C_m^{(1)} = x_0 \cos[(N-1)\omega_m] + x_1 \cos[(N-2)\omega_m] \tag{7.34}$$

Consider the trigonometric relations

$$\cos(a+b) = \cos a \cos b - \sin a \sin b \tag{7.35a}$$

$$\cos(a-b) = \cos a \cos b + \sin a \sin b \tag{7.35b}$$

Add (7.35a) and (7.35b) to get

$$\cos(a+b) = 2\cos a \cos b - \cos(a-b) \tag{7.35c}$$

It follows that the cosine function $\cos[(N-1)\omega_m]$ in (7.34) can be expressed as

$$\cos[(N-1)\omega_m] = \cos([(N-2)+1]\omega_m)$$
$$= 2\cos[(N-2)\omega_m]\cos\omega_m - \cos[(N-3)\omega_m]$$

Then (7.34) can be rewritten as

$$C_m^{(1)} = (2x_0 \cos\omega_m + x_1)\cos[(N-2)\omega_m] - x_0 \cos[(N-3)\omega_m] \tag{7.36a}$$

The next partial sum is also expressed as

$$C_m^{(2)} = (2x_0 \cos\omega_m + x_1)\cos[(N-2)\omega_m] + (x_2 - x_0)\cos[(N-3)\omega_m] \tag{7.36b}$$

Similarly

$$C_m^{(3)} = (2g_2 \cos\omega_m + x_2 - x_0)\cos[(N-3)\omega_m]$$
$$+ (x_2 - g_2)\cos[(N-4)\omega_m] \tag{7.36c}$$

The general expression is

$$C_m^{(i)} = g_i \cos[(N-i)\omega_m] + h_i \cos[(N-i-1)\omega_m] \quad i = 1, 2, \ldots, N-1 \quad (7.36d)$$

where

$$g_1 = x_0, \qquad\qquad h_1 = x_1$$
$$g_i = 2g_{i-1} \cos \omega_m + h_{i-1}, \quad h_i = x_i - g_{i-1}, \quad i = 2, 3, \ldots, N-1$$

The final cosine term is expressed as

$$C_m = C_m^{(N-1)} = g_{N-1} \cos \omega_m + h_{N-1} \qquad\qquad (7.37)$$

Similarly

$$\sin[(N-1)\omega_m] = \sin([(N-2)+1]\omega_m)$$
$$= 2\sin[(N-2)\omega_m]\cos \omega_m - \sin[(N-3)\omega_m]$$

The sine term defined in (7.33b) can be written as

$$S_m^{(i)} = g_i \sin[(N-i)\omega_m] + h_i \sin[(N-i-1)\omega_m] \quad i = 1, 2, \ldots, N-1$$

S_m can be computed by recursively solving for g_i as

$$S_m = S_m^{(N-1)} = g_{N-1} \sin \omega_m \qquad\qquad (7.38)$$

where g_i and h_i are defined in (7.36d). This method reduces the complexity down to $O(N)$. The total computation is $(2N+4)$ real multiplies and $(4N-2)$ real adds. If the input signal is real, cosine and sine terms (C_m and S_m) are also real. Thus the total computation reduces to $(N+2)$ real multiplies and $(2N-1)$ real adds.

Goertzel's algorithm can be implemented in MATLAB as follows (see Fig. 7.6).

```
function [Xndft, z] = ndft(x, omega, 'Goertzel')

% Input: x(n), ωm % Output: Xndft(zm), zm
% On the unit circle: Using the Goertzel algorithm
% with trigonometric series interpretation.
N = length(x); j = sqrt(-1);
for m = 1:N
   h = x(1); g = 0;
   for n = 2:N
      h_pre = h;
      h = x(n) - g;
      g = 2*g*cos(omega(m)) + h_pre;
   end
   C = g*cos(omega(m)) + h;
```

(continued)

```
    S = g*sin(omega(m));
    A(m) = C + j*S; % A(m) = Aₘ
end

X = exp(-j*(N-1)*omega).*A; Xndft = X; % Output
z = exp(j*omega); % zₘ = e^{jωₘ}, z(m) = zₘ

subplot(2,1,1);
stem(omega, abs(X)); title('Magnitude plot of the NDFT');
xlabel('Frequency \omega'); ylabel(' Magnitude');
axis([min(omega) max(omega) 0 max(abs(X))]);
subplot(2,1,2);
stem(omega, unwrap(angle(X))); title('Phase plot of the NDFT');
xlabel('Frequency \omega'); ylabel('Phase in radians');
axis([min(omega) max(omega) min(unwrap(angle(X)))
max(unwrap(angle(X)))]);
```

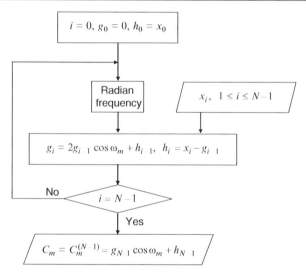

Fig. 7.6 A flow chart for Goertzel's algorithm. Only computation of C_m is shown

7.3.2 Inverse NDFT

In general there is no simple inversion formula for NDFT, since we have to derive an inverse matrix for the nonuniformly spaced frequency samples. Note that simple operation to determine the signals from a given NDFT vector $\underline{X}$ means $\underline{x} = [D]^{-1}\underline{X}$. The inverse matrix can be found by solving the linear system using Gaussian elimination [B27]. This involves $O(N^3)$ operations.

The inverse NDFT is analogous to the problem of polynomial interpolation techniques as the original signal vector on a uniform frequency grid is reconstructed from the nonuniform frequency grid. We deal with the following reconstruction or recovery problem. Let C denote the set of complex numbers. Given the values

$x_j \in C \, (j = 0, 1, \dots, N - 1)$ at nonequispaced points $p_j, j = 0, 1, \dots, N - 1$, the aim is to reconstruct a trigonometric polynomial $x(p_j)$ which is close to the original sample x_j as

$$x(p_j) = \sum_k X_k \exp\left(j2\pi p_j\right) \cong x_j, \quad \text{i.e.,} \quad [E]\,\underline{X} \cong \underline{x} \tag{7.39}$$

where $[E]$ represents the matrix form of inverse Fourier transform.

A standard method is to use the Moore–Penrose pseudoinverse solution [N6] which solves the general linear least squares problem

$$\text{Minimize}\,\|\underline{X}\|^2 \quad \text{subject to} \quad \|\underline{x} - [E]\,\underline{X}\|^2 = \text{minimum} \tag{7.40}$$

that minimizes the residual (error) norm $\|\cdot\|^2$. Of course, computing the pseudoinverse problem by the singular value decomposition is very expensive and there is no practical way at all.

As a more practical way, one can reduce the approximation error $\underline{r} = \underline{x} - [E]\,\underline{X}$, by using the weighted approximation problem

$$\min \|\underline{x} - [E]\,\underline{X}\|_{[W]} \tag{7.41}$$

where

$$\|\underline{x} - [E]\,\underline{X}\|_{[W]} = \left(\sum_{j=0}^{N-1} w_j \left|x_j - x(p_j)\right|^2\right)^{1/2}$$

where $[W]$ denotes a diagonal weighting matrix.

7.3.2.1 Lagrange Interpolation Method

The Lagrange interpolation technique is useful to convert the given NDFT coefficients into the corresponding Z-transform of the sequence [N26]. If this can be achieved, then the sequence (inverse transformed) can be identified as the coefficients of the Z-transform. Using the Lagrange polynomial of order $N - 1$, Z-transform coefficients can be expressed as

$$X(z) = \sum_{m=0}^{N-1} \frac{L_m(z)}{L_m(z_m)} X_{\text{ndft}}[m] \tag{7.42}$$

where $L_m(z) = \Pi_{i=0,\,i \neq m}^{N-1}(1 - z_i\,z^{-1}), m = 0, 1, \dots, N - 1$. Then, we only need to pick up the coefficients of $X(z)$ to obtain the original sample x_n. It gives reasonable results for shorter sequences, say, $N < 50$, but the interpolation error goes to infinity, when the sequence gets too large.

Let us illustrate with an example of a sequence of length $N = 3$. Assume the frequency points are denoted for simplicity as $\{z[k]\} = \{a, b, c\}$ and the NDFT sample sequence as $\{X_{\text{ndft}}[k]\} = \{\hat{X}[0], \hat{X}[1], \hat{X}[2]\}$. The Z-transform is then given as

$$X(z) = \frac{L_0(z)}{L_0(a)}\hat{X}[0] + \frac{L_1(z)}{L_1(b)}\hat{X}[1] + \frac{L_2(z)}{L_2(c)}\hat{X}[2] \tag{7.43}$$

Note that $L_k(z_i)$ is a constant value at a given data point. Thus we continue to expand (7.43) as

$$X(z) = (1 - bz^{-1})(1 - cz^{-1})\frac{\hat{X}[0]}{L_0(a)} + (1 - az^{-1})(1 - cz^{-1})\frac{\hat{X}[1]}{L_1(b)}$$
$$+ (1 - az^{-1})(1 - bz^{-1})\frac{\hat{X}[2]}{L_2(c)}$$

Expanding this into a second order polynomial gives

$$X(z) = \left(1 - (b + c)z^{-1} + bcz^{-2}\right)\frac{\hat{X}[0]}{L_0(a)} + \left(1 - (c + a)z^{-1} + caz^{-2}\right)\frac{\hat{X}[1]}{L_1(b)}$$
$$+ \left(1 - (a + b)z^{-1} + abz^{-2}\right)\frac{\hat{X}[2]}{L_2(c)}$$

The inverse NDFT $x[n]$ can be identified as the coefficients of this interpolating polynomial $X(z)$. Recall that $\{a[n]\} = \{a_0, a_1\} \leftrightarrow A(z) = a_0 + a_1 z^{-1}$. Thus $x[n]$ can be represented in matrix form as

$$\{x[n]\} = \begin{bmatrix} 1 & 1 & 1 \\ -(b + c) & -(c + a) & -(a + b) \\ bc & ca & ab \end{bmatrix} \begin{bmatrix} \hat{X}[0]/L_0(a) \\ \hat{X}[1]/L_1(b) \\ \hat{X}[2]/L_2(c) \end{bmatrix} \tag{7.44}$$

which yields three data points. However, this will in general lead to high complexity due to the "all combinations" property of the coefficients and data points.

MATLAB implementation of the inverse NDFT using Lagrange interpolation is as follows:

```
function x = indft (Xndft, z)

% Input: X_ndft(z_m), z_m  % Output: x(n)
% INDFT using Lagrange interpolation
format long g
N = length(Xndft); L = zeros(1, N);
vec_old = zeros(1, N);
for m = 1: N
    vec_new = 1;
```

(continued)

```
temp_vec = [z(1:m-1) z(m+1:N)];

% Computing the fundamental Lagrange polynomial Lm(Zm)
L(m) = prod(1-temp_vec/z(m));

% Computing the fundamental Lagrange polynomial Lm(Z|Zm)
for n = 1: N-1
    vec_new = conv(vec_new, [1-temp_vec(n)]);
end
vec_old = vec_old + vec_new * Xndft(m)/L(m);          .
end
x = vec_old;
format
```

Although the nonuniform sample positions can be determined arbitrarily, the condition number C of a matrix $[A]$ will greatly increase if any two points get too close, which is defined by

$$C([A]) = \|[A]\| \|[A]^{-1}\| \tag{7.45}$$

where $\|\cdot\|$ denotes the matrix norm. We have many different definitions on the matrix norm. In MATLAB, use the command norm(A, 'parameter') and cond (A, 'parameter') to check the results. If the parameter is '1', the maximum absolute column sum $\|[A]\|_1$ is calculated by $\|[A]\|_1 = \max\limits_{j} \sum\limits_{i=1}^{n} |a_{ij}|$. If it is '2', the square root of the maximum eigenvalue of $([A]^T)^* [A]$ (where $([A]^T)^*$ means the conjugate transpose of matrix $[A]$) is calculated by $\|[A]\|_2 =$ (maximum eigenvalue of $([A]^T)^* [A])^{1/2}$ which is often called the matrix norm. For example, given $[A] = \begin{bmatrix} 1 & 1 \\ 1 & -1 \end{bmatrix}$ which is 2×2 orthogonal matrix, $\|[A]\|_1 = 2$ and $\|[A]\|_2 = \sqrt{2}$. Condition number is obtained as

$$C(A) = norm(A, 2) * norm(inv(A), 2)$$

where $*$ denotes multiply in MATLAB. The matrix norm is greater than one for all matrices. Since a matrix $[A]$ and its inverse matrix are in counter-part relationship, the condition number is close to 1, saying $[A]$ is a well-conditioned matrix. A system is said to be singular if the condition number is infinite, and ill-conditioned if it is too large, where "too large" means roughly that $\log C$ is greater than the precision of matrix entries, where C denotes the condition number.

Thus a hierarchical sampling scheme is useful to determine the nonuniform sample locations by introducing interpolation and decimation, as depicted in Fig. 7.7. More significant regions are densely sampled, while the others are coarsely sampled.

Fig. 7.7 Hierarchical
interpolation by a factor of
two

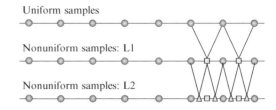

7.4 Two-Dimensional NDFT

As in the one-dimensional case, the two-dimensional NDFT can be calculated on
the actual sample locations in the time domain or in the frequency domain, when the
sampling is nonuniform. It is possible to detect exact frequency properties by
considering frequencies required by the designer. Assuming the nonuniform sampling is done in the frequency domain, the two-dimensional NDFT of a sequence
$x(n_1, n_2)$ of size $N_1 \times N_2$ is defined as

$$X(z_{1k}, z_{2k}) = \sum_{n_1=0}^{N_1-1} \sum_{n_2=0}^{N_2-1} x(n_1, n_2) z_{1k}^{-n_1} z_{2k}^{-n_2}, \quad k = 0, 1, ..., N_1 N_2 - 1 \qquad (7.46)$$

where (z_{1k}, z_{2k}) represent $N_1 N_2$ distinct points in the (z_1, z_2) space. These points can
be chosen arbitrarily but in such a way that the inverse transform exists.

The 2D sequence is considered as successive one-dimensional sequences, i.e.,
$N_1 \times N_2$ data sequence is rewritten to obtain $N_1 N_2$ points. Corresponding transform
matrix is of size $N_1 N_2 \times N_1 N_2$. We illustrate this by an example with $N_1 = N_2 = 2$.
Transform is expressed in a matrix form as

$$\underline{X} = [D]\,\underline{x} \qquad (7.47)$$

where

$$\underline{X} = \begin{bmatrix} X(z_{10}, z_{20}) \\ X(z_{11}, z_{21}) \\ X(z_{12}, z_{22}) \\ X(z_{13}, z_{23}) \end{bmatrix}, \quad \underline{x} = \begin{bmatrix} x(0,0) \\ x(0,1) \\ x(1,0) \\ x(1,1) \end{bmatrix}$$

and

$$[D] = \begin{bmatrix} 1 & z_{20}^{-1} & z_{10}^{-1} & z_{10}^{-1} z_{20}^{-1} \\ 1 & z_{21}^{-1} & z_{11}^{-1} & z_{11}^{-1} z_{21}^{-1} \\ 1 & z_{22}^{-1} & z_{12}^{-1} & z_{12}^{-1} z_{22}^{-1} \\ 1 & z_{23}^{-1} & z_{13}^{-1} & z_{13}^{-1} z_{23}^{-1} \end{bmatrix}$$

7.4.1 2D Sampling Structure

7.4.1.1 Uniform Rectangular Structure

This structure, as shown in Fig. 7.8a has long been adopted for most of frequency analysis and processing and the conventional uniform DFT is useful to treat the samples in this structure. The transform is basically separable.

7.4.1.2 Nonuniform Rectangular Structure

In this case, the sampling points lie at the vertices of a rectangular grid in the (z_1, z_2) space, as shown in Fig. 7.8b. The z_1 and z_2 coordinates are nonuniformly spaced but lines (row or column) are parallel to each other. Only the distances among lines are arbitrarily chosen, so long as they are distinct.

Since all points are defined on a straight line, the transform operation can be separable and can then be expressed in a simpler matrix form as

$$[X] = [D_1][x][D_2]^T \tag{7.48}$$

where

$$[X] = \begin{bmatrix} X(z_{10}, z_{20}) & X(z_{10}, z_{21}) & \cdots & X(z_{10}, z_{2, N_2-1}) \\ X(z_{11}, z_{20}) & X(z_{11}, z_{21}) & \cdots & X(z_{11}, z_{2, N_2-1}) \\ \vdots & \vdots & \ddots & \vdots \\ X(z_{1, N_1-1}, z_{20}) & X(z_{1, N_1-1}, z_{21}) & \cdots & X(z_{1, N_1-1}, z_{2, N_2-1}) \end{bmatrix}$$

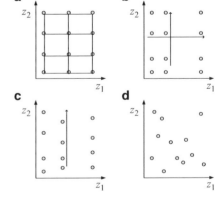

Fig. 7.8 Frequency sampling structure in 2D case.
a Uniform rectangular.
b Nonuniform rectangular.
c Nonuniform on parallel lines. d Arbitrary nonuniform structure

$$[X] = \begin{bmatrix} x(0,0) & x(0,1) & \cdots & x(0,N_2-1) \\ x(1,0) & x(1,1) & \cdots & x(1,N_2-1) \\ \vdots & \vdots & \ddots & \vdots \\ x(N_1-1,0) & x(N_1-1,1) & \cdots & x(N_1-1,N_2-1) \end{bmatrix}$$

$$[D_1] = \begin{bmatrix} 1 & z_{10}^{-1} & z_{10}^{-2} & \cdots & z_{10}^{-(N_1-1)} \\ 1 & z_{11}^{-1} & z_{11}^{-2} & \cdots & z_{11}^{-(N_1-1)} \\ \vdots & \vdots & \vdots & \ddots & \vdots \\ 1 & z_{1,N_1-1}^{-1} & z_{1,N_1-1}^{-2} & \cdots & z_{1,N_1-1}^{-(N_1-1)} \end{bmatrix}$$

$$[D_2] = \begin{bmatrix} 1 & z_{20}^{-1} & z_{20}^{-2} & \cdots & z_{20}^{-(N_2-1)} \\ 1 & z_{21}^{-1} & z_{21}^{-2} & \cdots & z_{21}^{-(N_2-1)} \\ \vdots & \vdots & \vdots & \ddots & \vdots \\ 1 & z_{2,N_2-1}^{-1} & z_{2,N_2-1}^{-2} & \cdots & z_{2,N_2-1}^{-(N_2-1)} \end{bmatrix}$$

Here $[D_1]$ and $[D_2]$ are called *Vandermonde* matrices of size $N_1 \times N_1$ and $N_2 \times N_2$, respectively. This type of matrix is sometimes called alternant matrix [N11]. The determinants of Vandermonde matrices have a particularly simple form as

$$\det[D_1] = \prod_{i \neq j, i > j} \left(z_{1i}^{-1} - z_{1j}^{-1} \right)$$

Since this structure yields separable transform, the matrix $[D]$ can be expressed in the Kronecker product form as

$$[D] = [D_1] \otimes [D_2] \tag{7.49}$$

Applying a property of the Kronecker product $\otimes$, the determinant of $[D]$ can be written as

$$\det[D] = (\det[D_1])^{N_2} (\det[D_2])^{N_1} \\ = \prod_{i \neq j, i > j} \left(z_{1i}^{-1} - z_{1j}^{-1} \right)^{N_2} \prod_{p \neq q, p > q} \left(z_{2p}^{-1} - z_{2q}^{-1} \right)^{N_1} \tag{7.50}$$

Thus the matrix $[D]$ is nonsingular provided $[D_1]$ and $[D_2]$ are nonsingular, i.e., if the points $z_{10}, z_{11}, \ldots, z_{1,N_1-1}$ and $z_{20}, z_{21}, \ldots, z_{2,N_2-1}$ are distinct. Then, degrees of freedom to calculate matrices reduces from the $N_1 N_2$ in the two-dimensional NDFT to $N_1 + N_2$. The inverse two-dimensional NDFT to obtain the two-dimensional data can be computed by solving two separate linear systems of sizes N_1 and N_2, respectively. This takes $O(N_1^3 + N_2^3)$ operations, instead of $O(N_1^3 N_2^3)$ operations.

7.4.1.3 Nonuniform on Parallel Lines

The separable operation may not be applied in this structure, since points are nonuniformly distributed on row directions, as shown in Fig. 7.8c. However, the two-dimensional NDFT matrix can be expressed as a generalized Kronecker product in this case.

$$[D] = \{[D_2]\} \otimes [D_1] \tag{7.51}$$

Here, $[D_1]$ is an $N_1 \times N_1$ Vandermonde matrix and $\{[D_2]\}$ is defined as a set of $N_1(N_2 \times N_2)$ Vandermonde matrices $[D_{2i}], i = 0, 1, \ldots, N_1 - 1$ represented as

$$\{[D_2]\} = \left\{ \begin{array}{c} [D_{20}] \\ [D_{21}] \\ \vdots \\ [D_{2,N_1-1}] \end{array} \right\} \tag{7.52}$$

where

$$[D_{2i}] = \begin{bmatrix} 1 & z_{20i}^{-1} & z_{20i}^{-2} & \cdots & z_{20i}^{-(N_2-1)} \\ 1 & z_{21i}^{-1} & z_{21i}^{-2} & \cdots & z_{21i}^{-(N_2-1)} \\ \vdots & \vdots & \vdots & \ddots & \vdots \\ 1 & z_{2,N_2-1,i}^{-1} & z_{2,N_2-1,i}^{-2} & \cdots & z_{2,N_2-1,i}^{-(N_2-1)} \end{bmatrix}, \quad i = 0, 1, \ldots, N_1 - 1$$

This means

$$[D] = \left\{ \begin{array}{c} [D_{20}] \otimes \underline{d}_0 \\ [D_{21}] \otimes \underline{d}_1 \\ \vdots \\ D_{2,N_1-1} \otimes \underline{d}_{N_1-1} \end{array} \right\} \tag{7.53}$$

where $\underline{d}_i$ denotes the ith row vector of matrix $[D_1]$. The determinant of $[D]$ can then be written as

$$\det [D] = (\det [D_1])^{N_2} \prod_{i=0}^{N_1-1} \det [D_{2i}] \tag{7.54}$$

Therefore, $[D]$ is nonsingular if the matrices $[D_1]$ and $[D_{2i}]$ are nonsingular. In this case the transform is partially separable.

7.4.1.4 Arbitrary Sample Structure

Since the samples are chosen arbitrarily in the frequency domain as shown in Fig. 7.8d, separable transform is not guaranteed. In this case, the successive one-dimensional transform should be applied. All the transform matrices should be larger than in the separable case, requiring higher complexity.

7.4.2 Example of Two-Dimensional Nonuniform Rectangular Sampling

The following are the MATLAB codes to implement the two-dimensional NDFT with nonuniform rectangular sampling defined in Fig. 7.8b.

```
function [Xk] = ndft2(xn, M, N, fs1, fs2)
% Computes NonUniform Discrete Fourier Transform
% - - - - - - - - - - - - - - - - - - - - - - - -
% xn = N-point 2D input sequence over 0 <= n <= N-1 (Uniform)
% Xk = 2D NDFT coefficient array (NonUniform)
% fs = nonuniform frequency vector (N-point)
% N, M = Length of NDFT
% Usage: Xk = ndft2d_r(xn, N, M, fs1, fs2)
%

n1 = 0: 1: M-1;                      % Index for input data
Wn = exp(-j*2*pi/M);                 % Twiddle factor
nk1 = fs1' * n1;                     % Creates an M × M matrix
DFTmtx1 = Wn .^ nk1;                 % DFT matrix (M × M)

n2 = 0: 1: N-1;                      % Index for input data
Wm = exp(-j*2*pi/N);                 % Twiddle factor
Nk2 = fs2' * n2;                     % Creates an N × N matrix
DFTmtx2 = Wm .^ nk2;                 % DFT matrix (N × N)
Xk = DFTmtx1 * xn * DFTmtx2.';
```

The example codes to test the two-dimensional NDFT of an image data are:

```
% 2D NonUniform DFT simulation
x = double(imread('image64.tif'));
figure,imshow(x, [])

t = 0 : 63;
fs1=[[0:0.5:15] [16:32] [33:2:63]];
fs2 = fs1;
Xf = ndft2(x, 64, 64, fs1, fs2);
figure,imshow(abs(Xf), [])

Xfc = fftshift(Xf);
figure,imshow(abs(Xfc), [])

Xfl = log(1+abs(Xfc));
figure,imshow(Xfl, [])

%nonuniform grid plotting
figure, fs11 = decimate(fs1,4); %64 pts -> 16 pts for showing purpose
[X,Y] = meshgrid(fs11); Z = zeros(16);
plot3(X, Y, Z, 'k.'); axis([0 64 0 64]) %Two-dim. nonuniform grids
displayed!
```

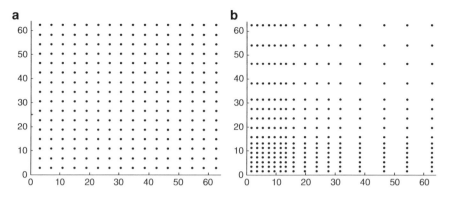

Fig. 7.9 Example of frequency sampling structure. **a** Uniform rectangular. **b** Nonuniform rectangular

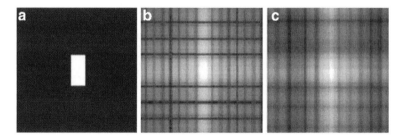

Fig. 7.10 Result of the two-dimensional NDFT. **a** Input image. **b** Obtained with uniform sampling. **c** Obtained with nonuniform rectangular sampling

In the simulation, choice of the nonuniform sampling structure is of main interest. For example, we design the low frequency emphasized structure as shown in Fig. 7.9b. Samples are densely populated at low frequency, while coarse at high frequency. The two-dimensional NDFT results are shown in Fig. 7.10, comparing to the uniform DFT. Central region representing low frequency components is expanded in Fig. 7.10b because of more samples in the region.

7.5 Filter Design Using NDFT

7.5.1 Low-Pass Filter Design

Consider the design of a linear-phase low-pass filter (LPF) with odd number of filter taps N. Then, we assume the impulse response of the filter is real and symmetric,

$$h[n] = h[N - 1 - n] \qquad n = 0, 1, \ldots, N - 1 \tag{7.55}$$

That is, $h[0] = h[N-1], h[1] = h[N-2], \ldots$, etc. The frequency response of the filter is represented in the form of discrete-time Fourier transform as

$$H(e^{j\omega}) = \sum_{n=0}^{N-1} h[n] e^{-j\omega n} \tag{7.56}$$

Since the filter is symmetric, this can be expanded as

$$H(e^{j\omega}) = h\left(\frac{N-1}{2}\right) e^{-j\omega\left(\frac{N-1}{2}\right)} + \sum_{n=1}^{(N-1)/2} h\left(\frac{N-1}{2}-n\right) \left[e^{-j\omega\left(\frac{N-1}{2}-n\right)} + e^{-j\omega\left(\frac{N-1}{2}+n\right)} \right]$$

$$= \left[h\left(\frac{N-1}{2}\right) + 2\sum_{n=1}^{(N-1)/2} h\left(\frac{N-1}{2}-n\right) \frac{1}{2} \left(e^{j\omega n} + e^{-j\omega n} \right) \right] e^{-j\omega\left(\frac{N-1}{2}\right)}$$

Thus

$$H(e^{j\omega}) = \sum_{k=0}^{(N-1)/2} r[k] \cos \omega k \; e^{-j\omega(N-1)/2} = R(\omega) \; e^{-j\omega(N-1)/2} \tag{7.57}$$

where the magnitude function $R(\omega)$ is a real, even, periodic function of ω, given by [N26]

$$R(\omega) = \sum_{k=0}^{(N-1)/2} r[k] \cos \omega k \tag{7.58}$$

and

$$\begin{aligned} r[0] &= h[(N-1)/2] \\ r[k] &= 2h[(N-1)/2 - k] \qquad k = 1, 2, \ldots, (N-1)/2 \end{aligned} \tag{7.59}$$

Now, we need to define the desired frequency response $R(\omega)$ in terms of the passband edge at ω_p, stopband edge at ω_s, and peak ripples δ_p and δ_s in the passband and stopband, respectively.

Chebyshev filters [B19, B40] have equiripples in the passband (type I) or in the stopband (type II) as shown in Fig. 7.11 with some acceptable tolerances. In addition, a transition band is specified between the passband and the stopband to permit the magnitude to drop off smoothly. The Chebyshev filters can be implemented in MATLAB as follows.

```
% Chebyshev type I and II filters
% Order of the filters
N = 5;
% Passband edge frequency
```

(continued)

```
Wp = 0.5;
% Ripple in the pass-band (dB)
RipplePass = 1;
% Ripple in the stop-band (dB)
RippleStop = 20;

w = 0:pi/255:pi;

[numer, denom] = cheby1(N, RipplePass, Wp);
chebyshev1 = abs(freqz(numer, denom, w));

[numer, denom] = cheby2(N, RippleStop, Wp);
chebyshev2 = abs(freqz(numer, denom, w));

subplot(2,1,1);
plot(w, chebyshev1); axis([0 3.14 -0.05 1.05]);
title('Chebyshev type I filter')
subplot(2,1,2);
plot(w, chebyshev2); axis([0 3.14 -0.05 1.05]);
title('Chebyshev type II filter')
```

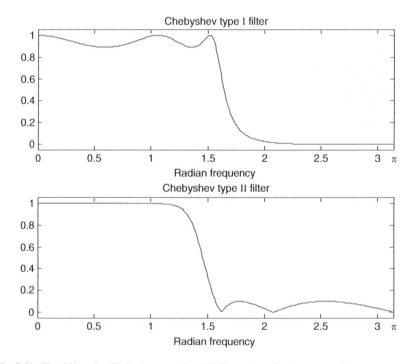

Fig. 7.11 The fifth order Chebyshev type I and II filters (magnitude responses)

7.5.1.1 Generation of the Desired Frequency Response

The real-valued amplitude response $R(\omega)$ is separated into two parts for the passband $R_p(\omega)$ and stopband $R_s(\omega)$ as

$$R(\omega) = \begin{cases} R_p(\omega), & 0 \le \omega \le \omega_p \\ R_s(\omega), & \omega_s \le \omega \le \pi \end{cases} \tag{7.60}$$

where $R_p(\omega)$ and $R_s(\omega)$ are defined as [N14]

$$R_p(\omega) = 1 - \delta_p T_P(X_p(\omega)) \tag{7.61}$$

and

$$R_s(\omega) = -\delta_s T_S(X_s(\omega)) \tag{7.62}$$

with $T_M(\cdot)$ denoting a Chebyshev polynomial of order M as

$$T_M(x) = \begin{cases} \cos(M\cos^{-1}(x)), & -1 \le x \le 1 \\ \cosh(M\cosh^{-1}(x)), & |x| > 1 \end{cases} \tag{7.63}$$

The Chebyshev polynomial yields equiripple in the frequency range which is less than or equal to ± 1 and monotone outside the range. The functions $X_p(\omega)$ and $X_s(\omega)$ are required to map the equiripple intervals of the Chebyshev polynomials $X_p(\omega)$ and $X_s(\omega)$ to the frequency range $0 \le \omega \le \omega_p$ in the passband and $\omega_s \le \omega \le \pi$ in the stopband, respectively. This mapping can be accomplished by choosing $X_p(\omega)$ and $X_s(\omega)$ as

$$\begin{aligned} X_p(\omega) &= A\cos(a\omega + b) + B \\ X_s(\omega) &= C\cos(c\omega + d) + D \end{aligned} \tag{7.64}$$

Here we need to obtain the eight parameters a, b, c, d, A, B, C, D by imposing appropriate constraints on the filter response $R_p(\omega)$ and $R_s(\omega)$ as follows:

1. Parameter b

Since the response $R(\omega)$ in (7.60) is even and symmetric with respect to $\omega = 0$, we have

$$R_p(\omega) = R_p(-\omega) \quad \text{and} \quad X_p(\omega) = X_p(-\omega) \tag{7.65}$$

By substituting (7.65) for $X_p(\omega)$ in (7.64), we obtain

$$\cos(a\omega + b) = \cos(-(a\omega - b)) = \cos(a\omega - b)$$

Thus b should be π, i.e.,

$$b = \pi \tag{7.66}$$

2. Parameter d

To derive other parameters, we put $R_s(\omega)$ be symmetric with respect to $\omega = \pi$. We have

$$R_s(\omega + \pi) = R_s(-\omega + \pi) \quad \text{and} \quad X_s(\omega + \pi) = X_s(-\omega + \pi) \quad (7.67)$$

By substituting (7.67) for $X_s(\omega)$ in (7.64), we obtain

$$\cos(c\omega + (c\pi + d)) = \cos(-c\omega + (c\pi + d))$$

Then, $(c\pi + d)$ should be π and

$$d = \pi(1 - c) \quad (7.68)$$

3. Parameter a

At the end of the passband, $\omega = \omega_p$, the filter response $R_p(\omega)$ in (7.61) should be equal to $1 - \delta_p$ to meet the transient response, i.e.,

$$R_p(\omega_p) = 1 - \delta_p \quad (7.69)$$

This implies that from (7.61)

$$T_P(X_p(\omega_p)) = 1 \quad (7.70)$$

From the definition of $T_M(\cdot)$ in (7.63), it follows that $X_p(\omega_p) = 1$. By substituting this for $X_p(\omega)$ in (7.64), we obtain

$$X_p(\omega_p) = A\cos(a\omega_p + b) + B = 1 \quad (7.71)$$

In (7.71), parameter b is known as in (7.66). By applying the trigonometric relations, we have

$$a = \frac{1}{\omega_p}\cos^{-1}\left(\frac{B - 1}{A}\right) \quad (7.72)$$

4. Parameter c

At the start of the stopband, $\omega = \omega_s$, the filter response $R_s(\omega)$ in (7.61) should be equal to δ_s to meet the transient response, i.e.,

$$R_s(\omega_s) = \delta_s \quad (7.73)$$

This implies that in (7.62) and (7.63)

$$T_S(X_s(\omega_s)) = 1 \quad \text{and} \quad X_s(\omega_s) = 1 \tag{7.74}$$

By substituting this for $X_s(\omega)$ in (7.64), we obtain

$$X_s(\omega_s) = C\cos(c\,\omega_s + d) + D = 1 \tag{7.75}$$

In (7.75), parameter d is known as in (7.68). By applying the trigonometric relations, we have

$$c = \frac{1}{(\omega_s - \pi)}\cos^{-1}\left(\frac{D-1}{C}\right) \tag{7.76}$$

5. Parameter B

An extremum of $R_p(\omega)$ in the Chebyshev filter with odd N occurs at $\omega = 0$. That is,

$$R_p(0) = 1 + \delta_p \tag{7.77}$$

From (7.61)

$$T_P(X_p(0)) = -1 \tag{7.78}$$

From the definition in (7.63)

$$\cos(M\cos^{-1}(X_p(0))) = -1$$
$$\cos^{-1}(X_p(0)) = \pi, \quad \text{if } M = 1$$
$$X_p(0) = -1$$

Substituting for $X_p(\omega)$ in (7.64) and using the value of b in (7.68), we obtain

$$D = C - 1 \tag{7.79}$$

6. Parameter D

An extremum of $R_s(\omega)$ in the Chebyshev filter with odd N occurs at $\omega = \pi$. That is,

$$R_p(\pi) = -\delta_s \tag{7.80}$$

From (7.62)

$$T_S(X_s(\pi)) = -1 \qquad (7.81)$$

From the definition in (7.63)

$$X_s(\pi) = -1$$

Substituting for $X_s(\omega)$ in (7.64) and using the value of d in (7.66), we obtain

$$B = A - 1 \qquad (7.82)$$

7. Parameter A

Minimum value of the passband response can be defined by

$$\min\left(R_p(\omega)\right) = -\delta_s \qquad (7.83)$$

that $R_p(\omega)$ has to be greater than the negative stopband ripple. We apply the constraint for (7.61) to get $T_p(X_p(\omega)) \rightarrow \max$, $\cos(M\cos^{-1}(x)) \rightarrow \max$, $\cos^{-1}(X_p(\omega)) \rightarrow \min$, $X_p(\omega) = A\cos(a\omega + b) + B \rightarrow \max$, $\cos(a\omega + b) = 1$ and $X_p(\omega) = A + B$. We also use the result in (7.82). Then

$$1 - \delta_p T_P(2A - 1) = -\delta_s$$
$$A = \frac{1}{2}\left[T_P^{-1}\left(\frac{1 + \delta_s}{\delta_p}\right) + 1\right] \qquad (7.84)$$

where $T_M^{-1}(\cdot)$ denotes an inverse Chebyshev function of order M.

8. Parameter C

Maximum value of the stopband response can be defined by

$$\max(R_s(\omega)) = 1 + \delta_p \qquad (7.85)$$

that $R_s(\omega)$ has to be smaller than the highest passband ripple. We apply the constraint for (7.62) to get

$$\delta_s T_S(C + D) = 1 + \delta_p$$
$$C = \frac{1}{2}\left[T_S^{-1}\left(\frac{1 + \delta_p}{\delta_s}\right) + 1\right] \qquad (7.86)$$

7.5.1.2 Location of Nonuniform Frequency Samples

Filters discussed here consist of N filter coefficients in the time domain. One way of designing the filter is to derive frequency response in the frequency domain and take N-point inverse transform. The desired frequency response is sampled at N nonuniformly spaced points on the unit circle in the z-plane [N17]. Then, an inverse NDFT of these frequency samples gives the filter coefficients. Optimal location of nonuniform samples is chosen on the extrema in the passband and stopband.

Since the impulse response of the filter is symmetric, the number of independent filter N_i is nearly half the total coefficient number N, which is given by

$$N_i = \frac{N+1}{2} \tag{7.87}$$

We have to locate N_i samples in the range $0 \le \omega \le \pi$ in a nonuniform manner. These samples are taken at the P extrema in the passband and the S extrema in the stopband. At the P extrema in the passband, amplitude response $R_p(\omega)$ is given by

$$R_p(\pi) = 1 \pm \delta_p \tag{7.88}$$

or

$$T_P\big(X_p(\omega)\big) = \pm 1$$

and

$$X_p(\omega) = \cos\left(\frac{k\pi}{P}\right), \quad k = 1, 2, \dots, N_p \tag{7.89}$$

We substitute (7.89) into (7.64) to obtain the optimal sample locations $\omega_k^{(p)}$ in the passband:

$$\omega_k^{(p)} = \frac{1}{a}\left\{ \cos^{-1}\left[\frac{\cos\left(\frac{k\pi}{P}\right) - B}{A} \right] - b \right\}, \quad k = 1, 2, \dots, N_p \tag{7.90}$$

Similarly, the optimal sample locations $\omega_k^{(s)}$ in the stopband is given as:

$$\omega_k^{(s)} = \frac{1}{c}\left\{ \cos^{-1}\left[\frac{\cos\left(\frac{k\pi}{S}\right) - D}{C} \right] - d \right\}, \quad k = 1, 2, \dots, N_s \tag{7.91}$$

Thus, P locations in the passband and S locations in the stopband can be defined by using the parameters in (7.90) and (7.91). The number of extrema $P + S$ equals the total number of alternations over the range $0 \le \omega \le \pi$. It also equals half the total length N excluding one center coefficient as

$$L = P + S = \frac{N-1}{2} \tag{7.92}$$

Thus, we still have to decide one more center location that can be placed either at the passband edge ω_p or at the stopband edge ω_s.

7.5.2 Example of Nonuniform Low-Pass Filter

We design a Chebyshev type I lowpass filter using the above parameters. We calculate extrema values in the ripples which will be used for obtaining impulse response of the filter with nonuniform samples in the frequency domain. The filter specification is given by:

Filter length $N = 37$,   Ripple ratio $\delta_p/\delta_s - 15$
Passband edge $\omega_p = 0.4\pi$,   Stopband edge $\omega_s = 0.5\pi$

The order of the filter polynomials for the passband and stopband are given by the number of passband order P and the stopband order S and the total alternations. In this example, we get $L = 18, P = 8$ and $S = 10$ in accordance with the relationships. First, the eight parameters are calculated by considering step 1 through step 8. Second, the passband function $X_p(\omega)$ illustrated in Fig. 7.12a and the stopband function $X_p(\omega)$ are calculated. The passband response of Chebyshev filter is shown

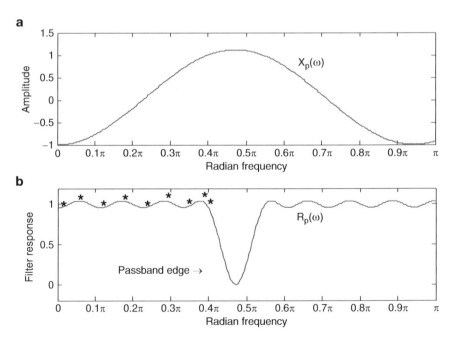

Fig. 7.12 Passband function (**a**) and passband response of the Chebyshev type I filter (**b**). $(P + 1)$ extrema will be used to design a filter with nonuniform samples

in Fig. 7.12b with four maxima and five minima which equals to $P + 1$ extrema. Design of the stopband response, calculation of extrema and inverse transform to obtain time-domain impulse response are left to the reader. A MATLAB program for the example is as follows.

```
% Chebyshev filter design using nonuniform samples
N = 37; % Filter length
wp = 0.4*pi; % Passband edge
ws = 0.5*pi; % Stopband edge
k = 15; % Ripple ratio delta_p/delta_s

Nextrema = (N-1)/2; % Number of extrema = P+S
P = (wp/(wp+(pi-ws)))*Nextrema; % Passband alternations
S = ((pi-ws)/(wp+(pi-ws)))*Nextrema; % Stopband alternations

delta_p = sqrt(k)*10^(-0.1162*(ws-wp)*(N-1)-0.65);
delta_s = delta_p/k;

% Rpw = Acos(aw+b)+B
b = pi;
% A = 0.5(aTp(rip)+1) where rip = (1+delta_s)/delta_p
rip = (1+delta_s)/delta_p;
temp = cos(acos(rip)/P); % Tp(x) = cos(P acos(x))
A = (temp+1)/2;
B = A-1;
a = acos((B-1)/A)/wp;

w = [0:0.01:pi];
Xpw = A*cos(a*w+b)+B;
Rpw = 1 - delta_p * cos(P*acos(Xpw));
subplot(2,1,1); plot(w, Xpw, 'LineWidth', 1.4); axis([0 pi -1 1.5]);
xlabel({'Radian Frequency';'(a)'}); ylabel('Amplitude');
text(2, 0.85, 'Xp(w)');
set(gca,'XLim',[0 pi],'XTick',[0:pi/10:pi]);
set(gca,'XTickLabel',{'0','0.1p','0.2p','0.3p','0.4p','0.5p','0.6p','0.7p',
'0.8p','0.9p','p'},'FontName','Symbol');
subplot(2,1,2); plot(w, Rpw, 'LineWidth', 1.4); axis([0 pi -0.2 1.2]);
xlabel({'Radian Frequency';'(b)'}); ylabel('Filter Response');
text(2, 0.85, 'Rp(w)');
set(gca,'XLim',[0 pi],'XTick',[0:pi/10:pi]);
set(gca,'XTickLabel',{'0','0.1p','0.2p','0.3p','0.4p','0.5p','0.6p','0.7p',
'0.8p','0.9p','p'},'FontName','Symbol');
pass = wp/pi*length(w);
Reduced = Rpw(1:pass);
[wmax,imax,wmin,imin] = extrema(Reduced);
text(w(imax), Rpw(imax), '*', 'FontSize', 20);
text(w(imin), Rpw(imin), '*', 'FontSize', 20);
text(0.5, 0.2, 'Passband edge \rightarrow');

function [wmax,imax,wmin,imin] = extrema(x)
%      EXTREMA Gets the extrema points from filter response
%      [XMAX,IMAX,XMIN,IMIN] = EXTREMA(W) returns maxima, index, minima,
%                                  index
%      XMAX - maxima points
%      IMAX - indice of the XMAX
%      XMIN - minima points
%      IMIN - indice of the XMIN

wmax = []; imax = []; wmin = []; imin = [];

% Vector input?
Nt = numel(x);
if Nt ~= length(x)
    error('Entry must be a vector.')
end
```

(continued)

```
% Not-a-Number?
inan = find(isnan(x));
indx = 1:Nt;
if ~isempty(inan)
    indx(inan) = [];
    x(inan) = [];
    Nt = length(x);
end

% Difference between subsequent elements
dx = diff(x);

% Flat peaks? Put the middle element:
a = find(dx~=0); % Indice where x changes
lm = find(diff(a)~=1) + 1; % Indice where a do not changes
d = a(lm) - a(lm-1); % Number of elements in the flat peak
a(lm) = a(lm) - floor(d/2); % Save middle elements
a(end+1) = Nt;

% Peaks?
xa = x(a); %
b = (diff(xa) > 0); % 1 => positive slopes (minima begin)
% 0 => negative slopes (maxima begin)
xb = diff(b); % -1 => maxima indice (but one)
% +1 => minima indice (but one)
imax = find(xb == -1) + 1; % maxima indice
imin = find(xb == +1) + 1; % minima indice
imax = a(imax);
imin = a(imin);

nmaxi = length(imax);
nmini = length(imin);

% Maximum or minumim on a flat peak at the ends?
if (nmaxi==0) && (nmini==0)
    if x(1) > x(Nt)
        wmax = x(1);
        imax = indx(1);
        wmin = x(Nt);
        imin = indx(Nt);
    elseif x(1) < x(Nt)
        wmax = x(Nt);
        imax = indx(Nt);
        wmin = x(1);
        imin = indx(1);
    end
    return
end

% Maximum or minumim at the ends?
if (nmaxi==0)
    imax(1:2) = [1 Nt];
elseif (nmini==0)
    imin(1:2) = [1 Nt];
else
    if imax(1) < imin(1)
        imin(2:nmini+1) = imin;
        imin(1) = 1;
    else
        imax(2:nmaxi+1) = imax;
        imax(1) = 1;
    end
    if imax(end) > imin(end)
        imin(end+1) = Nt;
    else
```

(continued)

```
        imax(end+1) = Nt;
    end
end
wmax = x(imax);
wmin = x(imin);
% Not-a-Number?
if ~isempty(inan)
    imax = indx(imax);
    imin = indx(imin);
end
```

7.6 Summary

This chapter has defined and developed nonuniform DFT (NDFT) including its properties. Having addressed every aspect of the DFT/FFT, it is only proper to focus on its applications in the concluding chapter (Chapter 8).

7.7 Problems

7.1 One of the advantages in the NDFT is the possibility of correct frequency detection residing in the signal, while the transform is in general not orthogonal. Inverse problem should be carefully considered. Rectangular structure is helpful to solve the problem, but the sample interval must be large enough to obtain the inverse matrix. Given a sound data input with 256 samples and sampling rate 8 kbps, derive the smallest sample interval in the nonuniform inverse transform.

7.2 Write a MATLAB program for the one-dimensional NDFT to compare with uniform DFT for "Windows XP Shutdown.wav" data. Plot original signal and transformed results. Refer to source code as:

```
% One-dimensional NDFT test
x = 1 : 256;
fid = fopen('Sound.wav','rb');
data = fread(fid, 256)';
subplot(311); plot(x, data)

fsd = 0 : 255;
Xfd = dft1d(data, 256);
subplot(312); plot(x, abs(Xfd))

fs = (0 : 255)/255;
fs1 = [[0:0.5:40] [41:1:174] [175:2:255]];
Xf = ndft1d(data, 256, fs1);
subplot(313); plot(x, abs(Xf))
```

7.3 Analyze and compare the fast NDFT algorithms: Horner's nested method [N7] and Goertzel's algorithm [N6].

7.4 Fast inverse transform is based on interpolation problem. Analyze and compare the Lagrange interpolation method [N26, B6] and Newton interpolation method [N26].

7.5 Write a MATLAB program for the two-dimensional NDFT with sampling structure of nonuniform grid on parallel lines in the column direction.

7.6 Derive the nonuniform sampling structure in the two-dimensional image data by considering human interested important region, say, edge information, in the spatial domain and its counterpart in the frequency domain.

7.7 Design a highpass filter based on nonuniform frequency samples.

7.8 Design an image compression algorithm by taking into account the two-dimensional NDFT that emphasizes important region or components, while removing less important components.

7.9 (a) Derive (7.37) from (7.33a). (b) Derive (7.38) from (7.33b).

Chapter 8
Applications

This chapter focuses on applications of FFT/IFFT in a number of diverse fields. In view of the extensive nature of their applications they are intentionally described in a conceptual form. The reader is directed to the references wherein the applications are described in detail along with the theoretical background, examples, limitations, etc. The overall objective is to expose the reader to the limitless applications of DFT in the general areas of signal/image processing.

8.1   Frequency Domain Downsampling

Assume the bandwidth of a signal is known. The sampling rate must be at least twice the highest frequency content in the signal. The Nyquist rate is two times of the bandwidth. Aliasing occurs when the bandlimited signal is sampled *below* the Nyquist rate!

Sampling a continuous signal below the Nyquist rate aliases energy at frequencies higher than half the sampling rate back into the baseband spectrum or below half the sampling rate.

This is equally true for decimation of samples. Thus aliasing that will be caused by downsampling a signal by the ratio of two to one, can be avoided by low-pass filtering (decimation filtering) the samples first so that the samples keep only less than half of the bandwidth (Fig. 8.1). To accurately implement decimation, this behavior should be preserved [F11, G2].

Consider downsampling by a factor of two. An N-point sequence (assume $N = 2^l$, l = integer) can be decomposed into two $N/2$-point sequences [one of even samples $x(2n)$ and another of odd samples $x(2n+1)$] (Fig. 8.2).

$$X^{\mathrm{F}}(k) = \sum_{n=0}^{N-1} x(n)\, W_N^{nk}, \qquad k, n = 0, 1, \ldots, N-1 \tag{2.1a}$$

K.R. Rao et al., *Fast Fourier Transform: Algorithms and Applications*,
Signals and Communication Technology,
DOI 10.1007/978-1-4020-6629-0_8, © Springer Science+Business Media B.V. 2010

Fig. 8.1 Decimator: Data/
time domain decimation

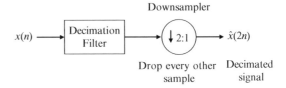

Downsampler

$x(n) \longrightarrow$ Decimation Filter $\longrightarrow$ $\downarrow$ 2:1 $\longrightarrow$ $\hat{x}(2n)$

Drop every other   Decimated
sample   signal

a   **b**

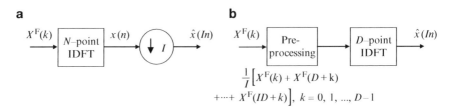

$\frac{1}{I}\big[X^{\mathrm{F}}(k) + X^{\mathrm{F}}(D+k)$
$+\cdots+ X^{\mathrm{F}}(ID+k)\big], \; k = 0, 1, ..., D-1$

Fig. 8.2 **a** Data domain downsampling. **b** DFT domain downsampling. ($\downarrow I$) represents a down-sampling by a factor of I. [F11] © 1999 IEEE

$$= \sum_{n=0}^{N/2-1} \left[x(2n)W_N^{2nk} + x(2n+1)W_N^{(2n+1)k}\right]$$

$$= \sum_{n=0}^{N/2-1} \left[x(2n)W_{N/2}^{nk} + x(2n+1)W_{N/2}^{nk}W_N^{k}\right] \tag{8.1}$$

where $W_N^{2nk} = W_{N/2}^{nk}$. Adding two corresponding coefficients from half DFT blocks results in downsampling in the data domain.

$$X^{\mathrm{F}}(k) + X^{\mathrm{F}}(k + N/2), \qquad k = 0, 1, \ldots, \frac{N}{2} - 1$$

$$= \sum_{n=0}^{N/2-1} \left[\left(x(2n)W_{N/2}^{nk} + x(2n+1)W_{N/2}^{nk}W_N^{k}\right) \right.$$
$$\left. + \left(x(2n)W_{N/2}^{n(k+N/2)} + x(2n+1)W_{N/2}^{n(k+N/2)}W_N^{k+N/2}\right) \right]$$

$$= \sum_{n=0}^{N/2-1} \left[x(2n)W_{N/2}^{nk} + x(2n+1)W_{N/2}^{nk}W_N^{k} + x(2n)W_{N/2}^{nk} - x(2n+1)W_{N/2}^{nk}W_N^{k}\right]$$

$$= 2 \sum_{n=0}^{N/2-1} x(2n)\, W_{N/2}^{nk}$$

$$= 2 \times N/2\text{-point DFT}\left[x(2n)\right] \tag{8.2}$$

where $W_N^{nN} = W_{N/2}^{n(N/2)} = 1$ and $W_N^{N/2} = -1$.

Thus when $N = I \times D$ where I is an integer, an N-point IDFT followed by downsampling by the factor I is equivalent to additions of I blocks of D DFT coefficients and a D-point IDFT.

Let $x(n)$ be an input sequence and $x(nI)$ be the downsampled sequence by a factor of I.

$$x(n) = \frac{1}{N} \sum_{k=0}^{N-1} X^F(k) W_N^{-nk}, \qquad n = 0, 1, \ldots, N-1 \qquad (2.1b)$$

$$x(nI) = \frac{1}{D} \sum_{k=0}^{D-1} \left[\frac{1}{I} \sum_{l=0}^{I-1} X^F(lD + k) \right] W_D^{-nk}$$

$$= D\text{-point IDFT}, \left[\frac{1}{I} \sum_{l=0}^{I-1} X(lD + k) \right], \quad n, k = 0, 1, \ldots, D-1 \qquad (8.3)$$

Example 8.1 Let $N = 8$, $D = 4$, $I = 2$ and a random vector $\underline{x}$ be $(1, 2, 3, 4, 5, 6, 7, 8)^T$ (see Fig. 8.3). Then the DFT of the vector $\underline{x}$ is

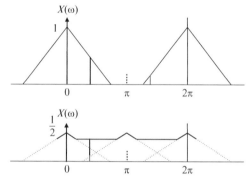

Analogy between continuous and discrete Fourier transforms

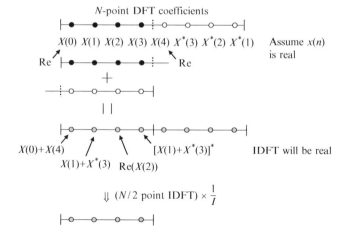

Fig. 8.3 DFT domain downsampling by a factor of 2:1. If N point IDFT is applied to the coefficients, zero insertion results. **a** Analogy between continuous and discrete Fourier transforms

$$\underline{X}^{\mathrm{F}} = \underline{X}_{\mathrm{r}} + j\underline{X}_{\mathrm{i}}$$

where

$$\underline{X}_{\mathrm{r}} = (36, -4, -4, -4, -4, -4, -4, -4)^T$$
$$\underline{X}_{\mathrm{i}} = (0, 9.657, 4, 1.657, 0, -1.657, -4, -9.657)^T$$

The DFT vector $\underline{X}^{\mathrm{F}}$ is partitioned into two blocks of four samples. Two samples from two associated blocks are added up.

$$\underline{X}^{\mathrm{F}}_{\mathrm{d}} = \frac{1}{2}[A]\,(\underline{X}_{\mathrm{r}} + j\underline{X}_{\mathrm{i}}) = (16, -4, -4, -4)^T + j(0, 4, 0, -4)^T$$

where

$$[A] = \begin{bmatrix} 1 & 0 & 0 & 0 & 1 & 0 & 0 & 0 \\ 0 & 1 & 0 & 0 & 0 & 1 & 0 & 0 \\ 0 & 0 & 1 & 0 & 0 & 0 & 1 & 0 \\ 0 & 0 & 0 & 1 & 0 & 0 & 0 & 1 \end{bmatrix} \tag{8.4}$$

Thus the data decimated by a factor of two is

$$\underline{x}_{\mathrm{d}} = D\text{-point IDFT }(\underline{X}^{\mathrm{F}}_{\mathrm{d}}) = (1, 3, 5, 7)^T$$

However, this is not actually two-to-one decimation. We need to apply a decimation filter to a input $x(n)$. All the samples of the input change. We take every other sample, or we delete every other sample. Thus the method described in (8.1) – (8.4) should follow a decimation filter as seen in Fig. 8.1 [D37, p. 220].

8.1.1 Frequency Domain Upsampling (Zero Insertion)

Similarly, upsampling (or zero insertion) in the data domain by a factor of I can be performed exactly in the frequency domain by repeatedly appending a DFT block $(I-1)$ times (Fig. 8.4) [IN5].

For upsampling by a factor of two, let $X^{\mathrm{F}}(k)$ be the DFT of $x(n)$, $n = 0, 1, \ldots, \frac{N}{2} - 1$. Hence

$$X^{\mathrm{F}}(k + N/2) = X^{\mathrm{F}}(k) \qquad k = 0, 1, \ldots, \frac{N}{2} - 1 \tag{8.5}$$

Then

$$x_{\mathrm{u}}(2m) = x(n) = N/2\text{-point IDFT of }[X^{\mathrm{F}}(\mathrm{k})] \quad k, m, n = 0, 1, \ldots, \frac{N}{2} - 1 \tag{8.6a}$$

a   **b**

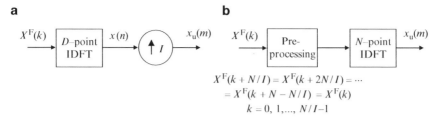

$$X^F(k + N/I) = X^F(k + 2N/I) = \cdots$$
$$= X^F(k + N - N/I) = X^F(k)$$
$$k = 0, 1, \ldots, N/I - 1$$

Fig. 8.4 a Data domain upsampling. ($\uparrow I$) represents upsampling by a factor of I. **b** DFT domain upsampling

Fig. 8.5 Interpolator: Data/ time domain interpolation

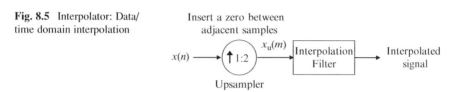

$$x_u(2m + 1) = x\left(n + \tfrac{1}{2}\right) = 0 \qquad m, n = 0, 1, \ldots, \tfrac{N}{2} - 1 \qquad (8.6b)$$

Example 8.2 Let a random vector $\underline{x}$ be $(1, 2, 3, 4)^T$ and $I = 2$. Then $D = 4$ and $N = 8$. The DFT of $\underline{x}$ is

$$\underline{X}^F = \underline{X}_r + j\underline{X}_i$$

where

$$\underline{X}_r = (10, -2, -2, -2)^T$$
$$\underline{X}_i = (0, 2, 0, -2)^T$$

Let the DFT vector $\underline{X}^F$ be extended by repeating it at the end such that its length is $N = D \times I$. Then the extended DFT sequence is

$$\underline{X}_u^F = [A]^T (\underline{X}_r + j\underline{X}_i)$$
$$= (10, -2, -2, -2, 10, -2, -2, -2)^T + j(0, 2, 0, -2, 0, 2, 0, -2)^T$$

where $[A]$ is defined in (8.4). Thus the data upsampled by a factor of two is

$$\underline{x}_u = N\text{-point IDFT}(\underline{X}_u^F) = (1, 0, 2, 0, 3, 0, 4, 0)^T$$

In case of one-to-two interpolation, put zeros in between samples in an input $x(n)$ as shown in Example 8.2. Apply an interpolation filter to all of them (Fig. 8.5). Samples from the input $x(n)$ stay same. Only in-between samples with the value zero change [D37, p. 220].

8.2   Fractal Image Compression [FR3, FR6]

Fractal or iterated function system (IFS) image coding [B23, IP34] can be accelerated by using FFTs for fast circular correlation.

The image to be coded is partitioned into a set of image blocks called *ranges* in fractal coding parlance. For each range, we search for another part of the image called a *domain* that gives a good approximation to the range when appropriately scaled in order to match the size of the range and transformed by a luminance transformation that provides for contrast and bright adjustment. The list of parameters specifying for each range, the corresponding domain and the affine luminance transformation together with the partition information is called a *fractal code*.

The goal in the encoding step is to find for each range a codebook block that gives the least L^2-error (Euclidean norm) when adjusted by an affine transformation. Thus a computationally expensive least-square optimization is required for each pair of range and codebook blocks in order to determine the optimal luminance transformation and the resulting approximation error.

Let an image to be coded be denoted by $[I] \in R^{N \times N}$ or a matrix $[I]$ of size $(N \times N)$, where N is a power of 2. The image is partitioned into non-overlapping range blocks $[g_r], r = 0, 1, \dots, N_r - 1$. Let domain block $[h]$ of size $(N/2 \times N/2)$ denote the downscaled version of the image $[I]$, i.e.,

$$h(m_1, m_2) = \frac{1}{4} \sum_{n_1=2m_1}^{2m_1+1} \sum_{n_2=2m_2}^{2m_2+1} I(n_1, n_2) \qquad 0 \le m_1, m_2 \le \frac{N}{2}$$

Each range block $[g_r]$ is then approximated as follows:

$$[\hat{g}_r] = s_r T_r([h]) + o_r \qquad r = 0, 1, \dots, N_r - 1 \qquad (8.7)$$

The parameters s_r, o_r are called *scaling factor* and *offset*, respectively. T_r is an isometric transformation, which shuffles pixels inside the decimated domain block $[h]$. Isometric transformations consist of circular shift and inversion of the domain block $[h]$. Thus codebook blocks $[c_l]$, $l = 0, 1, \dots, N_c - 1$, is generated from $[h]$. Each range block $[g_r]$ is then approximated by an affine transformation of a codebook block.

$$[\hat{g}_r] = s_r [c_l] + o_r \qquad r = 0, 1, \dots, N_r - 1 \qquad (8.8)$$

A range block is compared to all codebook blocks in order to determine the best codebook block giving the least approximation error.

For a given range, the corresponding canonical codebook consists of all image blocks in the downscaled version $[h]$ that have the same size and shape as the range. Note that the range can be of arbitrary shape such as a polygon. Using arbitrary but fixed scanning procedure, the range and codebook blocks are converted to vectors which are denoted by $\underline{R}, \underline{D}^R_{0,0}, \dots, \underline{D}^R_{N/2-1, N/2-1}$ and call blocks again. The number of codebook blocks in the canonical codebook for a given range is equal to $N^2/4$ since we allow codebook blocks to *wrap around* image borders. For simplicity this method does not consider isometric transformations of the downscaled image (rotations by multiple of $\pi/2$ and reflections). For better readability we simply write $\underline{D}$ instead of $\underline{D}^R_{m_1,m_2}$. The distortion function for the range vector $\underline{R}$ and a codebook vector $\underline{D}$ is a quadratic function of the parameters s, o of the affine luminance transformation:

$$d_{\underline{D},\underline{R}}(s,o) = \|\underline{R} - (s\underline{D} + o\underline{1})\|_2^2$$
$$= \langle\underline{D},\underline{D}\rangle s^2 + 2\langle\underline{D},\underline{1}\rangle so + no^2 - 2\langle\underline{R},\underline{D}\rangle s - 2\langle\underline{D},\underline{1}\rangle o + \langle\underline{R},\underline{R}\rangle \quad (8.9)$$

The constant block with unit intensity at the pixels of interest is converted to a vector $\underline{1}$. The symbol $\langle,\rangle$ denotes an inner product on a vector space of dimension n, where n is the number of pixels in the range block. As $\underline{R}$ and $\underline{D}$ are column vectors, $\langle\underline{R},\underline{D}\rangle = \underline{R}^T\underline{D}$.

Figure 8.6 illustrates the part of the algorithm that computes the array of $\langle\underline{D},\underline{1}\rangle$, $\langle\underline{D},\underline{D}\rangle$ and $\langle\underline{D},\underline{R}\rangle$ with the FFT-based approach. The products $\langle\underline{D},\underline{1}\rangle$ are obtained by the cross correlation of the downscaled image with a 'range' where all intensities are set to unity (called the *range shape matrix*). The sum of the squares $\langle\underline{D},\underline{D}\rangle$ is computed in the same way using the range shape matrix where all intensities in the downscaled image are squared before cross correlation.

This method has a strong potential in applications where an adaptive image partition provides for large irregularly shaped ranges and a fractal code is sought. The FFT-based approach handles the case of an irregular range shape with ease by zero padding the pixels that are not in the range. When ranges have different shapes, the computation of $\langle\underline{D},\underline{1}\rangle$ and $\langle\underline{D},\underline{D}\rangle$ cannot be done as Fig. 8.6 anymore as it is a uniform partition. This method can speed up a direct approach.

Example 8.3 Two-dimensional circular correlation is denoted as $\star$ and is shown for $N = 3$.

$$(h_1 \star h_2)(n_1,n_2) = \sum_{k_1=0}^{N-1}\sum_{k_2=0}^{N-1} h_1(k_1,k_2)h_2((k_1 - n_1) \bmod N, (k_2 - n_2) \bmod N)$$
$$(8.10)$$

$$(h_1 \star h_2)(0,0) = \text{sum of } \begin{bmatrix} h_1(0,0) & h_1(0,1) & h_1(0,2) \\ h_1(1,0) & h_1(1,1) & h_1(1,2) \\ h_1(2,0) & h_1(2,1) & h_1(2,2) \end{bmatrix} \circ \begin{bmatrix} h_2(0,0) & h_2(0,1) & h_2(0,2) \\ h_2(1,0) & h_2(1,1) & h_2(1,2) \\ h_2(2,0) & h_2(2,1) & h_2(2,2) \end{bmatrix}$$

$$(h_1 \star h_2)(0,1) = \text{sum of } \begin{bmatrix} h_1(0,0) & h_1(0,1) & h_1(0,2) \\ h_1(1,0) & h_1(1,1) & h_1(1,2) \\ h_1(2,0) & h_1(2,1) & h_1(2,2) \end{bmatrix} \circ \begin{bmatrix} h_2(0,2) & h_2(0,0) & h_2(0,1) \\ h_2(1,2) & h_2(1,0) & h_2(1,1) \\ h_2(2,2) & h_2(2,0) & h_2(2,1) \end{bmatrix}$$

$$(h_1 \star h_2)(0,2) = \text{sum of } \begin{bmatrix} h_1(0,0) & h_1(0,1) & h_1(0,2) \\ h_1(1,0) & h_1(1,1) & h_1(1,2) \\ h_1(2,0) & h_1(2,1) & h_1(2,2) \end{bmatrix} \circ \begin{bmatrix} h_1(0,1) & h_1(0,2) & h_1(0,0) \\ h_1(1,1) & h_1(1,2) & h_1(1,0) \\ h_1(2,1) & h_1(2,2) & h_1(2,0) \end{bmatrix}$$

$$(h_1 \star h_2)(1,0) = \text{sum of } \begin{bmatrix} h_1(0,0) & h_1(0,1) & h_1(0,2) \\ h_1(1,0) & h_1(1,1) & h_1(1,2) \\ h_1(2,0) & h_1(2,1) & h_1(2,2) \end{bmatrix} \circ \begin{bmatrix} h_1(2,0) & h_1(2,1) & h_1(2,2) \\ h_1(0,0) & h_1(0,1) & h_1(0,2) \\ h_1(1,0) & h_1(1,1) & h_1(1,2) \end{bmatrix}$$

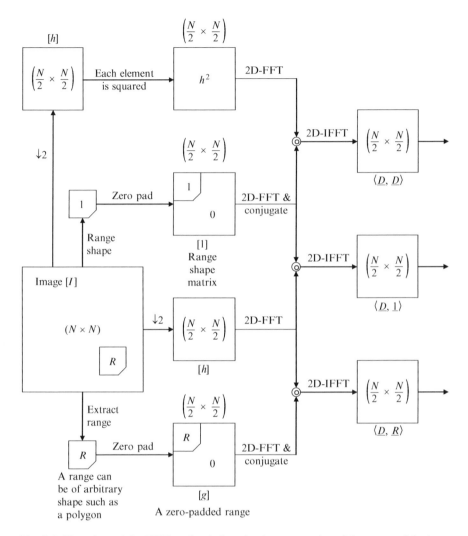

Fig. 8.6 Flow chart of the FFT-based technique for the computation of the arrays of the inner products $\langle \underline{D}, \underline{D} \rangle$, $\langle \underline{D}, \underline{1} \rangle$, $\langle \underline{D}, \underline{R} \rangle$. The symbol ∘ denotes the Hadamard product of two complex Fourier coefficient matrices. Here the range block need not be of *square shape* [FR6] (see also [FR3]). © 2000 Elsevier

To get an element of resultant matrix, circularly shift elements of the matrix $[h_2]$ horizontally and vertically, and then apply Hadamard product to the two matrices, and finally add all the elements. Hadamard product represents element-by-element multiplication of the two matrices.

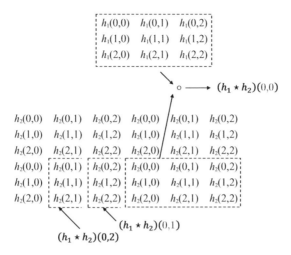

$$
\begin{array}{ccc}
h_1(0,0) & h_1(0,1) & h_1(0,2) \\
h_1(1,0) & h_1(1,1) & h_1(1,2) \\
h_1(2,0) & h_1(2,1) & h_1(2,2)
\end{array}
$$

$\circ \longrightarrow (h_1 \star h_2)(0,0)$

Example 8.4   Find 2-D circular correlation of $[h_1]$ and $[h_2]$

$$
[h_1] = \begin{bmatrix} 5 & 3 \\ 1 & 0 \end{bmatrix}, \qquad [h_2] = \begin{bmatrix} 1 & 2 \\ 3 & 4 \end{bmatrix}
$$

$(h_1 \star h_2)(0,0) = 14$

$$
\begin{aligned}
(h_1 \star h_2)(0,0) &= 5 \times 1 + 3 \times 2 + 1 \times 3 + 0 \times 4 = 14 \\
(h_1 \star h_2)(0,1) &= 5 \times 2 + 3 \times 1 + 1 \times 4 + 0 \times 3 = 17 \\
(h_1 \star h_2)(1,0) &= 5 \times 3 + 3 \times 4 + 1 \times 1 + 0 \times 2 = 28 \\
(h_1 \star h_2)(1,1) &= 5 \times 4 + 3 \times 3 + 1 \times 2 + 0 \times 1 = 31
\end{aligned}
$$

$$
[h_1] \star [h_2] = \text{2-D IDFT} \left([H_1]^* \circ [H_2] \right) = \begin{bmatrix} 14 & 17 \\ 28 & 31 \end{bmatrix}
$$

where $[H_1] = $ 2-D DFT of $[h_1]$ and $\circ$ is Hadamard product defined in (8.11).

FFT application (cross correlation via FFT)

$$[A] \circ [B] = \begin{bmatrix} a_{11} & a_{12} & \cdots & a_{1N} \\ a_{21} & a_{22} & & a_{2N} \\ \vdots & & \ddots & \vdots \\ a_{N1} & a_{N2} & \cdots & a_{NN} \end{bmatrix} \circ \begin{bmatrix} b_{11} & b_{12} & \cdots & b_{1N} \\ b_{21} & b_{22} & & b_{2N} \\ \vdots & & \ddots & \vdots \\ b_{N1} & b_{N2} & \cdots & b_{NN} \end{bmatrix} = [C] \quad (8.11)$$

where $\circ$ represents Hadamard product. Corresponding elements of $[A]$ and $[B]$ are multiplied to yield $[C]$.

8.3 Phase Only Correlation

Correlation between phase-only versions of the two images to be aligned is used for image registration [IP2].

Let $x(n_1, n_2)$ be the reference image and $y(n_1, n_2)$ be the translated image of $x(n_1, n_2)$ by (m_1, m_2). Then

$$y(n_1, n_2) = x(n_1 + m_1, n_2 + m_2) \quad n_1, n_2 = 0, 1, \ldots, N - 1$$
$$- (N - 1) \le m_1, m_2 \le N - 1 \quad (8.12)$$

From the Fourier shift property defined in (5.12), the DFT of (8.12) is

$$Y^{\mathrm{F}}(k_1, k_2) = X^{\mathrm{F}}(k_1, k_2) W_N^{-k_1 m_1} W_N^{-k_2 m_2} \quad (8.13)$$

The cross power spectrum of the phase only images is defined as

$$Z_{\mathrm{poc}}^{\mathrm{F}}(k_1, k_2) = \frac{X^{\mathrm{F}}(k_1, k_2) Y^{\mathrm{F}*}(k_1, k_2)}{|X^{\mathrm{F}}(k_1, k_2) Y^{\mathrm{F}*}(k_1, k_2)|} \quad k_1, k_2 = 0, 1, \ldots, N - 1 \quad (8.14)$$

$$\begin{aligned} Z_{\mathrm{poc}}^{\mathrm{F}}(k_1, k_2) &= \frac{X^{\mathrm{F}}(k_1, k_2) X^{\mathrm{F}*}(k_1, k_2) W_N^{k_1 m_1} W_N^{k_2 m_2}}{|X^{\mathrm{F}}(k_1, k_2) X^{\mathrm{F}*}(k_1, k_2) W_N^{k_1 m_1} W_N^{k_2 m_2}|} \\ &= \frac{X^{\mathrm{F}}(k_1, k_2) X^{\mathrm{F}*}(k_1, k_2) W_N^{k_1 m_1} W_N^{k_2 m_2}}{|X^{\mathrm{F}}(k_1, k_2) X^{\mathrm{F}*}(k_1, k_2)|} \\ &= \exp\left(-j2\pi \frac{k_1 m_1 + k_2 m_2}{N}\right) \end{aligned} \quad (8.15)$$

In the spatial domain this is equivalent to

$$z_{poc}(n_1, n_2) = \delta(n_1 - m_1, n_2 - m_2) \qquad n_1, n_2 = 0, 1, \ldots, N - 1 \qquad (8.16)$$

Thus an impulse centered on (m_1, m_2) is obtained with the range $0 \leq z_{poc}$ $(n_1, n_2) \leq 1$ (subscript poc stands for *phase only correlation*). This is shown in a block diagram format in Fig. 8.7.

Images shifted by (m_1, m_2) and $(m_1 - N, m_2 - N)$ relative to the original image have peaks at the same place with different peak values as shown in Fig. 8.8h and i. Thus every point except the origin represents two images shifted in the opposite directions of each other.

The Fourier phase correlation method is very robust against distortions. Only a small percentage of the original image is necessary to get good registration [IP19].

In terms of image matching, the phase correlation method is invariant to image shift/translation, occlusions and brightness change, and is robust against noise.

Equation (8.13) suggests a simpler way to isolate the term $\exp\left(-j2\pi \frac{k_1 m_1 + k_2 m_2}{N}\right)$ as

$$Z^{F}_{poc2}(k_1, k_2) = \frac{X^{F}(k_1, k_2)}{Y^{F}(k_1, k_2)} = \exp\left(-j2\pi \frac{k_1 m_1 + k_2 m_2}{N}\right) \qquad (8.17)$$

In the spatial domain this is equivalent to (8.16). However, the simpler method is less stable than the regular method, since X^{F}/Y^{F} blows up when Y^{F} comes close to zero whereas for the regular method the numerator and the denominator have the same magnitude [IP19].

Applications include finger print matching [IP16], waveform matching, iris recognition using phase based image matching, face recognition, palm print recognition, and detection of hard-cuts and gradual shot changes for archive film [IP17].

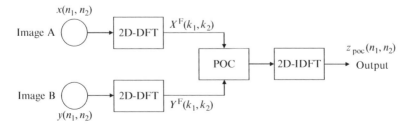

Fig. 8.7 Block diagram for implementation of phase only correlation (POC)

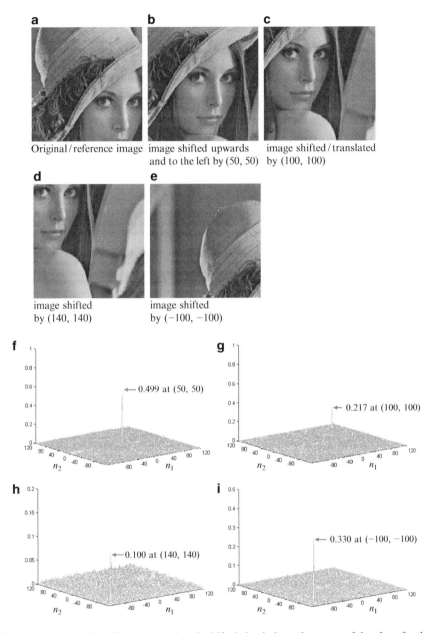

Fig. 8.8 Origin (0, 0) of the (n_1, n_2) plane is shifted circularly to the center of the plane for the phase correlation functions in (**f**)–(**i**). Here all the images are of size 240 × 240. Registration with two images shifted in the opposite directions of each other relative to the original image gives the same result (**h**) and (**i**) in terms of the actual coordinates of the peak at (−100, −100) as 140 − N = 100 with N = 240

8.4   Image Rotation and Translation Using DFT/FFT

This method is proposed by Cox and Tong [IP10], and is an application of the chirp z-algorithm (Section 3.14). Given the image $x(n_1, n_2)$ on an $N \times N$ square grid, we compute $x(n_1, n_2)$ on a grid rotated by an arbitrary angle θ in a counter-clockwise direction and shifted by an arbitrary translation (m_1, m_2).

First compute the 2-D FFT of $x(n_1, n_2)$, so that

$$X^{\mathrm{F}}(k_1, k_2) = \sum_{n_1=0}^{N-1} \sum_{n_2=0}^{N-1} x(n_1, n_2) W_N^{(n_1 k_1 + n_2 k_2)} \quad k_1, k_2 = 0, 1, \ldots, N-1 \quad (5.3a)$$

$$x(n_1, n_2) = \frac{1}{N^2} \sum_{k_1=0}^{N-1} \sum_{k_2=0}^{N-1} X^{\mathrm{F}}(k_1, k_2) W_N^{-(n_1 k_1 + n_2 k_2)} \quad n_1, n_2 = 0, 1, \ldots, N-1 \quad (5.3b)$$

where $W_N = \exp(-j2\pi/N)$.

Now compute the image $x(n_1, n_2)$ on the desired output grid.

$$x(n_1 \cos\theta - n_2 \sin\theta + m_1, n_1 \sin\theta + n_2 \cos\theta + m_2)$$

$$= \frac{1}{N^2} \sum_{k_1=0}^{N-1} \sum_{k_2=0}^{N-1} X^{\mathrm{F}}(k_1, k_2) \exp\left[\frac{j2\pi}{N}(k_1 m_1 + k_2 m_2)\right] \exp\left[\frac{j2\pi}{N}(k_1 n_1 + k_2 n_2)\cos\theta\right]$$

$$\times \exp\left[\frac{j2\pi}{N}(k_2 n_1 - k_1 n_2)\sin\theta\right] \quad (8.18)$$

To compute (8.18), we need to be able to compute the sum

$$g(n_1, n_2; \alpha, \beta) = \sum_{k_1=0}^{N-1} \sum_{k_2=0}^{N-1} G^{\mathrm{F}}(k_1, k_2) \exp(j2\pi[(k_1 n_1 + k_2 n_2)\alpha + (k_2 n_1 - k_1 n_2)\beta]) \quad (8.19)$$

for arbitrary α and β (here $\alpha = \dfrac{\cos\theta}{N}$ and $\beta = \dfrac{\sin\theta}{N}$), where

$$G^{\mathrm{F}}(k_1, k_2) = \frac{1}{N^2} X^{\mathrm{F}}(k_1, k_2) \exp\left[\frac{j2\pi}{N}(k_1 m_1 + k_2 m_2)\right] \quad (8.20)$$

The one-dimensional analog of (8.19) is as follows.

$$h(n; \alpha) = \sum_{k=0}^{N-1} H^{\mathrm{F}}(k) \exp(j2\pi k n \alpha) \quad (8.21)$$

This can be computed using the chirp-z algorithm, by expanding

$$2kn = k^2 + n^2 - (k-n)^2 \quad (8.22)$$

$$h(n; \alpha) = \exp\left(j\pi n^2 \alpha\right) \sum_{k=0}^{N-1} \left\{H^{\mathrm{F}}(k) \exp\left(j\pi k^2 \alpha\right)\right\} \exp\left[-j\pi(k-n)^2 \alpha\right]$$

$$= \exp\left(j\pi n^2 \alpha\right) \sum_{k=0}^{N-1} \hat{H}^{\mathrm{F}}(k) V(n-k)$$

$$= \exp\left(j\pi n^2 \alpha\right) \left\{\hat{H}^{\mathrm{F}}(k) * V(k)\right\} \tag{8.23}$$

where $\hat{H}^{\mathrm{F}}(k) = H^{\mathrm{F}}(k) \exp(j\pi k^2 \alpha)$ and $V(k) = \exp(-j\pi k^2 \alpha)$. This expansion is a multiplication, a convolution and another multiplication (Fig. 8.9). The convolution can be carried out quickly using two FFTs and one IFFT (Fig. 5.4).

Similarly, we can compute (8.19) efficiently. The integer expansion required is

$$2(k_1 n_1 + k_2 n_2) = k_1^2 + n_1^2 - k_2^2 - n_2^2 - (k_1 - n_1)^2 + (k_2 + n_2)^2 \tag{8.24a}$$

$$2(k_2 n_1 - k_1 n_2) = 2k_1 k_2 - 2n_1 n_2 - 2(k_1 - n_1)(k_2 + n_2) \tag{8.24b}$$

With this (8.19) can be represented as

$$g(n_1, n_2; \alpha, \beta) = Z^*(n_2, n_1) \sum_{k_1=0}^{N-1} \sum_{k_2=0}^{N-1} \left\{G^{\mathrm{F}}(k_1, k_2) Z(k_1, k_2)\right\} Z^*(k_1 - n_1, k_2 + n_2)$$

$$\tag{8.25}$$

where $Z(n_1, n_2) = \exp\left(j\pi\left[(n_1^2 - n_2^2)\alpha + 2n_1 n_2 \beta\right]\right)$, $G^{\mathrm{F}}(k_1, k_2)$ is defined in (8.20) and the superscript $*$ denotes complex conjugation. This implies two multiplications and a convolution, which can be done using three 2-D FFTs (Fig. 8.10). Matrix $\{Z(n_2, n_1)\}$ in (8.25) is the transpose of $\{Z(n_1, n_2)\}$. Another derivation of (8.25) is described in Problem 8.6 (Fig. 8.11).

Fig. 8.9 Block diagram for implementation of the chirp-z algorithm represented in (8.23), which is one-dimensional analog of (8.19)

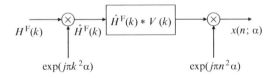

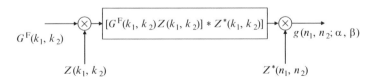

Fig. 8.10 Image rotation. Convolution can be done using three 2-D FFTs. In magnetic resonance imaging (MRI), the original data comes in Fourier space [IP10, W30]

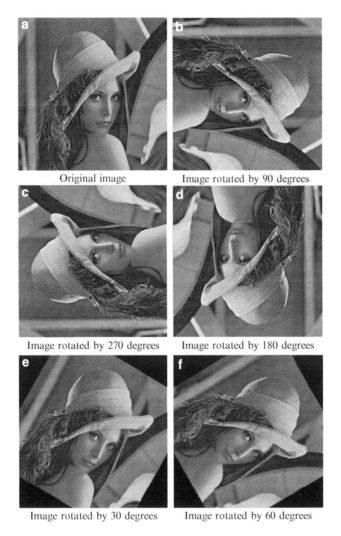

Fig. 8.11 a Original image, b Image rotated by 90°, c Image rotated by 270°, d Image rotated by 180°, e Image rotated by 30°, f Image rotated by 60°. b–d Images are rotated using the Cox-Tong method [IP10]. e–f Images are rotated in the spatial domain using the MATLAB command imrotate (I, −30, 'bilinear', 'crop')

This method only rotates images by 90°, 180° and 270° (see Fig. 8.11b–d) and extending this idea to other angles is a possible research topic that can be explored.

8.5 Intraframe Error Concealment [IP28]

In error prone environments such as wireless networks, a macroblock (MB) (16 × 16 size pels in video coding standards such as MPEG-1, 2, 4 and H.26*x* series [D37, IP28]) [IP29] can be lost or transmitted erroneously. One approach to address this is

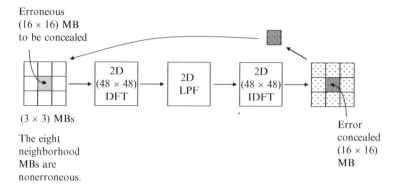

Fig. 8.12 Intraframe MB error concealment (DFT and IDFT are implemented by FFT and IFFT)

called error concealment where the corrupted MB is replaced with a neighborhood MB or weighted neighborhood MBs that are not damaged.

An example of intraframe error concealment is shown in Fig. 8.12. Consider the neighborhood MBs surrounding the (16×16) MB to be concealed which is initially filled with grey level values. This neighborhood amounts to (48×48) pels for (3×3) MBs. 2D FFT is then applied to this large area, followed by 2D low pass filtering that removes the discontinuity due to these inserted pels. This is followed by 2D IFFT, resulting in a replica of the input pels. The reconstructed pels as a result of low pass filtering are similar but not exactly the same as input pels. The cutoff frequency of the LPF influences the extent of dissimilarity. If the cutoff frequency is smaller, then the error concealed MB is strongly influenced by the surrounding pels and vice versa. The error affected MB (input) is replaced by the error concealed MB (output). The surrounding 8 MBs remain the same. The entire process, 2D-FFT, 2D-LPF and 2D-IFFT, is again repeated several times. During each repetition the cutoff frequency of the LPF is gradually increased. Based on the characteristics of the surrounding pels the LPF can be made directional thus improving the quality of error concealment. This repetition is interactive and is halted when the difference between the pels of the 2 MBs (error affected and error concealed) is less than a preset threshold.

8.6   Surface Texture Analysis

Various discrete orthogonal transforms such as the DFT, DCT, Walsh, phase-shift-invariant Walsh, BIFORE (BInary FOurier REpresentation) and Haar [G5, G11, G12, G13, G14, G15, G16, G17, G18, G19, G20, G21] are compared for their ability to characterize surface texture data. Both DFT and DCT are recommended for use in surface texture analysis in view of their rapid rates of convergence and also because of their ability to characterize the data and the machining peaks. Details of this comparison are described in [J3].

8.7   FFT-Based Ear Model

There are two versions for the objective measurements of perceived audio quality: basic version and advanced version [D46, D53]. The former includes *model output variables* (MOVs) that are calculated from the FFT-based ear model (Fig. 8.13). It uses a total of 11 MOVs for the prediction of the perceived basic audio quality. It uses a 2,048-point FFT.

The advanced version includes MOVs that are calculated from the filter bank-based ear model as well as the MOVs from the basic version. The spectrally adapted excitation patterns and the modulation patterns are computed from the filter bank-based part of the model only. The advanced version uses a total of five MOVs for the prediction perceived basic audio quality.

8.8   Image Watermarking

Watermarking of digital media for copyright protection and content verification has been extensively used to avoid/eliminate digital piracy by hiding a secret and personal message to protect the copyright of a product and to demonstrate its authenticity (content verification, data integrity and tamper proofing [E8]). The piracy includes illegal access, intentional tampering and copyright violation. The watermarked image must also withstand various operations such as filtering, dith-ering, photocopying, scanning, cropping, scaling, rotation, translation, JPEG com-pression [IP28], etc. In the watermark embedding scheme DFT phase spectrum [E1, E2] and magnitude spectrum [E4] have been used for watermark hiding (Fig. 8.14). Watermark embedding in the DFT magnitude spectrum is robust to elementary transformations such as rotation, scaling and translation [E4].

In [E2], it is shown that embedding a watermark in the phase spectrum of a 2D-DFT of real sequences such as images is very robust to tampering, and also to changes in image contrast. Also phase distortions deliberately introduced by an intruder to impede transmission of the watermark have to be large in order to be successful, resulting in severe degradation to image quality. Details of the water-mark embedding scheme in the phase spectrum of the image are described in [E2]. It is shown that the watermarked image has no visible artifacts and the watermark has survived up to 15:1 compression ratio when the watermarked image is subjected to JPEG encoder [IP28].

Integrating aspects of the human visual system into watermarking algorithms can be a research topic. In addition, a detailed study of the effects of image distortion on a watermark can be undertaken with a view to improving the

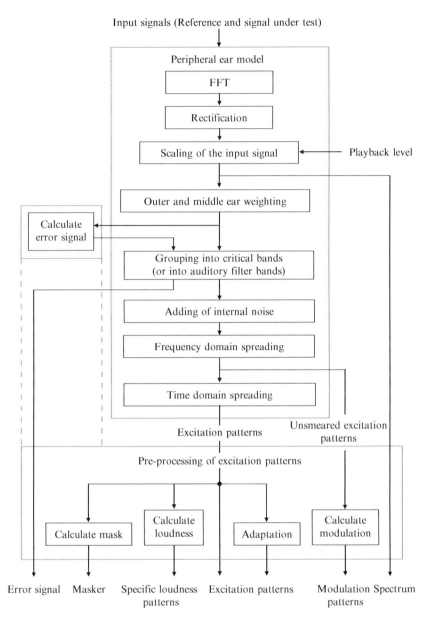

Input signals (Reference and signal under test)

Peripheral ear model

FFT

Rectification

Scaling of the input signal ◄──────── Playback level

Outer and middle ear weighting

Calculate error signal

Grouping into critical bands
(or into auditory filter bands)

Adding of internal noise

Frequency domain spreading

Time domain spreading

Excitation patterns   Unsmeared excitation patterns

Pre-processing of excitation patterns

Calculate mask   Calculate loudness   Adaptation   Calculate modulation

Error signal   Masker   Specific loudness patterns   Excitation patterns   Modulation Spectrum patterns

Fig. 8.13 Peripheral ear model and pre-processing of excitation patterns for the FFT-based part of the model. [D46] © 1998–2001 ITU-R

watermark detection. Novel techniques can be devised to make it possible to detect a watermark without requiring the original unmarked image [E2].

Ruanaidh and Pun [E4] have developed a method for embedding a watermark in an invariant domain by combining the DFT with log-polar map, which is robust to

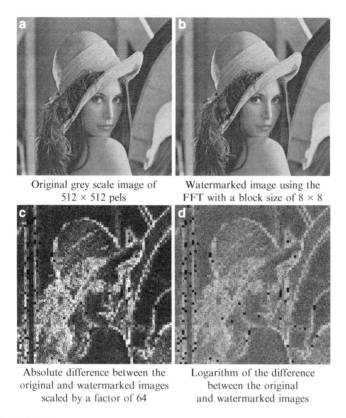

Fig. 8.14 Embedding a watermark in the phase spectrum of a 2D-DFT of images [E2]. **a** Original grey scale image of 512 × 512 pels. **b** Watermarked image using the FFT with a block size of 8 × 8. **c** Absolute difference between the original and watermarked images scaled by a factor of 64. **d** Logarithm of the difference between the original and watermarked images

rotation and scaling. The implementation is a compromise between the conflicting goals of watermark invisibility and its robustness. A prototype RST (rotation, scaling and translation) invariant watermarking scheme is shown in Fig. 8.15. As the log-polar mapping (LPM) and its inverse are lossy processes, an embedding process that avoids the need to pass the watermarked image through a log polar map is shown in Fig. 8.16. Only the 2D spread spectrum signal goes through the ILPM (inverse LPM). The scheme to extract the watermark from the stegoimage (watermarked image) is shown in Fig. 8.17.

8.9 Audio Watermarking

In [E15] M-band wavelet modulation is combined with code division multiple access CDMA technique to construct watermark signals (Fig. 8.18). The CDMA technique replaces the typical spread spectrum (SS) technique to improve both

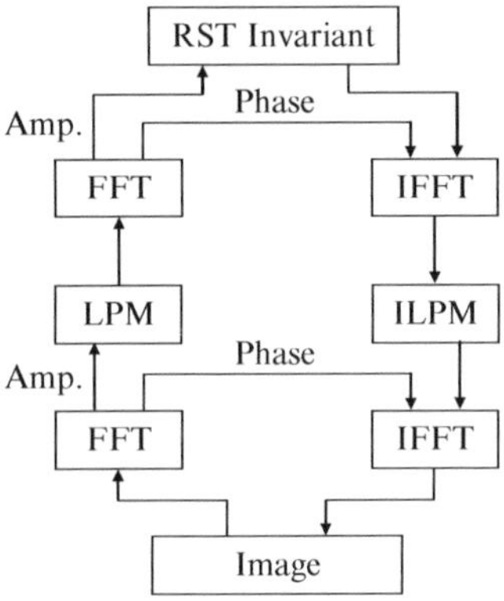

Fig. 8.15 A diagram of a
prototype RST-invariant
watermarking scheme. RST:
Rotation, Scale and
Translation. Amp:
Amplitude. LPM and ILPM
denote a log polar map and its
inverse. [E4] © 1998
Elsevier

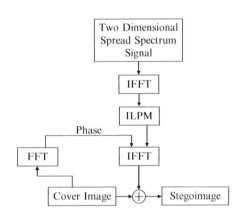

Fig. 8.16 A method of
embedding a watermark in an
image which avoids mapping
the original image into the
RST invariant domain. The
terms "cover image" and
"stego image" denote the
original image and
watermarked image,
respectively. [E4] © 1998
Elsevier

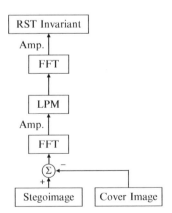

Fig. 8.17 A scheme to
extract a watermark from an
image. Amp: Amplitude. [E4]
© 1998 Elsevier

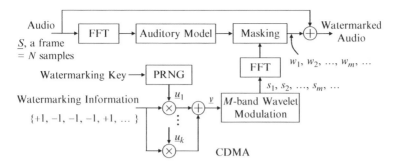

Fig. 8.18 Watermark embedding diagram. k is the number of CDMA carrier signals per frame. $M = 2^4$ when audio is sampled at 44.1 kHz. PRNG: Pseudorandom number generator. [E15] © 2002 IEICE

robustness and capacity requirements. In order to satisfy the orthogonality condition of CDMA carrier signals, the Gram–Schmidt orthogonalization process [B40] should be employed to modify the generated pseudo noise (PN) sequences $\underline{u}_i$. Let $b_i \in \{+1, -1\}$ denote a bit stream of watermarking information. The application of CDMA can be formulated as

$$\underline{v} = \sum_{i=1}^{k} b_i \underline{u}_i \tag{8.26}$$

where k is the number of CDMA carrier signals per frame, and k is limited so that the strength of watermark signal is within the perceptual constraints.

For the requirement of the acceptable quality, Ji et al. [E15] control the strength of the watermark signal by selecting appropriate masking thresholds. They use the frequency masking model defined in the MPEG-1 audio psychoacoustic model [E14].

8.9.1 Audio Watermarking Using Perceptual Masking

Swanson et al. [E14] have developed a watermarking procedure to embed copyright protection directly into digital audio exploiting the temporal and frequency perceptual maskings which guarantee that the embedded watermark is inaudible and robust to tampering and to various DSP operations. The audio watermarking system is described in a block diagram format in Fig. 8.19. The frequency masking model is based on audio psychoacoustic model 1 for MPEG-1 audio Layer I [D22, D26]. In Fig. 8.19, $S_i(k)$ is the log power spectrum of the ith audio block (16 ms segment of the audio signal sampled at 32 kHz, $N = 16$ ms $\times$ 32 kHz $= 512$ samples) weighted by the Hann window $h(n)$ i.e.,

$$h(n) = \frac{\sqrt{8/3}}{2} \left[1 - \cos\left(2\pi\frac{n}{N}\right) \right] \qquad (n = 0, 1, \dots, N-1) \tag{8.27}$$

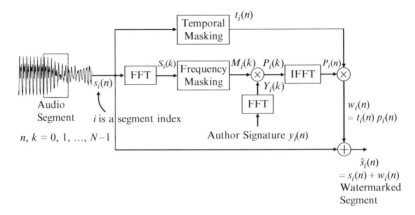

Fig. 8.19 Blockdiagram of audio watermarking procedure. [E14] © 1998 Elsevier

The power spectrum of the signal $s_i(n)$ is calculated as

$$S_i(k) = 10\log_{10}\left[\frac{1}{N}\left\|\sum_{n=0}^{N-1} s_i(n)h(n)\exp\left(\frac{-j2\pi nk}{N}\right)\right\|^2\right] \quad (k = 0, 1, \ldots, N-1)$$

(8.28)

The steps involved in watermarking on each audio segment with a unique noise-like sequence shaped by the masking phenomenon are outlined as follows:
For each audio segment $s_i(n)$:

1. Compute the power spectrum $S_i(k)$ of the audio segment $s_i(n)$ (Eq. 8.28).
2. Compute the frequency mask $M_i(k)$ of the power spectrum $S_i(k)$ (cf. Section 3.1 of [E14]).
3. Compute the FFT of author signature $y_i(n)$ which yields $Y_i(k)$.
4. Use the mask $M_i(k)$ to weight the noise-like author representation for that audio block, creating the shaped author signature $P_i(k) = Y_i(k)M_i(k)$.
5. Compute the inverse FFT of the shaped noise $p_i(n) = $ IFFT $[P_i(k)]$.
6. Compute the temporal mask $t_i(n)$ of $s_i(n)$ (cf. Section 3.2 of [E14]).
7. Use the temporal mask $t_i(n)$ to further shape the frequency shaped noise, creating the watermark $w_i(n) = t_i(n)p_i(n)$ of that audio segment.
8. Create the watermarked block $\hat{s}_i(n) = s_i(n) + w_i(n)$.

8.10 OFDM

OFDM/COFDM (orthogonal frequency domain multiplexing – coded OFDM) has been adopted in Europe for terrestrial digital television and HDTV broadcasting [O2]. Although FCC advisory committee for advanced television service (ACATS)

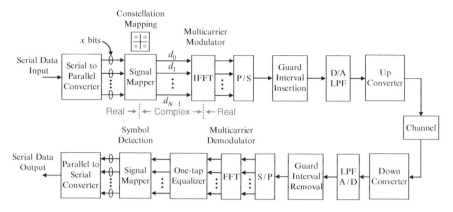

Fig. 8.20 FFT based OFDM system (P/S: Parallel to serial, S/P: Serial to parallel, D/A: Digital to analog, A/D: Analog to digital) [O2] © 1995 IEEE

has selected eight-VSB (vestigial sideband) digital modulation for terrestrial HDTV broadcasting, there is considerable debate over the use of COFDM versus VSB or QAM (quadrature amplitude modulation) for terrestrial HDTV broadcasting [O2, AP2]. The emphasis here is to describe the application of FFT in the OFDM (Fig. 8.20).

OFDM reduces the effect of frequency selective fading by reducing the data rate by splitting the data stream into several parallel blocks and then transmitting these blocks [O7]. By dividing and modulating the information among multiple carriers, it causes the signals to be resistant to ghosting and jamming [O11]. Other benefits of OFDM in wireless communication systems are high bandwidth efficiency, resistance to RF interference and robustness to multipath fading.

8.10.1 Signal Representation of OFDM Using IFFT/FFT

Denote the complex-valued signal points corresponding to the information signals in the $N/2$ subchannels as

$$\hat{d}_n = \hat{a}_n + j\hat{b}_n \qquad n = 0, 1, \ldots, \frac{N}{2} - 1 \qquad (8.29)$$

The symbols $\hat{a}_n$ and $\hat{b}_n$ take on values of $\pm 1, \pm 3, \ldots$ depending on the number of signal points in the signal constellations (Fig. 8.20). For example, $\hat{a}_n$ and $\hat{b}_n$ can be selected as $\{\pm 1, \pm 3\}$ for 16 QAM and $\{\pm 1\}$ for QPSK. The DFT of these information symbols $\{\hat{d}_n\}$ is a multicarrier OFDM signal $y(t)$ defined in (8.34). Since $y(t)$ must be real, we create N symbols from $N/2$ information symbols as (see the complex conjugate theorem in Section 2.3)

$$d_0 = \text{Re}\,(\hat{d}_0) \qquad (8.30a)$$

$$d_{N/2} = \mathrm{Im}(\hat{d}_0) \tag{8.30b}$$

$$d_n = \hat{d}_n \qquad n = 1, 2, \ldots, \frac{N}{2} - 1 \tag{8.30c}$$

$$d_{N-n} = \left(\hat{d}_n\right)^* \qquad n = 1, 2, \ldots, \frac{N}{2} - 1 \tag{8.30d}$$

The DFT of the complex data sequence $d_n = a_n + jb_n$, $(n = 0, 1, \ldots, N - 1)$ is

$$X^{\mathrm{F}}(k) = \sum_{n=0}^{N-1} d_n W_N^{nk} \qquad k = 0, 1, \ldots, N - 1$$

$$= \sum_{n=0}^{N-1} d_n \exp(-j2\pi f_n t_k) \tag{8.31}$$

where $f_n = \frac{n}{N\Delta t}$, $t_k = k\Delta t$, and Δt is an arbitrarily chosen symbol duration of d_n.

$$X^{\mathrm{F}}(k) = \sum_{n=0}^{N-1} (a_n + jb_n) \left[\cos(2\pi f_n t_k) - j\sin(2\pi f_n t_k)\right] \tag{8.32}$$

From the assumption of (8.30) imaginary terms of $X^{\mathrm{F}}(k)$ cancel out resulting in only the real part of $X^{\mathrm{F}}(k)$ as

$$Y^{\mathrm{F}}(k) = \mathrm{Re}[X^{\mathrm{F}}(k)] = \sum_{n=0}^{N-1} \left[a_n \cos(2\pi f_n t_k) + b_n \sin(2\pi f_n t_k)\right]$$

$$k = 0, 1, \ldots, N - 1 \tag{8.33}$$

LPF output of $Y^{\mathrm{F}}(k)$ at time intervals Δt closely approximates the FDM signal

$$y(t) = \sum_{n=0}^{N-1} \left[a_n \cos(2\pi f_n t_k) + b_n \sin(2\pi f_n t_k)\right] \qquad 0 \le t \le N\Delta t \tag{8.34}$$

Details of the various functional blocks in Fig. 8.20 are described in [O2]. The use of DFT for modulation and demodulation of multicarrier OFDM systems was proposed by Weinstein and Ebert [O1].

8.11 FFT Processors for OFDM

In multicarrier modulation, such as orthogonal frequency domain multiplexing (OFDM) and discrete multitone (DMT), data symbols are transmitted in parallel on multiple subcarriers. Multicarrier modulation techniques have been selected for communication standards, such as high speed transmission over telephone lines (such as DSL), wireless local area networks (WLAN), asymmetric digital subscriber lines (ADSL), very high speed digital subscriber lines (VDSL), digital audio broadcasting (DAB), digital video broadcasting (DVB) and powerline communications (PLC) [A31]. Multicarrier modulation-based transceivers involve real-time DFT computations (please see references listed in [A31]).

The FFT processor in [A31] uses the radix-4 DIF algorithm and in-place memory strategy. The processor can operate at 42 MHz and can compute a 256-point complex FFT in 6 µs.

A higher radix algorithm requires less computational cycles. The radix-2 algorithm requires four times more computational cycles than the radix-4 algorithm. However, the size of a radix-4 algorithm cannot be 128, 512, 2,048 or 8,192 since these are not powers of four. To compute FFTs that are not powers of four, the mixed-radix (MR) algorithm should be used.

The in-place algorithm minimizes the memory size since it stores butterfly outputs at the same memory locations used by inputs to the same butterfly.

The continuous-flow (CF) MR FFT processor in [A31] features the MR (radix-4/2) algorithm, an in-place strategy and the memory bank structure. The processor requires only two N-word memories. The memory is a dominant component in terms of hardware complexity and power consumption.

When the DFT length N can be decomposed into relatively prime factors, it can be implemented with Winograd Fourier transform algorithm (WFTA) [A3]. A systolic array for prime N-length DFT has been proposed and then combined with WFTA to control the increase in the hardware cost when the transform length is large [T1]. The performance of the systolic array for DFT is improved in terms of hardware, I/O cost and throughput.

A novel high performance 8k-point FFT processor architecture for OFDM utilized in DVB-T receiver (Fig. 8.21) [O5, O18] is developed based on radix-8 FFT algorithm. A novel distributed arithmetic-based non-butterfly multiplierless radix-8 FFT architecture is used in the 8,192 point FFT processor design. This involves in four radix-8 and one radix-2 FFT in stages ($8^4 \times 2 = 2^{12} \times 2 = 8,192$) (Fig. 8.22). The 8,192 point FFT can be implemented in 78 µs. Details on the gate

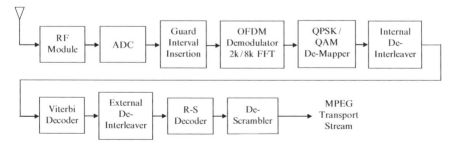

Fig. 8.21 DVB-T receiver. R-S: Reed Solomon. QPSK: Quadrature phase-shift keying. [O18] © 2007 IEEE

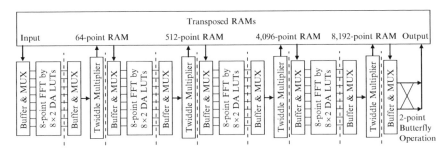

Fig. 8.22 Proposed 8k-point FFT architecture. DA: Distributed arithmetic. LUT: Lookup table. [O18] © 2007 IEEE

count, clock cycles, technology, speed throughput performance and area efficiency are described in [O18].

8.12 DF DFT-Based Channel Estimation Method [C16]

The discrete Fourier transform (DFT)-based channel estimation method derived from the maximum likelihood (ML) criterion is originally proposed for orthogonal frequency division multiplexing (OFDM) systems with pilot preambles [C15]. In order to save bandwidth and improve system performance, *decision-feedback* (DF) data symbols are usually exploited to track channel variations in subsequent OFDM data symbols, and this method is called DF DFT-based channel estimation. However, the working principle of this empirical method has not been explored from the viewpoint of Newton's method in previous studies. This paper derives the DF DFT-based channel estimation via Newton's method for *space-time block code* (STBC)/ OFDM systems (Fig. 8.23). In this way, the equivalence between the two methods is established. Their results indicate that both methods can be implemented through the same four components: a least-square (LS) estimator, an inverse DFT (IDFT) matrix, a weighting matrix and a DFT matrix, but with different connections. On the one hand, the gradient vector in Newton's method [W29] can be found by calculating the difference between an estimated channel frequency response and an LS estimate, followed by the IDFT operation. On the other hand, the inverse of the Hessian matrix in Newton's method [B27, W29] is just the weighting matrix operation in the DF DFT-based method.

8.12.1 DF DFT-Based Channel Estimation Method

As shown in Fig. 8.24b, the block diagram of the DF DFT-based channel estimation method is composed of an LS estimator, an IDFT matrix, a weighting matrix, and a

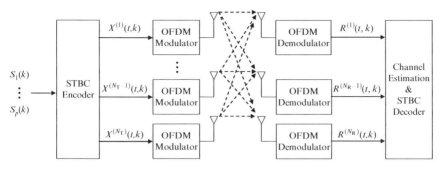

Fig. 8.23 STBC/OFDM system. $a = 1, 2, \ldots, N_T$; $b = 1, 2, \ldots, N_R$. [C16] © 2008 IEEE

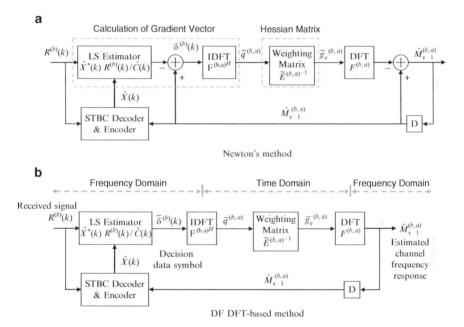

Fig. 8.24 Equivalence between **a** Newton's method and **b** the DF DFT-based method. D is a delay component. [C16] © 2008 IEEE

DFT matrix [C15]. The LS estimator exploits DF data symbols to produce an LS estimate, which is a noisy estimation of channel frequency response. After taking the IDFT to transfer the estimate to time domain, this estimate can be improved by using a weighting matrix which depends on the performance criterion chosen, either ML or minimum mean square error (MMSE) [C15]. Finally, the enhanced estimate is transformed back to the frequency domain to obtain a new estimate of the channel frequency response.

Ku and Huang [C16] developed the equivalence between Newton's method (Fig. 8.24a) and the DF DFT-based method for channel estimation in STBC/OFDM systems. The results can provide useful insights for the development of new algorithms.

8.13   The Conjugate-Gradient Fast Fourier Transform (CG-FFT)

The method of moments (MoM) is one of the effective methods to analyze antennas [K1, K2, K3, K4, B12].

The conjugate-gradient fast Fourier transform (CG-FFT) is successfully applied to the analysis of a large-scale periodic dipole array by speeding up the matrix-vector multiplication in the CG iterative solver for the MoM antenna analysis. Furthermore, an equivalent sub-array preconditioner is proposed to combine with the CG-FFT analysis to reduce iterative steps and the whole CPU-time of the iteration [K3].

The conjugate-gradient method with fast Fourier transform (CG-FFT) is very effective in dealing with the uniform rectangular array because its computational complexity can be reduced to $O(N \log_2 N)$ [K1, K4].

In [K4], CG-FMM-FFT is applied to a large-scale finite periodic array antenna composed of array elements with arbitrary geometry. Furthermore, the performance of the sub-array preconditioner is also compared with the near group preconditioner in the CG-FMM-FFT analysis of a large-scale finite periodic array antenna.

Fast multipole method-fast Fourier transform (FMM-FFT) together with pre-conditioners is developed and applied to analyzing large scale periodic antenna problems.

DFT (implementation via FFT) has been utilized in a multitude of speech coders, either in implementation of *time domain aliasing cancellation* (TDAC) [D1, D2] (via MDCT/MDST) or in developing a psychoacoustic model. Some of these are briefly reviewed.

8.14   Modified Discrete Cosine Transform (MDCT)

Several versions of the MDCT have been developed in the literature [D1, D2]. The MDCT and modified discrete sine transform (MDST) are employed in subband/transform coding schemes as the analysis/synthesis filter banks based on time domain aliasing cancellation (TDAC) [D1]. These are also called as "*TDAC transforms*". Princen, Bradley and Johnson [D1] defined two types of MDCT for evenly stacked and oddly stacked analysis/synthesis systems [D2, D20, D47, D49].

The *modulated lapped transform* (MLT) [D40] is used to implement block transform coding in video and audio compression (MPEG 1/2 audio and Dolby AC-3 (Figs. 8.26 and 8.27)). Several versions of the MLT have been developed and

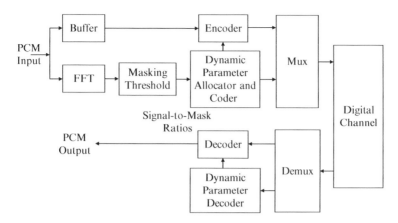

Fig. 8.25 Block diagram of perceptual-based coders. [D33] © 1995 IEEE

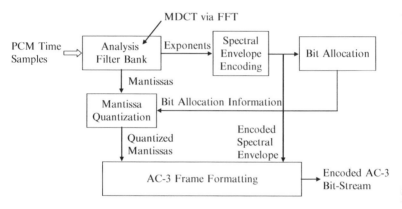

Fig. 8.26 The AC-3 (audio coder – 3) encoder (Dolby Labs). [D51] © 2006 IEEE

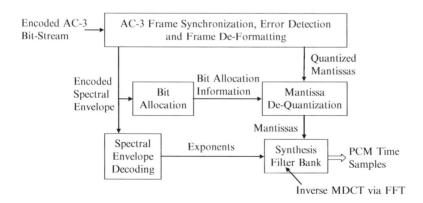

Fig. 8.27 The AC-3 (audio coder – 3) decoder (Dolby Labs). [D51] © 2006 IEEE

Table 8.1 Formulae and classification of CMFBs in audio coding standards. [D45] © 1999 AES

Class	MCT pair	CMFBs in standards
TDAC	$X_k = \sum\limits_{n=0}^{N-1} x_n \cos\left[\frac{\pi}{2N}\left(2n+1+\frac{N}{2}\right)(2k+1)\right]$ $x_n = \sum\limits_{k=0}^{N/2-1} X_k \cos\left[\frac{\pi}{2N}\left(2n+1+\frac{N}{2}\right)(2k+1)\right]$ for $k = 0, 1, \ldots, \frac{N}{2}-1$ and $n = 0, 1, \ldots, N-1$	MPEG-IV MPEG-II–AAC MPEG layer 3, second level AC-2 long transform
TDAC variant	$X_k = \sum\limits_{n=0}^{N-1} x_n \cos\left[\frac{\pi}{2N}(2n+1)(2k+1)\right]$ $x_n = \sum\limits_{k=0}^{N/2-1} X_k \cos\left[\frac{\pi}{2N}(2n+1)(2k+1)\right]$ for $k = 0, 1, \ldots, \frac{N}{2}-1$ and $n = 0, 1, \ldots, N-1$	AC-3 short transform 1
	$X_k = \sum\limits_{n=0}^{N-1} x_n \cos\left[\frac{\pi}{2N}(2n+1+N)(2k+1)\right]$ $x_n = \sum\limits_{k=0}^{N/2-1} X_k \cos\left[\frac{\pi}{2N}(2n+1+N)(2k+1)\right]$ for $k = 0, 1, \ldots, \frac{N}{2}-1$ and $n = 0, 1, \ldots, N-1$	AC-3 short transform 2
Polyphase filter bank	$X_k = \sum\limits_{n=0}^{N-1} x_n \cos\left[\frac{\pi}{N}\left(n-\frac{N}{4}\right)(2k+1)\right]$ $x_n = \sum\limits_{k=0}^{N/2-1} X_k \cos\left[\frac{\pi}{N}\left(n-\frac{N}{4}\right)(2k+1)\right]$ for $k = 0, 1, \ldots, \frac{N}{2}-1$ and $n = 0, 1, \ldots, N-1$	MPEG layers 1, 2 MPEG layer 3, first level

TDAC: Time domain aliasing cancellation
MCT: Modulated cosine transform
MLT: Modulated lapped transform
CMFB: Cosine modulated filter bank

are called as TDAC, MDCT and *cosine modulated filter banks* (CMFBs) (see Table 8.1). MPEG-1 audio layers 1–3 (Table 8.2, Figs. 8.28 and 8.29), MPEG-2 audio layers 1–4, MPEG-4 audio, MPEG-2 AAC audio[1] (Figs. 8.30 and 8.31) and Dolby AC-3 use the CMFBs to transform an audio sequence from time domain into subband or transform domain for compression (see [D37]).

The MDCT is the basic processing component in the international audio coding standards (MPEG series and H.262) [D37] and commercial audio products such as Sony MiniDisc/ATRAC/ATRAC2/SDDS digital audio coding systems (ATRAC: adaptive transform acoustic coding), AT&T Perceptual Audio Coder (PAC) or Lucent Technologies PAC/Enhanced PAC/Multichannel PAC for high quality compression (Tables 8.3 and 8.4) [D35].

The objective here is to focus on implementation of various revisions of MDCT and IMDCT (inverse MDCT) by FFT.

---

[1]AAC is used in MPEG-2 Part 7 and MPEG-4 Part 3.

Table 8.2 MPEG-1 layer specifications

	Layer I	Layer II	Layer III
Sampling frequency (kHz)	32, 44.1, 48	32, 44.1, 48	32, 44.1, 48
Minimum encoding/ decoding delay (ms)	19	35	59
Filter bank	MUSICAM filter bank (32 subbands)	MUSICAM filter bank (32 subbands)	MUSICAM filter bank and MDCT
Bandwidth of filter bank at 32 kHz sampling rate (Hz)	500	500	27.7 (assuming 18-point MDCT)
Psychoacoustics model	1 or 2	1 or 2	1 or 2 (frequency and temporal masking are used)
Masking threshold calculation	512 point FFT (coarse frequency resolution)	1,024 point FFT (finer frequency resolution)	1,024 point FFT psychoacoustic model 1; 1,024 and 256 for model 2
Bit allocation	Block of 12 samples from each of 32 subbands (= 384 input samples)	Block of 36 samples (three adjacent 12 sample blocks (= 3 × 384 = 1,152)	Adaptive block size to accommodate pre echo control
Quantization	Uniform	Uniform	Nonuniform
Entropy coding	No	No	Yes
Subjective test performance	Excellent at a stereo bit rate of 384 kbit/s	Excellent at a stereo bit rate of 256 kbit/s	An MOS increase of 0.6 over Layer II at a stereo bit rate of 128 kbit/s

MOS: Mean opinion score

The perceptual-based coder that exploits auditory masking is shown in Fig. 8.25. The amplitude resolution and hence the bit allocation and bit rate in each critical band is derived from the *signal-to-mask ratio* (SMR) versus the frequency. The SMR is determined for example from a 1,024 point FFT based spectral analysis of the audio block to be coded. Frequency domain coders with dynamic bit allocation to subbands or transform coefficients are described in detail in [D33].

The advanced television systems committee (ATSC) DTV standard includes digital high definition television (HDTV) and standard definition television (SDTV). The ATSC audio compression standard is the AC-3 (see [D51, D52]). Go to http://www.atsc.org/standards/to access and download the standards.

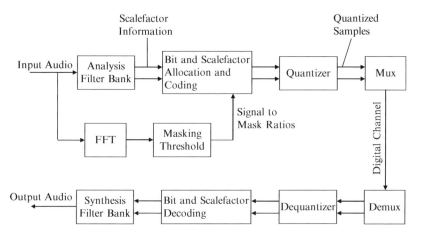

Fig. 8.28 Structure of MPEG-1 audio encoder and decoder (Layers I and II). [D33] ©1995 IEEE

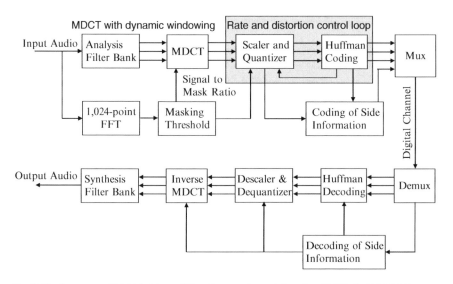

Fig. 8.29 Structure of MPEG-1 layer III audio encoder and decoder. [D33] © 1995 IEEE

Details of the gain control in Fig. 8.30 are shown in the encoder preprocessing module (Fig. 8.32). Details of the gain control in Fig. 8.31 are shown in the decoder postprocessing module (Fig. 8.33). Note that the 256 or 32 MDCT and IMDCT in Figs. 8.31 and 8.32 are implemented via FFT (Figs. 8.33 and 8.34).

8.15   Oddly Stacked TDAC

Odd stacked TDAC utilizes MDCT defined as [D23]

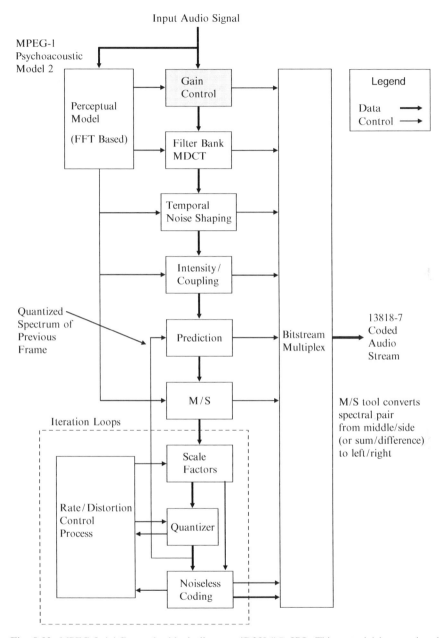

Fig. 8.30 MPEG-2 AAC encoder block diagram. [D39] "© ISO. This material is reproduced from ISO/IEC 13818-7:2006 with permission of the American National Standards Institute (ANSI) on behalf of the International Organization for Standardization (ISO). No part of this material may be copied or reproduced in any form, electronic retrieval system or otherwise or made available on the Internet, a public network, by satellite or otherwise without the prior written consent of the ANSI. Copies of this standard may be purchased from the ANSI, 25 West 43rd Street, New York, NY 10036(212) 642–4900, http://webstore.ansi.org"

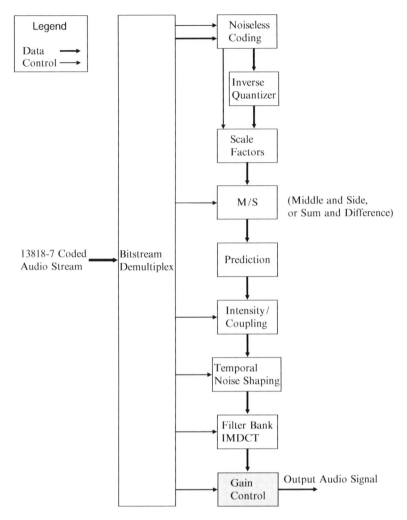

Fig. 8.31 MPEG-2 AAC decoder block diagram. [D39] "© ISO. This material is reproduced from ISO/IEC 13818-7:2006 with permission of the American National Standards Institute (ANSI) on behalf of the International Organization for Standardization (ISO). No part of this material may be copied or reproduced in any form, electronic retrieval system or otherwise or made available on the Internet, a public network, by satellite or otherwise without the prior written consent of the ANSI. Copies of this standard may be purchased from the ANSI, 25 West 43rd Street, New York, NY 10036 (212) 642–4900, http://webstore.ansi.org"

$$X^{\mathrm{MDCT}}(k) = \sum_{n=0}^{N-1} x(n) \cos\left[\frac{2\pi}{N}(n+n_0)\left(k+\frac{1}{2}\right)\right] \quad k = 0, 1, \ldots, N-1 \quad (8.35)$$

$x(n)$ is the quantized value of input signal $x(t)$ at sample n.

N = sample block length

$n_0 = \frac{(N/2)+1}{2}$ is the phase term required for aliasing cancellation.

$x(n) = n$th sample value

MDCT can be implemented via FFT as follows:

Table 8.3 Comparison of filter-bank properties. [D38] © 1997 AES

Feature	Layer 1	Layer 2	Layer 3	AC-2	AC-3	ATRAC[a]	PAC/ MPAC
Filter-bank type	PQMF	PQMF	Hybrid PQMF/ MDCT	MDCT/ MDST	MDCT	Hybrid QMF/ MDCT	MDCT
Frequency resolution at 48 kHz (Hz)	750	750	41.66	93.75	93.75	46.87	23.44
Time resolution at 48 kHz (ms)	0.66	0.66	4	1.3	2.66	1.3	2.66
Impulse response (LW)	512	512	1,664	512	512	1,024	2,048
Impulse response (SW)	–	–	896	128	256	128	256
Frame length at 48 kHz (ms)	8	24	24	32	32	10.66	23

[a]ATRAC is operating at a sampling frequency of 44.1 kHz. For comparison, the frame length and impulse response figures are given for an ATRAC system working at 48 kHz. (LW: Long window, SW: Short window)

Table 8.4 Comparison of currently available (as of 1997) audio coding systems (enc/dec: encoder/decoder, DCC: Digital compact cassette, MD: MiniDisc) [D38] © 1997 AES

	Bit rate	Quality	Complexity	Main applications	Available since
MPEG-1 layer 1	32–448 kb/s total	Good quality @ 192 kb/s/ch	Low enc/dec	DCC	1991
MPEG-1 layer 2	32–384 kb/s total	Good quality @ 128 kb/s/ch	Low decoder	DAB, CD-I, DVD	1991
MPEG-1 layer 3	32–320 kb/s total	Good quality @ 96 kb/s/ch	Low decoder	ISDN, satellite radio systems, Internet audio	1993
Dolby AC-2	128–192 kb/s/ch	Good quality @ 128 kb/s/ch	Low enc/dec	Point to point, cable	1989
Dolby AC-3	32–640 kb/s	Good quality @ 384 kb/s/ch	Low decoder	Point to multipoint, HDTV, cable, DVD	1991
Sony ATRAC	≈140 kb/s/ch		Low enc/dec	MD	1992
AT&T PAC			Low decoder		
APT-X100	Fixed compression 1:4		Very low enc/dec	Studio use	1989

1. Take FFT of $\left[x(n) \exp\left(-\dfrac{j\pi n}{N} \right) \right]$

$$= \hat{X}(k)$$
$$= R(k) + jQ(k) \qquad (8.36)$$

where $R(k)$ and $Q(k)$ are real and imaginary parts of $\hat{X}(k)$ respectively. Then

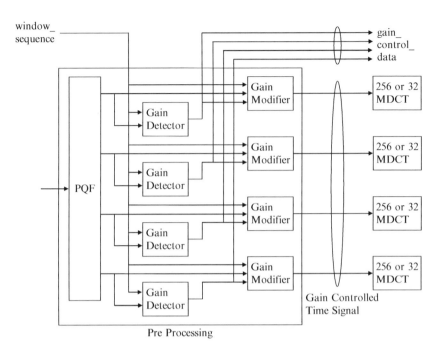

Fig. 8.32 Block diagram of the MPEG-2 AAC encoder preprocessing module. PQF: Polyphase quadrature filter. [D35] © 1997 AES

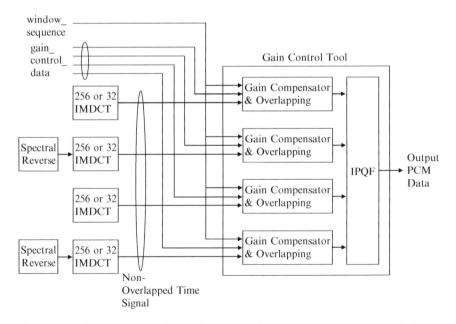

Fig. 8.33 Block diagram of the MPEG-2 AAC decoder postprocessing module. IPQF: Inverse polyphase quadrature filter. See also Fig. 8.34 for psychoacoustic model. [D35] © 1997 AES

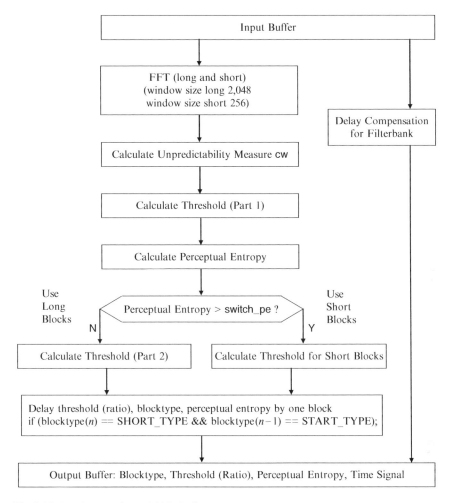

Fig. 8.34 Psychoacoustic model block diagram

$$X^{\mathrm{MDCT}}(k) = R(k) \cos\left[\frac{2\pi n_0}{N}\left(k + \frac{1}{2}\right)\right] + Q(k) \sin\left[\frac{2\pi n_0}{N}\left(k + \frac{1}{2}\right)\right] \qquad (8.37)$$

Proof:

$$\text{FFT of } \left[x(n) \exp\left(-\frac{j\pi n}{N}\right)\right]$$

$$= \sum_{n=0}^{N-1} x(n) \exp\left(-\frac{j\pi n}{N}\right) \exp\left(-\frac{j2\pi nk}{N}\right)$$

$$= \sum_{n=0}^{N-1} x(n) \exp\left[-\frac{j2\pi n}{N}\left(k + \frac{1}{2}\right)\right]$$

$$= \sum_{n=0}^{N-1} x(n) \cos\left[\frac{2\pi n}{N}\left(k + \frac{1}{2}\right)\right] - j \sum_{n=0}^{N-1} x(n) \sin\left[\frac{2\pi n}{N}\left(k + \frac{1}{2}\right)\right]$$

$$= R(k) + jQ(k) \qquad (8.38)$$

$$X^{\mathrm{MDCT}}(k) = R(k) \cos\left[\frac{2\pi n_0}{N}\left(k+\frac{1}{2}\right)\right] + Q(k) \sin\left[\frac{2\pi n_0}{N}\left(k+\frac{1}{2}\right)\right]$$

$$= \sum_{n=0}^{N-1} x(n)\left(\cos\left[\frac{2\pi n}{N}\left(k+\frac{1}{2}\right)\right]\cos\left[\frac{2\pi n_0}{N}\left(k+\frac{1}{2}\right)\right]\right.$$

$$\left. - \sin\left[\frac{2\pi n}{N}\left(k+\frac{1}{2}\right)\right]\sin\left[\frac{2\pi n_0}{N}\left(k+\frac{1}{2}\right)\right]\right)$$

$$= \sum_{n=0}^{N-1} x(n) \cos\left[\frac{2\pi}{N}(n+n_0)\left(k+\frac{1}{2}\right)\right] \tag{8.39}$$

which is (8.35). This is shown in block diagram format in Fig. 8.35.

IMDCT is defined as

$$x(n) = \frac{1}{N}\sum_{k=0}^{N-1} X^{\mathrm{MDCT}}(k) \cos\left[\frac{2\pi}{N}(n+n_0)\left(k+\frac{1}{2}\right)\right] \quad n=0,1,\ldots,N-1 \tag{8.40}$$

and can be implemented via IFFT as follows. Let

$$\hat{x}(n) = \mathrm{IFFT}\left[X^{\mathrm{MDCT}}(k) \exp\left(\frac{j2\pi k n_0}{N}\right)\right] \quad n=0,1,\ldots,N-1 \tag{8.41}$$

$$x(n) = r(n)\cos\left[\frac{\pi(n+n_0)}{N}\right] - q(n)\sin\left[\frac{\pi(n+n_0)}{N}\right] \tag{8.42}$$

where $r(n)$ and $q(n)$ are real and imaginary parts of $\hat{x}(n)$, respectively.

MDCT has been utilized in AC-3, AAC, MPEG-1 layer 3 second level, MPEG-4 audio and HDTV of ATSC [D16, D18, D20, D31, D32, D38, D39, D40, D43, D45, D47, D48, D49, D50]. Apart from the FFT implementation of the MDCT, various fast algorithms for efficient implementation of MDCT/MDST and their inverses also have been developed [D45, D48, D49]. An integer version of the MDCT called

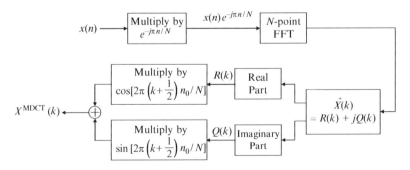

Fig. 8.35 Block diagram for implementation of oddly stacked TDAC via FFT

IntMDCT has been developed in [D47]. The later preserves most of the attractive properties of the MDCT. It provides perfect reconstruction, block overlapping, critical sampling, good frequency selectivity and fast algorithm. Added advantage is its application in lossless audio coding [D47].

See Fig. 11.15, the corresponding MATLAB simulation, and Fig. 11.18 in [M13] for waveform coding using the DCT and MDCT.

8.16 Perceptual Transform Audio Coder [D3, D4]

An estimate of the *perceptual entropy* (PE) of the audio signals is created from the combinations of several well known noise masking measures. These measures are combined with a heuristic method of tonality estimation of the short term frequency masking templates for audio stimuli. The perceptual entropy of each short term section of the audio stimuli is estimated as the number of bits required to encode the short term spectrum of the audio signal to the resolution required to inject noise at the masking template level. Details on perceptual entropy computation are shown in Fig. 8.36. Windowing and frequency transformation are implemented by a Hanning window followed by a real-complex FFT of length 2,048, retaining the first 1,024 transform coefficients (dc and the coefficient at $f_s/2$ are counted as one). Also critical band analysis is used in calculating the masking threshold. From the FFT the power spectrum is observed as $P(\omega) = \left[\mathrm{Re}^2(\omega) + \mathrm{Im}^2(\omega)\right]$ where $\mathrm{Re}(\omega) + j\,\mathrm{Im}(\omega)$ is the DFT coefficient. By summing the power spectra over various frequency bands, critical bands are obtained.

Details on suggesting that the PE measurement may well estimate a limit for transparent bit rate reduction for audio signals presented to the human ear are described in [D3, D4].

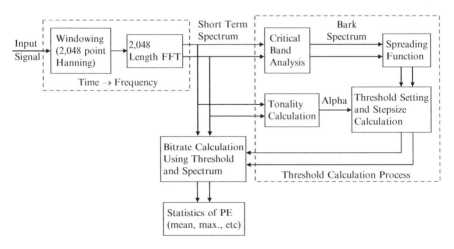

Fig. 8.36 Perceptual entropy (PE) calculation. [D4] © 1988 IEEE

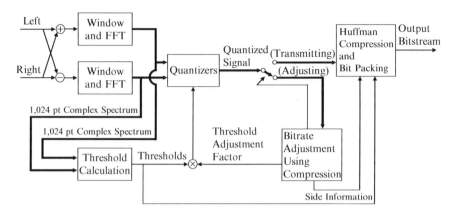

Fig. 8.37 Block diagram of SEPXFM coder (stereo-entropy-coded perceptual transform coder). [D6] © 1989 IEEE

The entropy-coded perceptual transform coder developed originally for mono-phonic signals [D3, D4], is extended to stereo signals [D6] (Fig. 8.37). The later case exploits both the redundancy in the stereo signal and the effects of acoustic mixing in the listening environment resulting in encoding the stereo signal at a bit rate much less than twice that for a monophonic signal.

8.17   OCF Coder

Optimum coding in the frequency domain (OCF) coder [D10] uses entropy coding of the spectral values for increased coding efficiency and flexibility of the coder. Both low-complexity adaptive transform coding (LC-ATC) and OCF use transform coding to get high redundancy removal and high adaptation to the perceptual requirements. Block diagram of the OCF coder is shown in Fig. 8.38. Decoder is shown in Fig. 8.39. The input signal is windowed and transformed via MDCT (modified DCT) which is used as a critically sampled filter bank. MDCT is implemented via the FFT. Inverse transformation (IMDCT) at the OCF-decoder is implemented via the IFFT.

8.18   NMR Measurement System

Another application of the FFT is in evaluating the audibility of quantization noise from an objective measurement based on noise-to-mask ratio (NMR) and masking flag [D10]. They use Hanning-windowed FFTs of 1,024 samples calculated every 512 samples (11.6 ms, sampling rate = 44.1 kHz). Block diagram of the NMR measurement system is shown in Fig. 8.40.

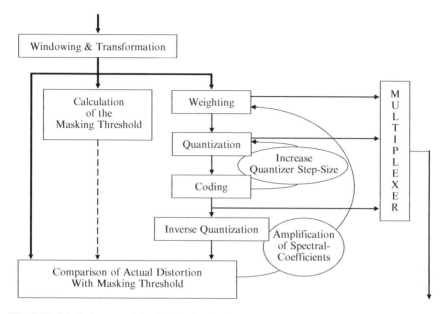

Fig. 8.38 Block diagram of the OCF coder. [D10] © 1990 IEEE

Fig. 8.39 Block diagram of the OCF decoder. [D10] © 1990 IEEE

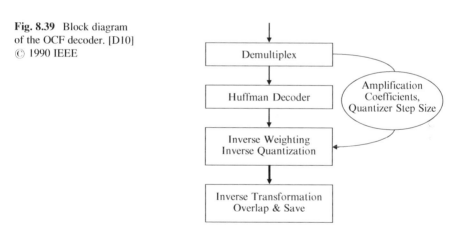

8.19 Audio Coder for Mobile Reception

A subband audio coding system working at a sampling rate of 48 kHz and on a COFDM (coded orthogonal FDM) broadcasting system has been developed by CCETT [D7]. This system addresses sound program production and bit rate reduction as well as channel coding and modulation. Block diagrams of the optimized subband coder and decoder are shown in Fig. 8.41.

The subband coder for a stereophonic program was implemented using DSPs on Eurocard boards compatible with a VME bus.

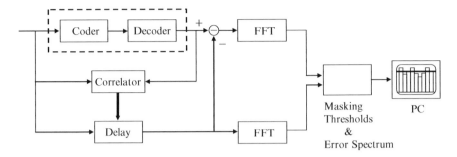

Fig. 8.40 Block diagram of the NMR measurement system. [D10] © 1990 IEEE

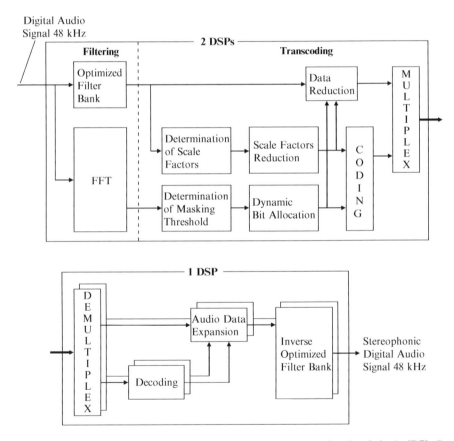

Fig. 8.41 Optimized stereophonic subband encoder (above) and decoder (below). [D7] © 1989 ITU

8.20   ASPEC (Adaptive Spectral Perceptual Entropy Coding of High Quality Music Signals)

As indicated in MUSICAM (masking pattern universal subband integrated coding and multiplexing) (see Section 8.23), ASPEC [D9, D17] is the other audio coding which was selected for extensive audio tests by ISO for possible application in MPEG audio. While MUSICAM (subband algorithm) was selected basically as MPEG audio, the psychoacoustic model of ASPEC was merged into the MPEG audio coding standard. ASPEC merges ideas from other high quality sound coding systems proposed by University of Erlangen/Fraunhofer Society, AT&T Bell Labs (two coders), University of Hannover/Thomson Consumer Electronics, and CNET [D17]. ASPEC has also met all the system requirements of the ISO proposal package description. Block diagrams of the single channel encoder (Fig. 8.42) and decoder (Fig. 8.43) are shown. Filter bank is implemented via the modified discrete cosine transform (MDCT). MDCT transforms $2n$ time domain samples to n frequency domain transform coefficients (downsampling) with a block size of n samples in time domain. Each sample is a member of two blocks. IMDCT (inverse MDCT) maps the n transform coefficients into $2n$ time domain samples. An overlap and add operation cancels the aliasing terms produced by the downsampling (TDAC).

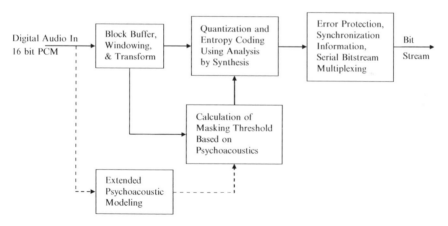

Fig. 8.42   Single-channel ASPEC encoder block diagram. [D17] © 1991 AES

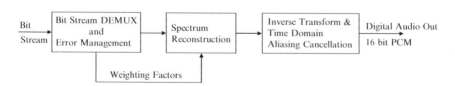

Fig. 8.43   Single-channel ASPEC decoder block diagram. [D17] © 1991 AES

MDCT

$$X_b(m) = \sum_{k=0}^{2n-1} f(k)x_b(k) \, \cos\left[\frac{\pi}{4n}(2k+1+n)(2m+1)\right]$$

(8.43)

$$m = 0, \, 1, \, \ldots, \, n-1$$

where $x_b(k)$ is the kth sample of block b with $x_b(k+n) = x_{b+1}(k)$ $k = 0, 1, \ldots,$ $n-1, f(k)$ $(k = 0, 1, \ldots, 2n-1)$ is the window function and $X_b(m)$ $(m = 0, 1, \ldots, n-1)$ is the mth transform coefficient. One of the window functions is

$$f(k) = \sin\left[\frac{\pi(2k+1)}{4n}\right] \qquad k = 0, \, 1, \, \ldots, \, 2n-1$$

(8.44)

The IMDCT is

$$y_b(p) = f(p) \sum_{m=0}^{n-1} X_b(m) \cos\left[\frac{\pi}{4n}(2p+1+n)(2m+1)\right]$$

(8.45)

$$p = 0, \, 1, \, \ldots, \, 2n-1$$

and

$$x_b(q) = y_{b-1}(q+n) + y_b(q) \quad q = 0, \, 1, \, \ldots, \, n-1$$

(8.46)

Among the several fast algorithms that have been developed for implementing MDCT and its inverse, one efficient algorithm uses FFT. The psychoacoustic model of the human auditory system for calculating the masking thresholds for bit allocation of the transform coefficients in the subbands is based on the FFT [D7].

8.21 RELP Vocoder (RELP: Residual Excited Linear Prediction)

Another application of FFT is in the RELP vocoder (Fig. 8.44). Details of this vocoder are described in [D30].

8.22 Homomorphic Vocoders

Homomorphic signal processing methods, such as homomorphic deconvolution, can be used both for vocal-tract characterization, as well as, for extraction of information relating to excitation [D30]. The basic idea in homomorphic vocoders is that the vocal tract and the excitation log-magnitude spectra can be combined additively to produce the speech log-magnitude spectrum.

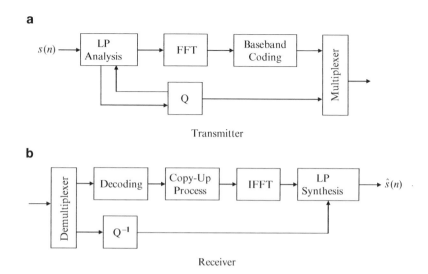

Fig. 8.44   FFT based RELP vocoder [D30]. © 1994 IEEE

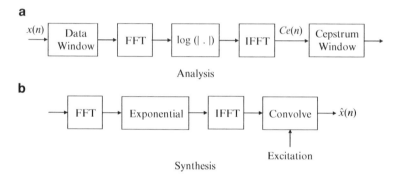

Fig. 8.45   A homomorphic speech analysis-synthesis system. [D30] © 1994 IEEE

A speech analysis-synthesis system that uses the cepstrum is shown in Fig. 8.45. The inverse Fourier transform of the log-magnitude spectrum of speech produces the cepstrum sequence $Ce(n)$. It can be shown that the ("que-frency") samples of the cepstrum that are near the origin are associated with the vocal tract. These coefficients can be extracted using a cepstral window. The length of the cepstral window must generally be shorter than the shortest possible pitch period. It can be also shown that for voiced speech the cepstral sequence has large samples at the pitch period. Therefore, the fundamental frequency can be estimated from the cepstrum.

The synthesizer takes the FFT of the cepstrum and the resulting frequency components are exponentiated. The IFFT of these components gives the impulse response of the vocal tract which is convolved with the excitation to produce synthetic speech. Although the cepstral vocoder did not find many applications at the time it was proposed, cepstrum-based methods for pitch and vocal-tract estimation found many other speech-processing applications. In addition, in contributions by Chung and Schafer [D5, D12] it was reported that good-quality speech at 4.8 kbits/s can be produced by combining homomorphic deconvolution with analysis-by-synthesis excitation modeling [D30].

8.23   MUSICAM (Masking-Pattern Universal Sub-Band Integrated Coding and Multiplexing)

In response to ISO, 14 companies have submitted proposals for developing a digital audio coding standard. Because of similarities between these coding proposals, a clustering into four development groups has been arranged. Finally two coding algorithms (MUSICAM – subband coder and ASPEC – transform coder, ASPEC is described in Section 8.20) have been tested extensively in Swedish Broadcasting Corpn., Stockholm. Based on the 11 performances and corresponding weighting factors (besides meeting the system requirements of the ISO proposal package description [D8]) MUSICAM achieved a higher score of nearly 6% over ASPEC (adaptive spectral entropy coding of high quality music signals) (Table 8.5) [D15].

MPEG audio [D50] resulted from collaboration of the two groups (MUSICAM and ASPEC) by combining the most efficient components of the two algorithms into one standard algorithm.

The sub-band coding scheme MUSICAM developed by French, Dutch and German engineers [D11] is based on properties of the human sound perception, i. e., temporal and spectral masking effects of the ear (Fig. 8.46).

Filter bank yields subbands of the audio signal (useful for temporal masking effects). FFT is carried out in parallel to the filtering of the audio signal for the calculation of the dynamic bit allocation to the subbands. By combining the high frequency resolution of the FFT and the high time resolution of the scale factors, human ears masking thresholds can be accurately estimated both in time and frequency.

8.24   AC-2 Audio Coder

Dolby Labs developed the digital audio coder called *Dolby AC-2* at sampling rates of 32, 44.1 and 48 kHz for both mono and stereo at compression ratios of 5.4:1, 5.6:1 and 6.1:1 respectively for 16 bit PCM [D13, D14, D19, D21, D24].

Table 8.5 Scoring of subjective and objective tests. [D15] © 1990 IEEE	Algorithm	ASPEC	MUSICAM
	Subjective tests	3,272	2,942
	Objective tests	4,557	5,408
	Total	7,829	8,350

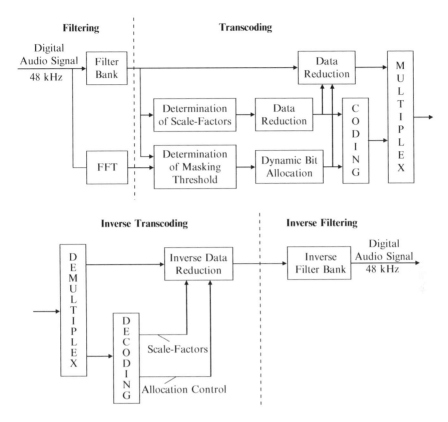

Fig. 8.46 Block diagram of the MUSICAM encoder (above) and decoder (below) [D11]. © 1990 IEEE

A windowed overlap-add process is used in the analysis/synthesis filter bank and has been efficiently implemented via the FFT. Encoder/decoder block diagrams are shown in Figs. 8.47 and 8.48 respectively. As mentioned earlier, TDAC and subband decomposition (critical bands) are implemented via alternating MDCT/ MDST on overlapping blocks. This is the evenly stacked TDAC.

MDCT/MDST are defined as

$$X^C(k) = \frac{1}{N} \sum_{n=0}^{N-1} x(n) \cos\left[\frac{2\pi k}{N}(n+n_0)\right] \qquad k = 0, 1, \ldots, N-1 \qquad (8.47)$$

and

$$X^S(k) = \frac{1}{N} \sum_{n=0}^{N-1} x(n) \sin\left[\frac{2\pi k}{N}(n+n_0)\right] \qquad k = 0, 1, \ldots, N-1 \qquad (8.48)$$

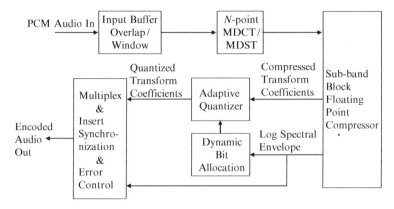

Fig. 8.47 AC-2 digital audio encoder family block diagram. [D21] © 1992 IEEE

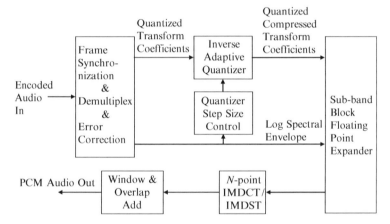

Fig. 8.48 AC-2 digital audio decoder family block diagram. [D21] © 1992 IEEE

where $n_0 = \left(\frac{N}{2}+1\right)\big/2$. In the TDAC each new N-sample block overlaps the previous block by 50% of the block length. Critical sampling of the transform filter bank is achieved as each MDCT or MDST generates only $N/2$ unique nonzero transform coefficients.

The IMDCT/IMDST convert these transform coefficients into N aliased time domain audio samples.

$$\hat{x}^C(n) = \sum_{k=0}^{N-1} X^C(k) \cos\left[\frac{2\pi k}{N}(n+n_0)\right] \quad n = 0, 1, \ldots, N-1 \tag{8.49}$$

$$\hat{x}^S(n) = \sum_{k=0}^{N-1} X^S(k) \sin\left[\frac{2\pi k}{N}(n+n_0)\right] \quad n = 0, 1, \ldots, N-1 \tag{8.50}$$

FFT techniques have been applied efficiently in implementing MDCT/MDST and their inverses. By considering MDCT and MDST as real and imaginary components of a single complex FFT, a single FFT can be used to implement both the transforms. These are described in detail in Sections 8.25 and 8.26.

8.25 IMDCT/IMDST Implementation via IFFT

The N aliased time domain samples $\hat{x}^C(n)$ and $\hat{x}^S(n)$ result from IMDCT/IMDST which are implemented via IFFT as follows:

$$
\begin{aligned}
\hat{x}^C(n) &= \text{Re} \sum_{k=0}^{N-1} X^C(k) \exp\left[\frac{j2\pi k}{N}(n+n_0) \right] \\
&= \text{Re}\left(\sum_{k=0}^{N-1} \left[X^C(k) \exp\left(\frac{j2\pi k n_0}{N} \right) \right] \left[\exp\left(\frac{j2\pi k n}{N} \right) \right] \right)
\end{aligned}
\tag{8.51}
$$

which is the real part of the IFFT of $\left[N X^C(k) \exp\left(\frac{j2\pi k n_0}{N}\right) \right]$. Similarly the IMDST is the imaginary part of the IFFT of $\left[N X^S(k) \exp\left(\frac{j2\pi k n_0}{N}\right) \right]$, or

$$
\hat{x}^S(n) = \text{Im}\left(\text{IFFT}\left[N X^S(k) \exp\left(\frac{j2\pi k n_0}{N} \right) \right] \right)
\tag{8.52}
$$

where $n_0 = \left(\frac{N}{2}+1\right)/2$, $X^C(k)$. and $X^S(k)$ are MDCT and MDST of $x(n)$ respectively (see (8.47) and (8.48)). As $\left[X^C(k) \exp\left(\frac{j2\pi k n_0}{N}\right) \right]$ is conjugate symmetric,[2] the imaginary part of the IFFT in (8.51) is zero. Similarly as $\left[X^S(k) \exp\left(\frac{j2\pi k n_0}{N}\right) \right]$ is conjugate antisymmetric,[3] the real part of the IFFT in (8.52) is zero. As the two IFFT outputs in (8.51) and (8.52) are purely real and imaginary respectively, the two N-point IFFTs can be combined into a single N-point complex output IFFT which reduces the multiply-add count of computing (8.51) and (8.52) respectively by a factor of two. This is shown in block diagram format in Fig. 8.49.

Proof: For $N = 8$

$$
X^C(N-k) = -X^C(k) \quad k = 1, 2, \ldots, \frac{N}{2}-1
\tag{8.53a}
$$

$$
X^C(N/2) = 0
\tag{8.53b}
$$

---

[2]The conjugate symmetric sequence for $N = 8$ is defined as
$\{X^F(0), X^F(1), \ldots, X^F(7)\} = \{X^F(0), X^F(1), X^F(2), X^F(3), X^F(4), [X^F(3)]^*, [X^F(2)]^*, [X^F(1)]^*\}$
[3]The conjugate antisymmetric sequence for $N = 8$ is defined as
$\{X^F(0), X^F(1), \ldots, X^F(7)\} = \{X^F(0), X^F(1), X^F(2), X^F(3), X^F(4), -[X^F(3)]^*, -[X^F(2)]^*, -[X^F(1)]^*\}$

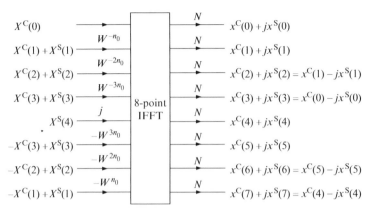

Fig. 8.49 IMDCT/IMDST implementation via IFFT for $N = 8$, $n_0 = \left(\frac{N}{2} + 1\right)/2$ and $W = e^{-j2\pi/N}$ (here $W = W_8$)

$$\left\{\exp\left(\frac{j2\pi k n_0}{N}\right)\right\}_{k=0}^{N-1} = (1, d_1, d_2, d_3, j, -d_3^*, -d_2^*, -d_1^*) \qquad (8.54)$$

From (8.53) and (8.54)

$$\left[X^C(k)\exp\left(\frac{j2\pi k n_0}{N}\right)\right]_{k=0}^{N-1} = \{X^C(0), X^C(1)d_1, X^C(2)d_2, X^C(3)d_3,$$
$$0, X^C(3)d_3^*, X^C(2)d_2^*, X^C(1)d_1^*\} \qquad (8.55)$$

where $X^C(k)$ is always real. Thus $\left[X^C(k)\exp\left(\frac{j2\pi k n_0}{N}\right)\right]$ is conjugate symmetric.

$$\hat{x}^C(n) = \mathrm{Re} \sum_{k=0}^{N-1} \left[X^C(k)\exp(j2\pi k n_0/N)\right] W_N^{-kn} \qquad (8.51a)$$

$$\begin{aligned}
\hat{x}^C(n) &= \mathrm{Re}\big[X^C(0) + X^C(1)d_1 e^{j2n\pi/N} + X^C(2)d_2 e^{j4n\pi/N} + X^C(3)d_3 e^{j6n\pi/N} \\
&\quad + X^C(3)d_3^* e^{j10n\pi/N} + X^C(2)d_2^* e^{j12n\pi/N} + X^C(1)d_1^* e^{j14n\pi/N}\big] \\
&= \mathrm{Re}\big[X^C(0) + X^C(1)d_1 e^{j2n\pi/N} + X^C(2)d_2 e^{j4n\pi/N} + X^C(3)d_3 e^{j6n\pi/N} \\
&\quad + X^C(3)d_3^* e^{-j6n\pi/N} + X^C(2)d_2^* e^{-j4n\pi/N} + X^C(1)d_1^* e^{-j2n\pi/N}\big] \\
&= \mathrm{Re}\big[X^C(0) + (c_1 + c_1^*) + (c_2 + c_2^*) + (c_3 + c_3^*)\big]
\end{aligned} \qquad (8.56)$$

where $c_k = X^C(k)\, d_k\, e^{j2\pi nk/N}$, $k = 1, 2, \ldots, \frac{N}{2} - 1$ and n is an integer. Sum of the terms inside the square brackets of (8.56) is real.

Similarly for MDST

$$X^S(0) = 0 \qquad (8.57a)$$

$$X^S(N-k) = X^S(k) \quad k = 1, 2, \ldots, \tfrac{N}{2} - 1 \tag{8.57b}$$

From (8.57) and (8.54)

$$\left[X^S(k)\exp\left(\frac{j2\pi kn_0}{N}\right)\right]_{k=0}^{N-1} = \{0, X^S(1)d_1, X^S(2)d_2, X^S(3)d_3, \tag{8.58}$$
$$jX^S(4), -X^S(3)d_3^*, -X^S(2)d_2^*, -X^S(1)d_1^*\}$$

Thus $\left[X^S(k)\exp\left(\frac{j2\pi kn_0}{N}\right)\right]$ is conjugate antisymmetric.

$$\hat{x}^S(n) = \mathrm{Im}\sum_{k=0}^{N-1}\left[X^S(k)\exp(j2\pi kn_0/N)\right]W_N^{-kn} \tag{8.59}$$

$$\begin{aligned}
\hat{x}^S(n) &= \mathrm{Im}\big[X^S(1)d_1e^{j2n\pi/N} + X^S(2)d_2e^{j4n\pi/N} + X^S(3)d_3e^{j6n\pi/N} \\
&\quad + jX^S(4) - X^S(3)d_3^*e^{j10n\pi/N} - X^S(2)d_2^*e^{j12n\pi/N} - X^S(1)d_1^*e^{j14n\pi/N}\big] \\
&= \mathrm{Im}\big[X^S(1)d_1e^{j2n\pi/N} + X^S(2)d_2e^{j4n\pi/N} + X^S(3)d_3e^{j6n\pi/N} \\
&\quad + jX^S(4) - X^S(3)d_3^*e^{-j6n\pi/N} - X^S(2)d_2^*e^{-j4n\pi/N} - X^S(1)d_1^*e^{-j2n\pi/N}\big] \\
&= \mathrm{Im}\big[(c_1 - c_1^*) + (c_2 - c_2^*) + (c_3 - c_3^*) + jX^S(4)\big]
\end{aligned} \tag{8.60}$$

where $c_k = X^S(k)d_ke^{j2\pi kn/N}, k = 1, 2, \ldots, \tfrac{N}{2} - 1$ and n is an integer. Sum of the terms inside the square brackets of (8.60) is imaginary.

The complex conjugate theorem is shown in (2.14). Similarly when $x(n)$ is an imaginary sequence, the N-point DFT is conjugate antisymmetric as

$$X^F\left(\tfrac{N}{2} + k\right) = -X^{F^*}\left(\tfrac{N}{2} - k\right), \quad k = 0, 1, \ldots, \tfrac{N}{2} \tag{8.61}$$

This implies that both $X^F(0)$ and $X^F(N/2)$ are imaginary. Given $x(n) \Leftrightarrow X^F(k)$ then $x^*(n) \Leftrightarrow X^{F^*}(-k)$. The DFT symmetry properties are summarized as follows.

Data domain $\Leftrightarrow$ DFT domain

Real sequence $\Leftrightarrow$ Conjugate symmetric   [see (2.14) and (8.56)]

$$\frac{1}{2}[x(n) + x^*(n)] \quad \Leftrightarrow \quad \frac{1}{2}\left[X^F(k) + X^{F^*}(-k)\right] \quad \text{from linearity} \tag{8.62}$$

where $x(n)$ is a complex number.

Imaginary sequence $\Leftrightarrow$ Conjugate antisymmetric   [see (8.60) and (8.61)]

$$\frac{1}{2}[x(n) - x^*(n)] \quad \Leftrightarrow \quad \frac{1}{2}\left[X^F(k) - X^{F^*}(-k)\right] \tag{8.63}$$

where $x(n)$ is a complex number.

From the duality of the DFT relations

Conjugate symmetric $\Leftrightarrow$ Real

$$\frac{1}{2}[x(n) - x^*(N - n)] \qquad \Leftrightarrow \qquad \frac{1}{2}\left[X^F(k) + X^{F^*}(k)\right] \qquad (8.64)$$

e.g. (0, 1, 2, 3, 2, 1) $\qquad\qquad$ (9, $-$ 4, 0, $-$ 1, 0, $-$ 4)

Conjugate antisymmetric $\Leftrightarrow$ Imaginary

$$\frac{1}{2}[x(n) - x^*(N - n)] \qquad \Leftrightarrow \qquad \frac{1}{2}\left[X^F(k) - X^{F^*}(k)\right] \qquad (8.65)$$

Thus

Symmetric real $\Leftrightarrow$ Symmetric real

$$x\left(\frac{N}{2} + n\right) = x\left(\frac{N}{2} - n\right) \qquad \text{or} \qquad x(n) = x(N - n) \quad n = 0, 1, \ldots, \frac{N}{2} \quad (8.66)$$

e.g. $\underline{x} = (3, 4, 2, 1, 2, 4)^T$

Symmetric imaginary $\Leftrightarrow$ Symmetric imaginary $\qquad\qquad\qquad$ (8.67)

e.g. $(j3, j, j2, j4, j2, j)$ $\qquad\qquad$ $(j13, -j2, j4, j, j4, -j2)$

Symmetric complex $\Leftrightarrow$ Symmetric complex

(from linear combination of (8.66) and (8.67))

Assume $x(0) = x\left(\frac{N}{2}\right) = 0$. Then more antisymmetry properties of the DFT and its inverse are as follows ([IP19], p. 242 and [B23], p. 49).

Antisymmetric real $\Leftrightarrow$ Antisymmetric imaginary $\qquad\qquad$ (8.68)

e.g. (0, 1, 0, -1) $\qquad\qquad$ (0, $-j2$, 0, $j2$)

Antisymmetric imaginary $\Leftrightarrow$ Antisymmetric real $\qquad\qquad$ (8.69)

e.g. (0, j, 0, $-j$) $\qquad\qquad\qquad$ (0, 2, 0, -2)

Antisymmetric complex $\Leftrightarrow$ Antisymmetric complex $\qquad\qquad$ (8.70)

(from linear combination of (8.68) and (8.69))

e.g. (0, 1$+ j$, 0, $- (1+ j)$) $\qquad$ (0, 2$-j2$, 0, $- (2- j2)$)

Note $X^F(0) = X^F\left(\frac{N}{2}\right) = 0$ for (8.68)–(8.70).

8.26   MDCT/MDST Implementation via IFFT

By considering MDCT and MDST as real and imaginary components of a single complex IFFT, a single IFFT can be used to implement both the transforms. From the definitions of MDCT/MDST in (8.47) and (8.48), respectively

$$
\begin{aligned}
X^C(k) + jX^S(k) &= \frac{1}{N}\sum_{n=0}^{N-1} x(n)\left(\cos\left[\frac{2\pi k}{N}(n+n_0)\right] + j\sin\left[\frac{2\pi k}{N}(n+n_0)\right]\right) \\
&= \frac{1}{N}\sum_{n=0}^{N-1} x(n)\exp\left[\frac{j2\pi k}{N}(n+n_0)\right] \\
&= W_N^{-k\,n_0}\frac{1}{N}\sum_{n=0}^{N-1} x(n)\,W_N^{-k\,n} \\
&= W_N^{-k\,n_0}\,\mathrm{IFFT}\left[x(n)\right]
\end{aligned}
\tag{8.71}
$$

where $n_0 = \left(\frac{N}{2}+1\right)/2$. This is shown in block diagram format in Fig. 8.50, for $N=8$.

Summarizing, FFT and its inverse have been extensively utilized in implementing MDCT/MDST as filter banks and in developing psychoacoustic models for audio coders that have been adapted by international standards bodies. Other applications include RELP vocoder/homomorphic vocoder, OCF coder, perceptual transform audio coder, NMR measurement, etc.

8.27   Autocorrelation Function and Power Density Spectrum

In this section, we compute the autocorrelation and the power density spectrum of a sequence of random variables.

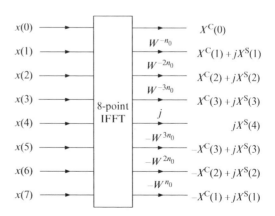

Fig. 8.50 MDCT/MDST implementation via IFFT for $N=8$, $n_0 = \left(\frac{N}{2}+1\right)/2$ and $W = e^{-j2\pi/N}$ (Here $W = W_8$)

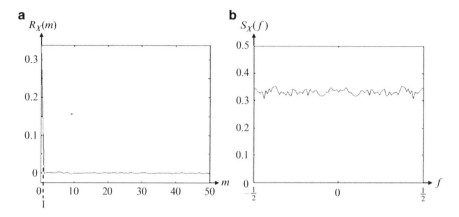

Fig. 8.51 a The autocorrelation and **b** the power density spectrum of a sequence of random variables. Assume $f_s = 1$

1. Generate a discrete-time sequence $\{x_n\}$ of $N = 2,000$ statistically independent and identically distributed random variables, which are selected from a uniform distribution over the interval $(-1, 1)$. The unbiased estimate of the autocorrelation of the sequence $\{x_n\}$ is defined as

$$R_X(m) = \left(\frac{1}{N-m}\right) \sum_{n=1}^{N-m} x_n x_{n+m} \quad m = 0, 1, \ldots, M$$

$$= \left(\frac{1}{N-|m|}\right) \sum_{n=|m|}^{N} x_n x_{n+m} \quad m = -1, -2, \ldots, -M \quad (8.72)$$

where $M = 50$. Compute $R_X(m)$ and plot it (see Fig. 8.51a).

2. Determine and plot the power density spectrum of the sequence $\{x_n\}$ by computing the DFT of $R_X(m)$ (Fig. 8.51b), which is defined as

$$S_X(f) = S_X^F(k) = (2M+1)\text{-point DFT}\,[R_X(m)] = \sum_{m=-M}^{M} R_X(m) \exp\left[\frac{-j2\pi mk}{(2m+1)}\right]$$

$$f_s = 1 \quad \tfrac{1}{2} \leq f \leq \tfrac{1}{2} \quad -M \leq k, m \leq M \quad (8.73)$$

DFT is implemented via the FFT.

8.27.1 Filtered White Noise

A white random process $x(t)$ with the power density spectrum $S_X(f) = 1$ for all f excites a linear filter with the impulse response (Fig. 8.52)

Fig. 8.52 Filtered white
noise

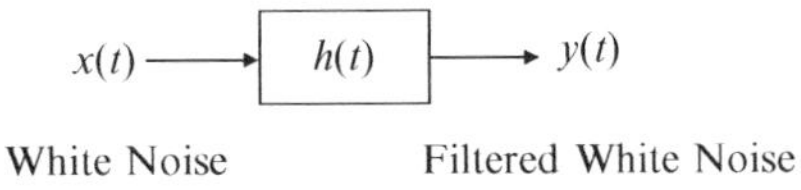

$$x(t) \longrightarrow \boxed{h(t)} \longrightarrow y(t)$$

White Noise   Filtered White Noise

$$h(t) = \begin{cases} e^{-t/4}, & t \geqslant 0 \\ 0 & \text{otherwise} \end{cases} \tag{8.74}$$

1. Determine and plot the power density spectrum $S_Y(f)$ of the filter output.

$$H(f) = \int_{-\infty}^{\infty} \left(e^{-t/4} \right) e^{-j2\pi ft} dt = \int_{0}^{\infty} e^{-\left(\frac{1}{4} + j2\pi f \right)t} dt = \frac{1}{\frac{1}{4} + j2\pi f} \tag{8.75}$$

$$f_s = 1, \quad -\frac{1}{2} \leqslant f \leqslant \frac{1}{2}$$

$$S_Y(f) = S_X(f)H(f)H^*(f) = S_X(f)|H(f)|^2 = \frac{1}{\frac{1}{16} + (2\pi f)^2} \tag{8.76}$$

Thus $S_Y(f) = 16$ for $f = 0$.

2. By using the inverse FFT on samples of $S_Y(f)$, compute and plot the autocorrelation function of the filter output $y(t)$ (see Fig. 8.53).

Approximation errors between the Fourier transform and DFT are shown in Fig 8.53. However, if we let $h(t) = e^{-t/40}$ for $t \geqslant 0$, the two transforms show similar results.

8.28   Three-Dimensional Face Recognition

In [LA7] various projection-based features using DFT or DCT are applied on registered 3D scans of faces for face recognition. Feature extraction techniques are applied to three different representations of registered faces i.e., 3D point clouds, 2D depth images and 3D voxel (volume element) (Table 8.6).

Using 3D-RMA face database [LA8], Dutagaci, Sankur and Yemez [LA7] have conducted the recognition performance of the various schemes (Table 8.7). The database contains face scans of 106 subjects. Details on training and test data and number of sessions per person (also suggestions for improving the recognition performance) are described in [LA7].

New precise measurement method of electric power harmonics based on FFT is described by Wu and Zhao [LA9].

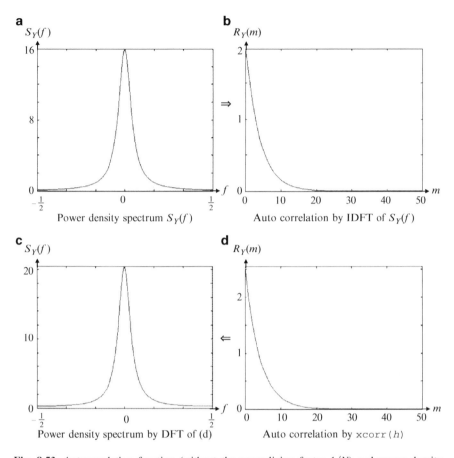

Fig. 8.53 Autocorrelation function (without the normalizing factor $1/N$) and power density spectrum. The more slowly $R(m)$ goes down, the sharper $S_Y(f)$ gets. Assume $f_s = 1$. **a** Power density spectrum $S_Y(f)$. **b** Auto correlation by IDFT of $S_Y(f)$. **c** Power density spectrum by DFT of (**d**). **d** Auto correlation by `xcorr(h)`

Table 8.6 Representation schemes and features used for 3D face recognition. [LA7] © 2006 SPIE

Representation	Features
3D Point Cloud	▪ 2 D DFT (Discrete Fourier Transform)
	▪ ICA (Independent Component Analysis)
	▪ NNMF (Nonnegative Matrix Factorization)
2D Depth Image	▪ Global DFT
	▪ Global DCT
	▪ Block-based DFT (Fusion at feature level)
	▪ Block-based DCT (Fusion at feature level)
	▪ Block-based DFT (Fusion at decision level)
	▪ Block-based DCT (Fusion at decision level)
	▪ ICA (Independent Component Analysis)
	▪ NNMF (Nonnegative Matrix Factorization)
3D Voxel Representation	▪ 3D DFT (Discrete Fourier Transform)

Table 8.7 Recognition performances and number of features. [LA7] © 2006 SPIE

Representation	Features	Number of features	Recognition accuracy in %
3D Point Cloud	2D DFT	$2 \times 400 - 1 = (799)$	95.86
	ICA	50	99.79
	NNMF	50	99.79
2D Depth Image	Global DFT	$2 \times 8 \times 8 - 1 = (127)$	98.24
	Global DCT	$11 \times 11 = (121)$	96.58
	Block-based DFT (Fusion at feature level)	20×20 blocks (12 blocks) $2 \times 2 \times 2 - 1$ for each block (84)	98.76
	Block-based DCT (Fusion at feature level)	20×20 blocks (12 blocks) 3×3 for each block (108)	98.24
	Block-based DFT (Fusion at decision level)	20×20 blocks (12 blocks) $4 \times 4 - 1$ for each block (180)	98.13
	Block-based DCT (Fusion at decision level)	20×20 blocks (12 blocks) 6×6 for each block (432)	97.82
	ICA	50	96.79
	NNMF	50	94.43
3D Voxel Representation	3D DFT	$2 \times 4 \times 4 \times 4 - 1 = (127)$	98.34

8.29 Two-Dimesional Multirate Processing

In this section, we review the basic concepts of two-dimensional multirate systems [F28]. For signals that are functions of time, the term "multirate" refers to systems in which the sampling rate of signals at different points in the system is different. For two-dimensional signals, we use "multirate" to refer to systems in which the lattices (to be defined shortly) on which the signals are defined differ, regardless of the physical significance of the lattice axes.

The *integer lattice* Λ is defined to be the set of all integer vectors $\mathbf{n} = (n_1, n_2)^T$. The sampling sublattice $\Lambda_{[D]}$ generated by the sampling matrix $[D]$ is the set of integer vectors $\mathbf{m}$ such that $\mathbf{m} = [D]\mathbf{n}$ for some integer vector $\mathbf{n}$. Consider, for example, the sampling matrix $[D]$ given by

$$[D] = [D_3] = \begin{pmatrix} 2 & 1 \\ -1 & 1 \end{pmatrix} \qquad (8.79)$$

The lattices Λ and $\Lambda_{[D]}$, are shown in Fig. 8.55e. The lattice Λ is light and dark circles. The sublattice $\Lambda_{[D]}$ is dark circles. In order to properly define a sublattice, a sampling matrix must be nonsingular with integer-valued entries. There are an infinite number of sampling matrices which generate a given sublattice. Each of these can be obtained from any of the others by postmultiplication by an integer matrix whose determinant is ± 1. A coset of a sublattice is the set of points obtained by shifting the entire sublattice by an integer shift vector $\mathbf{k}$. There are exactly

$D = |\det([D])|$ distinct cosets of $\Lambda_{[D]}$, and their union is the integer lattice Λ. We refer to the vector **k** associated with a certain coset as a coset vector.

8.29.1 Upsampling and Interpolation

For the case of upsampling by a factor of D, the upsampled version $y(\mathbf{n})$ of $x(\mathbf{n})$ is

$$y(\mathbf{n}) = \begin{cases} x\left([D]^{-1}\mathbf{n}\right), & \text{if } [D]^{-1}\mathbf{n} \in \Lambda \\ 0, & \text{otherwise} \end{cases} \qquad \mathbf{n} = \begin{pmatrix} n_1 \\ n_2 \end{pmatrix} \qquad (8.80)$$

Then the DFT of the upsampled signal $y(\mathbf{n})$ is

$$Y^F(\boldsymbol{\omega}) = X^F\left([D]^T\boldsymbol{\omega}\right) \qquad \boldsymbol{\omega} = \begin{pmatrix} \omega_1 \\ \omega_2 \end{pmatrix} \qquad (8.81)$$

$$Y(\mathbf{z}) = X\left(\mathbf{z}^{[D]}\right) \qquad \mathbf{z} = \begin{pmatrix} z_1 \\ z_2 \end{pmatrix} \qquad (8.82)$$

The result of upsampling on the Fourier transform of a signal is a reduction of the size of features (such as passbands) and a skewing of their orientation. As indicated by the movement of the dark square in Fig. 8.54, one entire period of $X^F(\boldsymbol{\omega})$, i.e., the unit frequency cell $\{\boldsymbol{\omega} \in [-\pi, \pi] \times [-\pi, \pi]\}$, is mapped to the baseband region

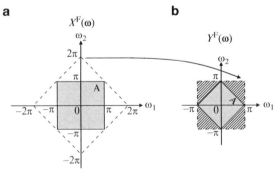

Dark square is a unit period.

Striped areas are images caused by upsampling and need to be filtered out. There are exactly D complete images $(D = |\det[D]| = 2)$.

Fig. 8.54 Movement of a unit period in frequency domain to baseband and images for one-to-two *upsampling* with $[D_2]^T = \begin{pmatrix} 1 & -1 \\ 1 & 1 \end{pmatrix}$. **a** *Dark square* is a unit period. **b** Striped areas are images caused by upsampling and need to be filtered out. There are exactly D complete images $(D = |\det[D]| = 2)$

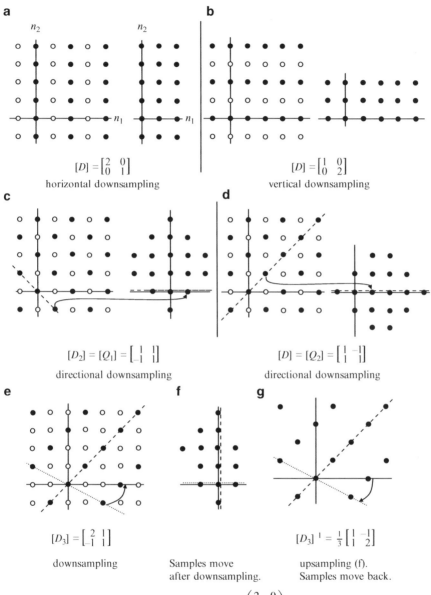

Fig. 8.55 a Horizontal downsampling $[D] = \begin{pmatrix} 2 & 0 \\ 0 & 1 \end{pmatrix}$. **b** Vertical downsampling $[D] = \begin{pmatrix} 1 & 0 \\ 0 & 2 \end{pmatrix}$. **c** Directional downsampling $[D_2] = [Q_1] = \begin{pmatrix} 1 & 1 \\ -1 & 1 \end{pmatrix}$. **d** Directional downsampling $[D] = [Q_2] = \begin{pmatrix} 1 & -1 \\ 1 & 1 \end{pmatrix}$. **e** Downsampling $[D_3] = \begin{pmatrix} 2 & 1 \\ -1 & 1 \end{pmatrix}$. **f** Samples move after downsampling. **g** Upsampling. **f** Samples move back $[D_3]^{-1} = \frac{1}{3} \begin{pmatrix} 1 & -1 \\ 1 & 2 \end{pmatrix}$. **a–d** Downsampling ratios are 2:1; **f–g** Downsampling and upsampling ratios are 3:1

$$\left\{ [D]^T \omega : \omega \in [-\pi,\pi] \times [-\pi,\pi] \right\} \tag{8.83}$$

For example, if $[D] = [D_2]$ then

$$[D]^T\boldsymbol{\omega} = [D_2]^T \qquad \boldsymbol{\omega} = \begin{pmatrix} 1 & -1 \\ 1 & 1 \end{pmatrix} \qquad \overset{\text{(NEW)}}{\begin{pmatrix} \pi \\ \pi \end{pmatrix}} = \overset{\text{(OLD)}}{\begin{pmatrix} 0 \\ 2\pi \end{pmatrix}}$$

$$[D]^T\boldsymbol{\omega} = [D_3]^T \qquad \boldsymbol{\omega} = \begin{pmatrix} 2 & -1 \\ 1 & 1 \end{pmatrix} \qquad \overset{\text{(NEW)}}{\begin{pmatrix} \frac{2}{3}\pi \\ \frac{1}{3}\pi \end{pmatrix}} = \overset{\text{(OLD)}}{\begin{pmatrix} \pi \\ \pi \end{pmatrix}}$$

which is centered at the origin. Images of the unit cell are mapped to regions surrounding the baseband. There are exactly D ($D = 2$ for our example: the dark square and the rest are shown in Fig. 8.54b) complete images of one period of $X^F(\boldsymbol{\omega})$ in $Y^F(\boldsymbol{\omega})$ [F28].

An upsampler followed by a filter that passes one of the images and stops the others is called the interpolator.

Mapping of (8.83) for upsampling can be alternatively expressed as follows [F29]. (Both expressions are useful.) From the inspection of the Fourier relationship of (8.80), it is seen that the dark square spectral region in Fig. 8.54a:

$$\{-\pi \leqslant \omega_1 \leqslant \pi\} \cap \{-\pi \leqslant \omega_2 \leqslant \pi\} \quad \text{Unit frequency cell} \tag{8.84}$$

is mapped to the dark square (parallelogram shaped) region

$$\{-\pi \leqslant d_{11}\omega_1 + d_{21}\omega_1 \leqslant \pi\} \cap \{-\pi \leqslant d_{12}\omega_1 + d_{22}\omega_1 \leqslant \pi\} \tag{8.85}$$

where

$$[D] = \begin{pmatrix} d_{11} & d_{12} \\ d_{21} & d_{22} \end{pmatrix} \tag{8.86}$$

8.29.2 Downsampling and Decimation

The $[D]$-fold downsampled version $y(\mathbf{n})$ of $x(\mathbf{n})$ is expressed as

$$y(\mathbf{n}) = x([D]\mathbf{n}) \qquad \mathbf{n} = \begin{pmatrix} n_1 \\ n_2 \end{pmatrix} \tag{8.87}$$

where $[D]$ is called the *sampling matrix* and is a nonsingular 2×2 matrix with integer entries. It is easy to verify that the downsampling factor is $D = |\det[D]|$. The reciprocal of D is the sampling density (sampling rate). Then the DFT of the downsampled signal $y(\mathbf{n})$ is

$$Y^F(\boldsymbol{\omega}) = \frac{1}{D} \sum_{l=0}^{D-1} X^F \left(\left([D]^T \right)^{-1} (\boldsymbol{\omega} - 2\pi \mathbf{k}_l) \right) \qquad (8.88)$$

There are D coset vectors $\mathbf{k}$ associated with the cosets of $[D]^T$.

Similar to (8.82), $X^F(\boldsymbol{\omega} - 2\pi \mathbf{d}_l)$, $l = 0, D - 1$ are mapped to the same region:

$$\left\{ [D]^T \boldsymbol{\omega} : \boldsymbol{\omega} \in \text{passband} \right\} \qquad (8.89)$$

Example: For $X^F(\boldsymbol{\omega} - 2\pi \mathbf{d}_0) = X^F(\boldsymbol{\omega})$

$$[D]^T \boldsymbol{\omega} = [D_2]^T \boldsymbol{\omega} = \begin{pmatrix} 1 & -1 \\ 1 & 1 \end{pmatrix} \begin{pmatrix} 0 \\ -\pi \end{pmatrix} = \begin{pmatrix} \pi \\ -\pi \end{pmatrix}$$

$$[D]^T \boldsymbol{\omega} = [D_3]^T \boldsymbol{\omega} = \begin{pmatrix} 2 & -1 \\ 1 & 1 \end{pmatrix} \begin{pmatrix} 0 \\ -\pi \end{pmatrix} = \begin{pmatrix} \pi \\ -\pi \end{pmatrix}$$

Consider the sampling matrix $[D_3] = \begin{pmatrix} 2 & 1 \\ -1 & 1 \end{pmatrix}$ (Figs. 8.56 and 8.57). We first select a complete set of coset vectors $\mathbf{k}_l$ associated with $\Lambda_{[D]^T}$. Since $|\det[D]^T| = 3$, there are three distinct cosets. One choice for the $\mathbf{k}_l$ is

$$\mathbf{k}_0 = \begin{pmatrix} 0 \\ 0 \end{pmatrix}, \quad \mathbf{k}_1 = \begin{pmatrix} 1 \\ 0 \end{pmatrix}, \quad \mathbf{k}_2 = \begin{pmatrix} 2 \\ 0 \end{pmatrix} \qquad (8.90)$$

The next step is to determine the aliasing offsets of $2\pi \left([D]^T \right)^{-1} \mathbf{k}_l$.

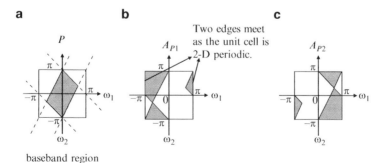

Fig. 8.56 Baseband region P for $[D_3] = \begin{pmatrix} 2 & 1 \\ -1 & 1 \end{pmatrix}$ and the two aliases of it, AP_1 and AP_2. [F28] © 1991 IEEE

Fig. 8.57 A set of three coset vectors for the sampling matrix
$$[D_3] = \begin{pmatrix} 2 & 1 \\ -1 & 1 \end{pmatrix} = (\mathbf{d}_1, \mathbf{d}_2).$$
When the coset vectors are used in a polyphase context, they are referred to as polyphase shift vectors for $[D]$

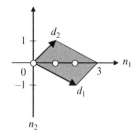

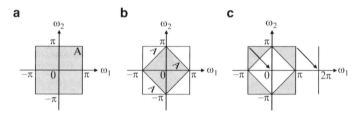

Fig. 8.58 Movement of unit period to baseband and images under upsampling for $[D_2]^T = \begin{pmatrix} 1 & -1 \\ 1 & 1 \end{pmatrix}$. **c** Images (*dark areas*) are created by upsampling and filtered out

$$\left([D_3]^T\right)^{-1} = \begin{pmatrix} 1 & 1 \\ -1 & 2 \end{pmatrix} \tag{8.91}$$

It is generally true that the frequency regions defined by AP_1 and AP_2 are admissible passbands (Fig. 8.58).

Frequency partition for alias free for two-to-one decimation, the square rotate clockwise while the area shrinks to half. Dark areas are passbands. We have already seen the frequency partition in Fig. 5.36a for DFT geometrical zonal filtering.

$$[D_2] = \begin{pmatrix} 1 & 1 \\ -1 & 1 \end{pmatrix} \quad \Rightarrow \quad \left([D_2]^T\right)^{-1} = \begin{pmatrix} 1 & 1 \\ -1 & 1 \end{pmatrix} \tag{8.92}$$

The following fundamental property defines an admissible passband. Let P denotes the set of frequencies in passband, and AP_l, $l = 1, \ldots, D-1$ denote the sets obtained by shifting P by the aliasing offsets $2\pi \left([D]^T\right)^{-1} \mathbf{k}_l$:

$$P = \left\{ \left([D]^T\right)^{-1} \boldsymbol{\omega} : \boldsymbol{\omega} \in [-\pi,\pi] \times [-\pi,\pi] \right\}$$
$$\text{e.g.} \quad \left([D]^T\right)^{-1} \boldsymbol{\omega} = \left([D]^T\right)^{-1} \begin{pmatrix} \pi \\ \pi \end{pmatrix} = \begin{pmatrix} \pi \\ 0 \end{pmatrix} \tag{8.93}$$

$$AP_l = \left\{ \boldsymbol{\omega} : \boldsymbol{\omega} + 2\pi \left([D]^T\right)^{-1} \mathbf{k}_l \in P \right\} \qquad l = 1, \ldots, D-1 \tag{8.94}$$

Then, the intersection of P and any of the AP_l must be empty, i.e.,

$$P \cap AP_l = \emptyset \qquad l = 1, \ldots, D-1 \qquad (8.95)$$

Otherwise, there exists some frequency in the passband that aliases onto another frequency. Intuitively, if two frequencies that are separated by an aliasing offset are both in P, the passband is inadmissible because those two frequencies will alias to the same frequency during downsampling. Regardless of shape, an admissible passband can be no larger than $(1/D)$th the size of the unit frequency cell.

Let the symbol $\Leftrightarrow$ denotes the correspondence between one- and two-dimensional expressions.

$$y(\mathbf{n}) = x([D]\mathbf{n}) \qquad \Leftrightarrow \qquad y(n) = x(2n) \qquad (8.96)$$

$$s_{[D]}(\mathbf{n}) = \begin{cases} 1, & \text{if } \mathbf{n} \in \Lambda_{[D]} \\ 0, & \text{otherwise} \end{cases}$$

$$s_{[D]}(\mathbf{n}) = \frac{1}{D}\sum_{l=0}^{D-1} e^{j2\pi\mathbf{k}_l^T[D]^{-1}\mathbf{n}} \qquad \Leftrightarrow \qquad \frac{1}{2}\left(1 + e^{\frac{j2\pi n}{2}}\right) \qquad (8.97)$$

$$\mathbf{e}_{[D]}(\boldsymbol{\omega}) = \mathbf{e}_{[D]}\begin{pmatrix} \omega_1 \\ \omega_2 \end{pmatrix} = \begin{pmatrix} e^{-j2\pi\boldsymbol{\omega}^T \mathbf{d}_1} \\ e^{-j2\pi\boldsymbol{\omega}^T \mathbf{d}_2} \end{pmatrix} \qquad (8.98)$$

$$[D] = (\mathbf{d}_1, \mathbf{d}_2), \qquad \mathbf{d}_1 = \begin{pmatrix} d_{11} \\ d_{21} \end{pmatrix}, \qquad \mathbf{d}_2 = \begin{pmatrix} d_{12} \\ d_{22} \end{pmatrix} \qquad (8.99)$$

where $\mathbf{d}_k$ is the kth column of $[D]$.

$$s_{[D]}(\mathbf{n}) = \frac{1}{D}\sum_{l=0}^{D-1}\left[\mathbf{e}_{[D]}{}^{-1}(2\pi\mathbf{k}_l)\right]^{-\mathbf{n}} \qquad (8.100)$$

The downsampled signal $y(\mathbf{n})$ of (8.96) can be obtained by a two-step process in which $x(\mathbf{n})$ is first multiplied by $s_{[D]}(\mathbf{n})$ to form an intermediate signal

$$w(\mathbf{n}) = x(\mathbf{n})\, s_{[D]}(\mathbf{n}) \qquad (8.101)$$

and intermediate signal is then downsampled. This two-step process [F28] is an extension of the one-dimensional case [IP34, p. 441, F30, p. 91].

$$w(\mathbf{n}) = \frac{1}{D}\sum_{l=0}^{D-1} x(\mathbf{n})\left[\mathbf{e}_{[D]}{}^{-1}(2\pi\mathbf{k}_l)\right]^{-\mathbf{n}}$$

Taking z-transforms yields

$$W(\mathbf{z}) = \frac{1}{D}\sum_{l=0}^{D-1} X\left[\mathbf{e}_{[D}{}^{-1]}(2\pi\mathbf{k}_l)\mathbf{z}\right] \qquad (8.102)$$

Since $w(\mathbf{n})$ is nonzero only on $\Lambda_{[D]}$ and equals $x(\mathbf{n})$ there,

$$Y(\mathbf{z}) = \sum_{\mathbf{n}\in\Lambda} y(\mathbf{n})\mathbf{z}^{-\mathbf{n}} = \sum_{\mathbf{n}\in\Lambda} x([D]\mathbf{n})\mathbf{z}^{-\mathbf{n}} = \sum_{\mathbf{n}\in\Lambda} w([D]\mathbf{n})\mathbf{z}^{-\mathbf{n}} \qquad (8.103)$$

Substituting $\mathbf{m} = [D]\mathbf{n}$,

$$Y(z) = \sum_{\mathbf{m}\in\Lambda} w(\mathbf{m})\mathbf{z}^{-[D]^{-1}\mathbf{m}} = \sum_{\mathbf{n}\in\Lambda} w(\mathbf{n})\mathbf{z}^{-[D]^{-1}\mathbf{n}} \tag{8.104}$$

$$= \sum_{\mathbf{n}\in\Lambda} w(\mathbf{n})\left(\mathbf{z}^{[D]^{-1}}\right)^{-\mathbf{n}} = w\left(\mathbf{z}^{[D]^{-1}}\right) \tag{8.105}$$

Therefore, from (8.102)

$$Y(\mathbf{z}) = \frac{1}{D}\sum_{l=0}^{D-1} X\left(e_{[D]^{-1}}(2\pi\mathbf{k}_l)\,\mathbf{z}^{[D]^{-1}}\right) \tag{8.106}$$

$$Y(\mathbf{z}) = \frac{1}{D}\sum_{l=0}^{D-1} X\left(e^{-j2\pi\mathbf{k}_l^T[D]^{-1}}\mathbf{z}^{[D]^{-1}}\right) \tag{8.107}$$

$$Y^{\mathrm{F}}(\boldsymbol{\omega}) = \frac{1}{D}\sum_{l=0}^{D-1} X^{\mathrm{F}}\left(\left([D^{-1}]^T\right)\boldsymbol{\omega} - 2\pi\left([D^{-1}]^T\right)\mathbf{k}_l\right) \tag{8.108}$$

Multirate identity (often called *noble identity*) is a useful tool to analyze multi-dimensional multirate systems. Its analysis side is illustrated in Fig. 8.59 where $H(\boldsymbol{\omega})$ and $[D]$ are a two-dimensional filter and a 2×2 downsampling matrix, respectively. With multirate identity, the order of a filter and a downsampling matrix can be interchanged. This is because any system whose z-transform is a function of $\mathbf{z}^{[D]}$ has an impulse response that is nonzero only on $\Lambda_{[D]}$. The synthesis one can be inferred similarly [F28, F32].

Definition 8.1 Let the lth column of the matrix $[D]$ be given by $\mathbf{d}_l$. The quantity $\mathbf{z}^{[D]}$ is defined to be a vector whose lth element is given by

$$z_l = \mathbf{z}^{\mathbf{d}_l} \tag{8.109}$$

Example: Let $\mathbf{z} = \begin{pmatrix} z_1 \\ z_2 \end{pmatrix} = \begin{pmatrix} e^{j\omega_1} \\ e^{j\omega_2} \end{pmatrix}$, $\boldsymbol{\omega} = \begin{pmatrix} \omega_1 \\ \omega_2 \end{pmatrix}$, $[D] = (\mathbf{d}_1, \mathbf{d}_2) = \begin{pmatrix} 2 & 2 \\ -1 & 1 \end{pmatrix}$.

Then elements of the vector $\mathbf{z}^{[D]}$ are computed as

$$z_1 = \mathbf{z}^{\mathbf{d}_1} = e^{j\boldsymbol{\omega}^T\mathbf{d}_1} = e^{j(2\omega_1-\omega_2)} = z_1^2 z_2^{-1} \tag{8.110}$$

Fig. 8.59 Multirate identity. [F32] © 2009 IEEE

$$z_2 = \mathbf{z}^{\mathbf{d}_2} = e^{j\boldsymbol{\omega}^T \mathbf{d}_2} = e^{j(2\omega_1 + \omega_2)} = z_1^2 z_2^1 \tag{8.111}$$

$$\mathbf{z}^{[D]} = \begin{pmatrix} z_1^2 z_2^{-1} \\ z_1^2 z_2^1 \end{pmatrix} \tag{8.112}$$

$$H\left(\mathbf{z}^{[D]}\right) = H\left(\begin{matrix} z_1^2 z_2^{-1} \\ z_1^2 z_2^1 \end{matrix} \right) \tag{8.113}$$

Notice that $\mathbf{z}^{[D]}$ is mapped to a vector as $[D]$ is a matrix, and $\mathbf{z}^{\mathbf{d}_l}$ is mapped to a complex number as $\mathbf{d}_l$ is a vector.

Definition 8.2 A *generalized quincunx sampling matrix* is a matrix whose entries are ± 1 with determinant two. Typical quincunx sampling matrices are

$$[Q_1] = \begin{pmatrix} 1 & 1 \\ -1 & 1 \end{pmatrix} \qquad [Q_2] = \begin{pmatrix} 1 & -1 \\ 1 & 1 \end{pmatrix} \tag{8.114}$$

Since $[Q_1]$ is the most commonly used of theses, we will use it by default unless specifically mentioned otherwise. Quincunx downsampling results in a down-sampled rotated representation.

We need to start from upsampling when we figure out a frequency representation of passband for downsampling (Fig. 8.60).

A resampling matrix is unimodular. A unimodular matrix has only 0, $+1$ or -1 entries. Its determinant is ± 1. Its inverse matrix is also unimodular.

$$[R_1] = \begin{pmatrix} 1 & 1 \\ 0 & 1 \end{pmatrix} \qquad [R_2] = \begin{pmatrix} 1 & -1 \\ 0 & 1 \end{pmatrix}$$
$$[R_3] = \begin{pmatrix} 1 & 0 \\ 1 & 1 \end{pmatrix} \qquad [R_4] = \begin{pmatrix} 1 & 0 \\ -1 & 1 \end{pmatrix} \tag{8.115}$$

where $[R_i], i = 1, \dots, 4$ are called *resampling matrices* and *diamond-conversion matrices*. Apply each matrix to the diamond-shape passband to obtain the associated parallelogram passband (Fig. 8.61).

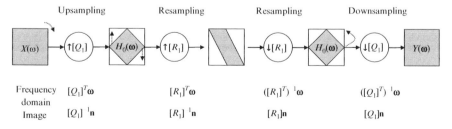

Fig. 8.60 Downsampling this bandlimited signal stretches its transform to the unit cell defined in (8.84), with no aliasing

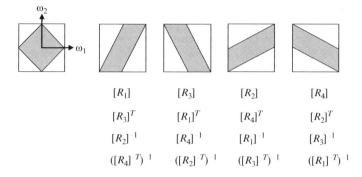

$[R_1]$	$[R_3]$	$[R_2]$	$[R_4]$
$[R_3]^T$	$[R_1]^T$	$[R_4]^T$	$[R_2]^T$
$[R_2]^{-1}$	$[R_4]^{-1}$	$[R_1]^{-1}$	$[R_3]^{-1}$
$([R_4]^T)^{-1}$	$([R_2]^T)^{-1}$	$([R_3]^T)^{-1}$	$([R_1]^T)^{-1}$

Fig. 8.61 A *diamond-shaped* passband and four parallelogram passbands by $[R_i], i = 1, \ldots, 4$. Three equivalent relations are listed below each matrix. [F31] © 2004 IEEE

8.30   Fast Uniform Discrete Curvelet Transform (FUDCuT)

As wavelet analysis is very effective at representing objects with isolated point singularities, ridgelet analysis can be very effective at representing objects with changes or singularities along lines [F25]. One can loosely view ridgelets as a way of concatenating one-dimensional wavelets along lines. Hence the motivation for using ridgelets in image processing tasks is very appealing as singularities are often joined together along edges or contours in images [F34].

Since in two-dimensional signal, points and lines are related via the Radon transform [B6] (a point in the Radon transform represents a line in a real-life image), the wavelet and ridgelet transforms are linked via the Radon transform. Ridgelets can effectively deal with *linelike* phenomena in the two-dimensional case, where wavelets do not effectively deal with singularities along lines and curves.

8.30.1   The Radon Transform

The continuous Radon transform of a function $f(x, y)$, denoted as $R_p(t, \theta)$, is defined as its line integral along a line inclined at an angle θ from the y-axis in the spatial domain (x, y) and at a distance t from the origin [B6, F34]. Mathematically, it is written as

$$R_p(t, \theta) = \int \int_{-\infty}^{\infty} f(x, y) \delta(x \cos \theta + y \sin \theta - t) dxdy \qquad (8.116)$$

$$-\infty < t < \infty, \quad 0 \le \theta < \pi$$

where $\delta(t)$ is the Dirac delta function. In the digital domain this is the addition of all pixels that lie along a particular line defined by intercept point t and slope θ and lying on the (x, y) plane (`radon`, `iradon`).

It is instructive to note that the application of a one-dimensional Fourier transform to the Radon transform $R_p(t, \theta)$ along t results in the two-dimensional Fourier transform in polar coordinates. More specifically, let $F^F(\omega_1, \omega_2)$ be the two-dimensional Fourier transform of $f(x, y)$. In polar coordinates we write $F_p^F(\xi, \theta) = F^F(\xi \cos \theta, \xi \sin \theta)$. Then we have

$$F^F(\xi \cos \theta, \xi \sin \theta) = \text{1-D FT}\left[R_p(t, \theta)\right] = \int_{-\infty}^{\infty} R_p(t, \theta) \exp(-j\xi t) dt \qquad (8.117)$$

This is the famous *projection-slice* theorem and is used in image reconstruction from projection methods [F36, B6]. Similarly, an inverse one-dimensional Fourier transform of the two-dimensional Fourier of an image yields a Radon transform of the image. Different from a regular inverse, this inverse one-dimensional Fourier transform is defined as its line integral along a line inclined at an angle θ from the ω_2-axis in the Fourier domain (ω_1, ω_2), where the inverse is applied to a line for each θ to obtain the two-dimensional Radon transformed data. The relation will be used to derive the ridgelet and thus curvelet transforms which come next in this section (see Fig. 8.62).

8.30.2 The Ridgelet Transform

The ridgelet transform [F36] is precisely the application of a one-dimensional wavelet transform to the slices of the Radon transform where the angular variable θ is constant and t is varying. To complete the ridgelet transform, we must take a one-dimensional wavelet transform along the radial variable in Radon space.

Fig. 8.62 Relations between the transforms. The Radon transform is the application of 1-D Fourier transform to the slices of the 2-D Fourier transform and the ridgelet transform is the application of 1-D wavelet transform to the slices of the Radon transform. It is noted that the two-dimensional Fourier transform is used to compute the Radon transform in this section rather than a direct computation of the Radon transform. [F36] © IEEE 2003

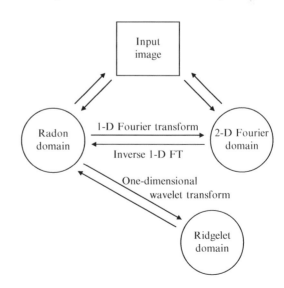

$$RI(a, b, \theta) = \int_{-\infty}^{\infty} \psi_{a,b}(t) R_{\mathrm{p}}(t, \theta) \, dt \qquad (8.118)$$

where one-dimensional wavelets are defined as

$$\psi_{a,b}(t) = a^{-\frac{1}{2}} \psi\left(\frac{t-b}{a}\right) \qquad (8.119)$$

where $a = 2^m$ represents dilation, $b = k2^m$ represents translation and m is a level index. m and k are integer. The ridgelet transform is optimal to detect lines of a given size, which is the block size.

The ridgelet transform is precisely the application of a one-dimensional wavelet transform to the slices of the Radon transform. Thus the finite ridgelet transform (FRIT) [F33] is the application of the finite Radon transform (FRAT) [F33] to the entire image followed by the wavelet transform on each row. The inverse FRIT is the entire operation in reverse. Apply the inverse wavelet transform to the rows, and then use the inverse FRAT on the entire image (Fig. 8.63).

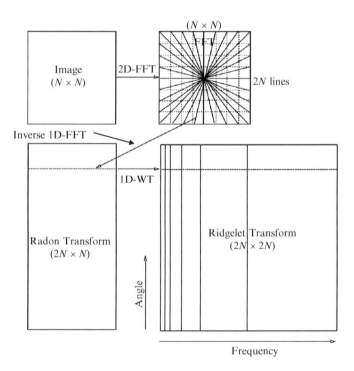

Fig. 8.63 Ridgelet transform flowgraph. Each of the $2N$ radial lines in the Fourier domain is processed separately. The one-dimensional inverse FFT is calculated along each radial line followed by a one-dimensional nonorthogonal wavelet transform. In practice, the one-dimensional wavelet coefficients are directly computed in the Fourier space. [F35] © IEEE 2002

8.30.3 The Curvelet Transform

The curvelet transform [LA25] begins with the application of the à trous wavelet [F33], and then is followed by the repeated application of the ridgelet transform.

The à trous subband filtering algorithm is especially well-adapted to the needs of the digital curvelet transform. The algorithm decomposes an $N \times N$ image I as a superposition of the form

$$I(x, y) = c_j(x, y) + \sum_{m=1}^{M} w_m(x, y) \tag{8.120}$$

where c_j is a coarse or smooth version of the original image I and w_m represents the *details* of I at scale 2^{-m}. Thus the algorithm outputs $M + 1$ subband arrays of size $N \times N$. Scale $m = 1$ corresponds to the finest scale i.e., high frequencies.

The algorithm is most simply stated as [F35]:

1. Apply the à trous algorithm with M scales to an input image (M subband arrays of size $N \times N$) (see Fig. 8.64).
2. Set block size $B_1 = B_{min}$ (e.g. $B_{min} = 16$).
3. For subband index $m = 1, 2, \ldots, M$
   (a) Partition the subband with a block size B_m and apply the discrete ridgelet transform to each block
   (b) If m modulo $2 = 1$ (or m is even), then the block size of the next subband is $B_{m+1} = 2B_m$
   (c) Else $B_{m+1} = B_m$.

Example: For an image of size $N \times N$, each subband also has the same size of $N \times N$. Let the number of fine scales $M = 5$. Set the size of block B_1. For $m = 1$, divide the subband $m = 1$ with block size B_1 and set $B_2 = 2B_1$. For $m = 2$, $B_3 = B_2$. For $m = 3$, $B_4 = 2B_2$. For $m = 4$, $B_5 = B_4$.

The first generation of curvelet transform was updated in 2003 [F24]. The use of the ridgelet transform was discarded, thus reducing the amount of redundancy in the transform and increasing the speed considerably. In this new method called the *second-generation curvelet*, an approach of curvelets as tight frames is taken. Using tight frames, an individual curvelet has a frequency support in a parabolic-wedge area of the frequency domain as seen in Fig. 8.65a [F37].

The main idea of this construction is to decompose any two-dimensional function into spaces that are strictly bandpass in frequency domain. The support shapes of these function spaces are concentric wedges (see Fig. 8.65a). The curvelet coefficients for each scale-direction wedges are estimated as the inner product of the given two-dimensional function and the bandlimited curvelet function centered on a grid that is inversely proportional to the wedge-shape support of the curvelet in the frequency domain (see Fig. 8.65b). Assume that we have two smooth functions W and V, satisfying the *admissibility* condition [F24, F38]

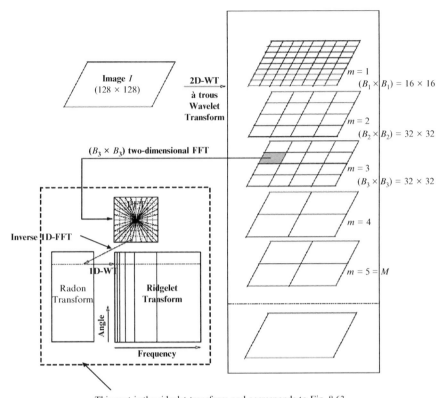

This part is the ridgelet transform and corresponds to Fig. 8.63.

Fig. 8.64 Curvelet transform flowgraph. The figure illustrates the decomposition of the original image into subbands followed by the spatial partitioning of each subband. The ridgelet transform is then applied to each block. [F35] © IEEE 2002

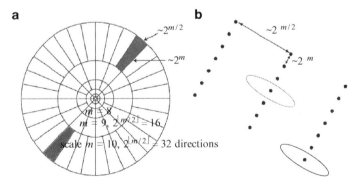

Fig. 8.65 Curvelet tiling of space and frequency. **a** Tiling of the frequency plane (an individual curvelet has frequency support in a pair of parabolic-wedge areas of the frequency domain) and **b** the spatial grid of the curvelet at a given scale m and orientation l. [F24] Copyright © 2006 Society for Industrial and Applied Mathematics. Reprinted with permission all rights reserved

$$\sum_{m=-\infty}^{\infty} W^2(2^{-m}r) = 1 \tag{8.121}$$

$$\sum_{l=-\infty}^{\infty} V^2(t-1) = 1 \tag{8.122}$$

The frequency window $U_m(r, \theta)$ for the curvelet transform is defined with r and θ polar coordinates in the Fourier domain by

$$U_m(r, \theta) = 2^{-3m/4} W(2^{-m}r) V\left(\frac{2\lfloor m/2\rfloor \theta}{2\pi}\right) \tag{8.123}$$

where $\lfloor m/2\rfloor$ is the integer part of $m/2$. The support of $U_m(r, \theta)$ is one of the two dark concentric wedges (or rectangles) in Fig. 8.65a. To obtain real-valued curvelets, a symmetric window function $\hat{\varphi}_m(r, \theta)$ in the frequency domain is defined as

$$\hat{\varphi}_m(r, \theta) = U_m(r, \theta) + U_m(r, \theta + \pi) \tag{8.124}$$

(Recall the complex conjugate theorem of (2.14).) This window is used to define the curvelet function at the first direction of scale m, $\hat{\varphi}_{m,1}(\omega) = \hat{\varphi}_m(r, \theta)$. In the frequency plane, each curvelet direction is indexed by $L = (m, l)$. At each scale m, the number of directions is $2^{\lfloor m/2\rfloor}$. At scale m, the curvelet direction l is generated by rotating the curvelet window function $\hat{\varphi}_m(r, \theta)$ by a rotation angle $\theta_l = (l-1)2\pi 2^{-\lfloor m/2\rfloor}$, with $l = 1, 2, \ldots, 2^{-\lfloor m/2\rfloor}$.

The coarse scale curvelet function is defined by polar window $W(r)$ as follows $\hat{\varphi}_0(\omega)W(r)$. From the definition of $W(r)$ and $V(\theta)$ functions in (8.121) and (8.122), it can be shown that the function $\hat{\varphi}_0(\omega)$ and scaled version of $\hat{\varphi}_{m,l}^2(\omega)$ are summed up to one for all ω, where $\omega = (\omega_1, \omega_2)$ is the two-dimensional frequency variable.

The curvelet transform [F24] has become popular in image processing since it is proved to be an essentially optimal representation of two variable functions which are smooth except at discontinuities along a C^2 curve (a function of the class C^2 is twice continuously differentiable), and has been used and performed very well especially for images with edges in many image processing applications, such as denoising, deconvolution, astronomical imaging, component separation (see [F25]). The implementation of its discrete version which is called the fast discrete curvelet transform (FUDCuT) is based on the FFT.

The fast uniform discrete curvelet transform (FUDCuT) is an invertible multi-resolution directional transform which is basically a filter bank implementation in the frequency domain of the continuous curvelet transform. The implementation processes of the forward transform of FUDCuT are as follows. First, the two-dimensional data (image) is transformed into the frequency domain via FFT. By windowing using the curvelet windows, we then obtain the curvelet coefficients in the frequency domain. Applying IFFT, we obtain the curvelet coefficients in the time domain. The inverse transform is simply obtained by reversing these steps. The transform has one-sided frequency support, which results in the complex-valued coefficients. An example of the FUDCuT decomposition is shown in Fig. 8.66, where the Lena image is decomposed into two levels with 6 and twelve directional subbands.

The implementation processes of the forward and inverse transforms of FUD-CuT are as follows [F25].

1. The 2-D image $x(n_1, n_2)$ is transformed into the frequency domain via 2-D FFT

$$X^{\mathrm{F}}(k_1, k_2) = 2\text{-D FFT}[x(n_1, n_2)]$$

2. The frequency domain $X^{\mathrm{F}}(k_1, k_2)$ is windowed by $U_{m,l}(k_1, k_2)$ to obtain the coefficients of scale m, direction l in the frequency domain

$$C_{m,l}^{\mathrm{F}}(k_1, k_2) = U_{m,l}(k_1, k_2) X^{\mathrm{F}}(k_1, k_2)$$

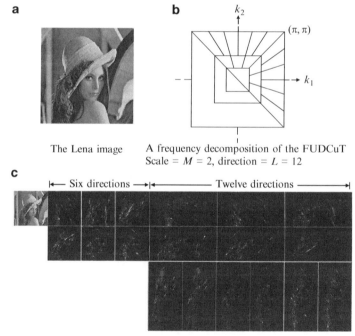

a

The Lena image

b

A frequency decomposition of the FUDCuT Scale $= M = 2$, direction $= L = 12$

c

$\longmapsto$ Six directions $\longrightarrow\!\longleftarrow$ Twelve directions $\longrightarrow$

FUDCuT coefficients of the Lena image

Fig. 8.66 An example of the fast uniform discrete curvelet transform (FUDCuT). The number of directions increases with frequency. Each bandpass region can have 3×2^m (scale $m = 1, 2$) directions. [F26] © 2009 IEEE

where the window $U_{m,l}(k_1, k_2)$ is defined as the product of two windows:

$$U_{m,l}(k_1, k_2) = W_m(k_1, k_2)V_{m,l}(k_1, k_2)$$

The window $W_m(k_1, k_2)$ is defined as

$$W_m(k_1, k_2) = \sqrt{\phi_{m+1}^2(k_1, k_2) - \phi_m^2(k_1, k_2)}$$

where

$$\Phi_m(k_1, k_2) = \phi(2^{-m}k_1)\phi(2^{-m}k_2)$$

The function $\phi(k)$ is the Meyer window, which is given by

Number '3' means period π has three sections.

$$\phi(k) = \begin{cases} 1 & \text{if } |\omega| \leq \frac{2\pi}{3} \\ \cos\left[\frac{\pi}{2}v\left(\frac{3|\omega|}{2\pi} - 1\right)\right] & \text{if } \frac{2\pi}{3} \leq |\omega| \leq \frac{4\pi}{3} \\ 0 & \text{otherwise} \end{cases} \qquad (8.125)$$

Number '2' means a flat line of the window stretches for two sections from origin and then a slope stretches for two sections.

Figure 8.67 illustrates the Meyer window. On the other hand, the window $V_{m,l}(k_2, k_2)$ is defined as

$$V_{m,l}(k_1, k_2) = \phi\left(2^{\lfloor m/2 \rfloor}\frac{k_1}{k_2} - l\right)$$

where $\lfloor m/2 \rfloor$ is the integer part of $m/2$ (e.g. $\lfloor 5/2 \rfloor = 2$).

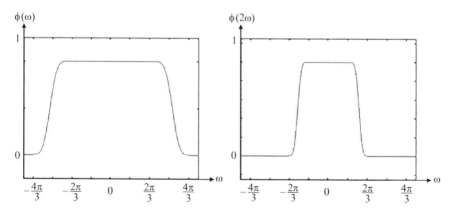

Fig. 8.67 Meyer window

3. For $m = 1, \ldots, M$ and $l = 1, \ldots, L$, the curvelet coefficient $c_{m,l}(n_1, n_2)$ in scale m, direction l, in the time domain is obtained by the 2-D inverse FFT (IFFT) of $C_{m,l}^{F}(k_1, k_2)$

$$c_{m,l}(n_1, n_2) = \text{2-D IFFT}\left[C_{m,l}^{F}(k_1, k_2)\right]$$

The inverse FUDCuTis obtained as follows.

1. For each scale m and direction l, the curvelet coefficient $c_{m,l}(n_1, n_2)$ is transformed into the frequency domain via 2-D FFT.

$$C_{m,l}^{F}(k_1, k_2) = \text{2-DFFT}\left[c_{m,l}(n_1, n_2)\right]$$

2. For each scale m and direction l, $C_{m,l}^{F}(k_1, k_2)$ is multiplied by the window $U_{m,l}(k_1, k_2)$ to obtain $X_{m,l}^{F}(k_1, k_2)$.

$$X_{m,l}^{F}(k_1, k_2) = U_{m,l}(k_1, k_2) C_{m,l}^{F}(k_1, k_2)$$

3. Add $X_{m,l}^{F}(k_1, k_2)$ all together i.e.,

$$X^{F}(k_1, k_2) = \sum_{m=1}^{M} \sum_{l=1}^{M} X_{m,l}^{F}(k_1, k_2)$$

4. Finally, apply 2-D IFFT to obtain $x(n_1, n_2)$.

$$x(n_1, n_2) = \text{2-D IFFT}\left[X^{F}(k_1, k_2)\right]$$

This is true since the windows $U_{m,l}(k_1, k_2)$ are designed such that

$$U_0^2(k_1, k_2) + \sum_{m=1}^{M} \sum_{l=1}^{L} U_{m,l}^2(k_1, k_2) = 1$$

The equivalent of the two-level FUDCuT filter bank is shown in Fig. 8.68. Function $v(x)$ is an auxiliary window function (or a smooth function) for Meyer wavelets with the property (Figs. 8.69 and 8.70)

$$v(x) + v(1 - x) = 1 \qquad\qquad (8.126)$$

Symbol d is a degree of the polynomial defining $v(x)$ on $[0, 1]$ (for software see [F24]).

$$v(x) = 0 \qquad \text{for } x < 0 \qquad\qquad (8.127)$$

a

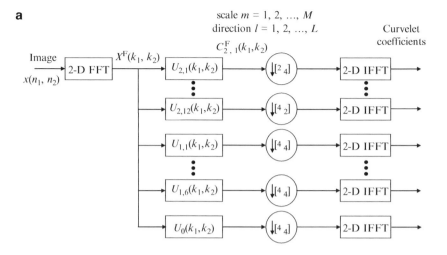

scale $m = 1, 2, ..., M$
direction $l = 1, 2, ..., L$

$C_{2,1}^{F}(k_1, k_2)$

Curvelet coefficients

Image $x(n_1, n_2)$ — 2-D FFT — $X^F(k_1, k_2)$

Forward transform

b

Curvelet coefficients

$C_{2,1}^{F}(k_1, k_2)$

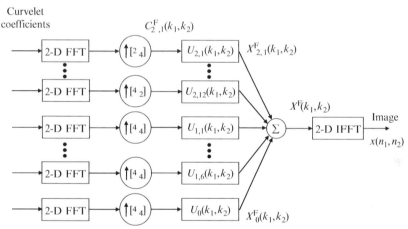

Inverse transform

Fig. 8.68 Fast uniform discrete curvelet transform (FUDCuT). The symbol ($\downarrow[^2{}_4]$) indicates downsampling or decimation by two in the row direction and four in the column direction, and the symbol ($\uparrow[^2{}_4]$) indicates upsampling or interpolation by similar rates. $[D_0] = \text{diag}(2^J, 2^J) = \text{diag}(4,4)$ for the maximum scale $J = 2$ [F25]

$$v(x) = x \qquad \text{for } 0 \leqslant x \leqslant 1 \text{ and } d = 0 \qquad (8.128)$$

$$v(x) = x^2(3 - 2x) \qquad \text{for } 0 \leqslant x \leqslant 1 \text{ and } d = 1 \qquad (8.129)$$

$$v(x) = x^3\left(10 - 15x + 6x^2\right) \qquad \text{for } 0 \leqslant x \leqslant 1 \text{ and } d = 2 \qquad (8.130)$$

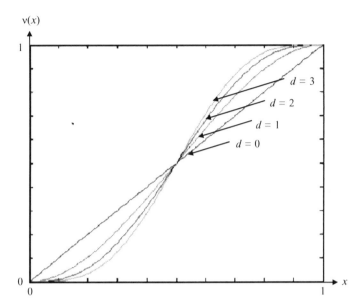

Fig. 8.69 An auxiliary window function for Meyer wavelets

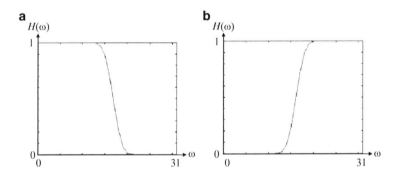

Fig. 8.70 a Lowpass and **b** highpass Meyer windows with $L = 4$, $d = 3$

$$v(x) = x^4\left(35 - 84x + 70x^2 - 20x^3\right) \qquad \text{for } 0 \leqslant x \leqslant 1 \text{ and } d = 3 \qquad (8.131)$$

$$v(x) = 1 \qquad \text{for } x > 0 \qquad (8.132)$$

Dyadic points are $1, 2^L$.

$$
\begin{aligned}
H &= 1 & 1 \leqslant \omega &\leqslant 2^L - \left\lfloor \frac{2^L}{3} \right\rfloor \\
H(\omega) &= \cos\left[\frac{\pi}{2} v\left(\frac{3\omega}{2^{L+1}} - 1\right)\right] & 2^L - \left\lfloor \frac{2^L}{3} \right\rfloor + 1 \leqslant \omega &\leqslant 2^L + \left\lfloor \frac{2^L}{3} \right\rfloor + 1 & (8.133) \\
H &= 0 & 2^L + \left\lfloor \frac{2^L}{3} \right\rfloor + 2 \leqslant \omega &\leqslant 2^{L+1}
\end{aligned}
$$

$$H = 0 \qquad\qquad\qquad 1 \leqslant \omega \leqslant 2^L - \left\lfloor \tfrac{2^L}{3} \right\rfloor$$

$$H(\omega) = \sin\left[\tfrac{\pi}{2} v\left(\tfrac{3\omega}{2^{L+1}} - 1\right)\right] \quad 2^L - \left\lfloor \tfrac{2^L}{3} \right\rfloor + 1 \leqslant \omega \leqslant 2^L + \left\lfloor \tfrac{2^L}{3} \right\rfloor + 1 \qquad (8.134)$$

$$H = 1 \qquad\qquad 2^{L+1} - \left(2^L - \left\lfloor \tfrac{2^L}{3} \right\rfloor - 1\right) + 1 \leqslant \omega \leqslant 2^{L+1}$$

8.31 Problems

(Section 8.1: 8.1–8.3)

8.1 When $X^F(k), k = 0, 1, \ldots, N - 1$ is given, derive an equation for a decimation by a factor of three. Assume that three divides N, i.e. $N = 3, 9, 27, \ldots$.

8.2 Show that upsampling by a factor of two can be performed exactly in the frequency domain by appending a DFT block after the same. See (8.6).

8.3 Given a random vector $\underline{x} = (1, 2, 3, 4)^T$, show frequency domain interpolation by a factor of three by using the MATLAB.

8.4 Create a random input sequence, $x(n)$ of length 18. The impulse response of a low pass filter or antialiasing filter is $h(n) = \{\tfrac{1}{4}, \tfrac{1}{2}, \tfrac{1}{4}\}$. Use MATLAB to implement 2:1 time domain decimation in the frequency domain as part of fast convolution filtering. Use the overlap-save method with three segments of length $(L + M - 1) = 8$, $L = 6$, $M = 3$. Thus the lengths of forward and inverse FFTs are eight and four, respectively.

(Section 8.4: 8.5–8.6)

8.5 Derive (8.25) from (8.19).

8.6 The spatial domain is described by the two-point vector $(n_1, n_2)^T$ which is denoted as $\underline{n}$. This leads to compact expressions for the 2-D DFT and its inverse.

$$X(\underline{k}) = \sum_{\underline{n}} x(\underline{n}) \exp\left\{\frac{-j2\pi}{N} \langle \underline{k}, \underline{n} \rangle\right\} \quad \underline{k} = \begin{pmatrix} k_1 \\ k_2 \end{pmatrix} \quad \underline{n} = \begin{pmatrix} n_1 \\ n_2 \end{pmatrix} \qquad (P8.1a)$$

$$x(\underline{n}) = \frac{1}{N^2} \sum_{\underline{k}} X^F(\underline{k}) \exp\left\{\frac{j2\pi}{N} \langle \underline{k}, \underline{n} \rangle\right\} \qquad (P8.1b)$$

where $\langle \underline{k}, \underline{n} \rangle = \underline{k}^T \underline{n}$ denotes an inner product in R^2. Lawton's chirp algorithm [IP3, IP9] is generalized for a 2-D rotation. We assume in this problem that $[A]$ is a symmetric matrix. Use the following matrix $[A]$ for parts (a) and (b).

$$[A] = \begin{pmatrix} -\cos\theta & \sin\theta \\ \sin\theta & \cos\theta \end{pmatrix} = \begin{pmatrix} -\alpha & \beta \\ \beta & \alpha \end{pmatrix} \qquad (P8.2)$$

(a) Let

$$Z(\underline{n}) = \exp\left\{\frac{-j\pi}{N}\langle \underline{n}, [A]\underline{n}\rangle\right\} = \exp\left(\frac{j\pi}{N}\left[(n_1^2 - n_2^2)\alpha - 2n_1 n_2 \beta\right]\right) \qquad \text{(P8.3)}$$

Starting from

$$x([A]\underline{n}) = x(-n_1 \cos\theta + n_2 \sin\theta, \, n_1 \sin\theta + n_2 \cos\theta) \qquad .$$

$$= \frac{1}{N^2} \sum_{k_1=0}^{N-1} \sum_{k_2=0}^{N-1} X^{\mathrm{F}}(k_1, k_2) \exp\left(\frac{j2\pi}{N}\left[(-k_1 n_1 + k_2 n_2)\alpha + (k_1 n_2 + k_2 n_1)\beta\right]\right)$$

$$\text{(P8.4)}$$

show that

$$x([A]\underline{n}) = Z(\underline{n}) \sum_{\underline{k}} \{X^{\mathrm{F}}(\underline{k})Z(\underline{k})\}Z^*(\underline{k} - \underline{n}) \qquad \text{(P8.5)}$$

(b) Let

$$Z(\underline{n}) = \exp\left\{\frac{j\pi}{N}\langle \underline{n}, [A]\underline{n}\rangle\right\} \qquad \text{(P8.6)}$$

Different from part (a), starting from

$$X^{\mathrm{F}}([A]\underline{k}) = X^{\mathrm{F}}(-k_1 \cos\theta + k_2 \sin\theta, \, k_1 \sin\theta + k_2 \cos\theta)$$

$$= \sum_{n_1=0}^{N-1} \sum_{n_2=0}^{N-1} x(n_1, n_2) \exp\left(\frac{-j2\pi}{N}\left[(-k_1 n_1 + k_2 n_2)\alpha + (k_1 n_2 + k_2 n_1)\beta\right]\right)$$

$$\text{(P8.7)}$$

show that

$$X^{\mathrm{F}}([A]\underline{k}) = Z(\underline{k}) \sum_{\underline{n}} \{x(\underline{n})Z(\underline{n})\}Z^*(\underline{k} - \underline{n}) \qquad \text{(P8.8)}$$

Use the following matrix $[A]$ for parts (c) and (d).

$$[A] = \begin{pmatrix} \sin\theta & \cos\theta \\ \cos\theta & -\sin\theta \end{pmatrix} = \begin{pmatrix} \alpha & \beta \\ \beta & -\alpha \end{pmatrix} \qquad \text{(P8.9)}$$

(c) Repeat part (a) by letting

$$Z(\underline{n}) = \exp\left\{\frac{j\pi}{N}\langle \underline{n}, [A]\underline{n}\rangle\right\} \tag{P8.10}$$

In other words, start from the following equation to get (P8.5).

$$x([A]\underline{n}) = x(n_1 \sin\theta + n_2 \cos\theta, n_1 \cos\theta - n_2 \sin\theta)$$

$$= \frac{1}{N^2} \sum_{k_1=0}^{N-1}\sum_{k_2=0}^{N-1} X^F(k_1, k_2) \exp\left(\frac{j2\pi}{N}[(k_1 n_1 - k_2 n_2)\alpha + (k_1 n_2 + k_2 n_1)\beta]\right)$$

$$\tag{P8.11}$$

(d) Repeat part (b) by letting

$$Z(\underline{n}) = \exp\left\{\frac{-j\pi}{N}\langle \underline{n}, [A]\underline{n}\rangle\right\} \tag{P8.12}$$

In other words, start from the following equation to get (P8.8).

$$X^F([A]\underline{k}) = X^F(k_1 \sin\theta + k_2 \cos\theta, k_1 \cos\theta - k_2 \sin\theta)$$

$$= \sum_{n_1=0}^{N-1}\sum_{n_2=0}^{N-1} x(n_1, n_2) \exp\left(\frac{-j2\pi}{N}[(k_1 n_1 - k_2 n_2)\alpha + (k_1 n_2 + k_2 n_1)\beta]\right)$$

$$\tag{P8.13}$$

8.7 Derive (8.33) from (8.29).

8.8 Show (8.41) yields (8.42).

8.9 Draw a block diagram similar to that for implementation of oddly stacked TDAC via FFT (Fig. 8.34) for implementing IMDCT via IFFT (see Problem 8.8).

8.10 Derive the following equations.

(a) (8.62)	(b) (8.63)	(c) (8.64)	(d) (8.65)

8.11 Let a sampling matrix $[D_3] = \begin{bmatrix} 2 & 1 \\ -1 & 1 \end{bmatrix}$ (see (8.79)). Similar to Fig. 8.56, to show the movement of a unit period in frequency domain to baseband and images, draw the baseband region and images after upsampling. Show details and results of both mapping methods.

Hint: see [F28].

8.32   Projects

8.1 Image registration by Fourier phase correlation method
   (a) Repeat the simulation shown in Fig. 8.7. Register an image with itself and plot the result. Compare those phase correlation functions with ordinary/ standard correlation functions (See Fig. 1 in [IP16]). Use the Lena and Girl images, and the MATLAB code in Appendix H.2. See [IP20]
   (b) Add random noise to the translated images. Register noisy images
   (c) Register the original image with nthe following translated, re-worked images
      (1) Noisy, blurred images.
      (2) Multiply all the pixel values by a number.
      (3) Take the square root of the pixel values.
   (d) Repeat the simulation for the simpler method defined in (8.17).
8.2 Implement phase based watermarking on an image and show the original and watermarked images (see Figs. 3 and 4 in [E2]). Show also an image similar to Fig. 5 in [E2]. Implement JPEG encoder [IP27] on the watermarked image and show that it survives up to 15:1 compression ratio (CR).

See [E4] for Projects 8.3–8.7.

8.3 Implement and display the 512×512 Lena image (watermarked) using the Fourier-Mellin transform (see Fig. 9 in [E4]). Watermark is "The watermark", in ASCII code.
8.4 Rotate the image (Fig. 9 in [E4]) by $43°$, scale by 75% and compress the image (JPEG [IP27, IP28], quality factor 50%) and display the image (Fig. 10 in [E4]). The quality factor is explained in Appendix B.
8.5 Recover the watermark using the scheme described in (Fig. 8.16) Fig. 8 in [E4].
8.6 Obtain and display the log polar map of the Lena image similar to Fig. 11 in [E4].
8.7 Obtain and display the reconstructed Lena image by applying inverse log-polar map to the result of Problem 8.6 (similar to Fig. 12 in [E4]).
8.8 The authors of [E14] indicate that they are updating the frequency masking model to the psychoacoustic model specified in MPEG-1 audio Layer III [D22, D26]. Embed the watermarking using this model (i.e., replace frequency masking based on MPEG-1 audio Layer I – see Fig. 8.28) and implement the watermarking process (see Fig. 8.19). Simulate and test this system. See Figs. 7–13 in [E14].
8.9 *Image Rotation*
   (a) Rotate the Lena image by $60°$ in the spatial domain. Use the MATLAB command imrotate [IP19].
   (b) Reduce the Lena image to the size of 64×64 to reduce the computation time. Implement (8.25) as it is and then rotate the Lena image by $90°$, $180°$ and $270°$. (see Example 5.1)

(c) Implement (8.25) by using the chirp-z algorithm instead of the convolution computation. Rotate the 512×512 Lena image by $90°$, $180°$ and $270°$ [IP10]. Compare calculation/elapsed times of (b) and (c). Use tic and toc.

(d) Given the DFT of an image, rotate the 512×512 Lena image by $180°$ in the DFT domain by using the DFT permutation property (see Example 2.1). Flip the image along left/right and along up/down directions respectively using the same method. (use the commands fliplr and flipud)

(e) Image rotation by $270°$ in the spatial domain can be accomplished in the DFT domain:

I = imread ('lena.jpg '),	A = fftshift (fft2 (I));	B = transpose (A);
C = fftshift (B);	D = ifft2 (C);	imshow (real (D), [ ])

See Section 8.27 for Projects 8.10 and 8.11.

8.10 Implement the steps shown in Fig. 8.51.

8.11 Implement the steps shown in Fig. 8.53.

(a) Use $N = 301$ frequency samples to plot the power density spectrum of $S_Y(f)$ defined in (8.76).

(b) Use the inverse FFT on $S_Y(f)$, to plot the auto correlation function of the filter output $y(t)$.

(c) Use $M = 151$ time samples to implement the decreasing exponential function defined in (8.74).

(d) Compute the auto correlation function $R(m)$, $m = 0, 1, \ldots, 2M - 1$ of (8.74). ($M = 151$) (see Examples 2.5 and 2.6).

(e) Use the FFT on $R(m)$ to plot $S_Y(k)$.

(f) As an alternative to step (e), compute $|H(k)|^2$ where $H(k) = \mathrm{DFT}[h(n)]$, $0 \le k, n \le 300$.

8.12 In [LA12] Raičević and Popović (please access the paper) apply adaptive directional filtering in the frequency domain for fingerprint image enhancement and as well denoising. Block diagram for the fingerprint enhancement is shown in Fig. 4. Note that this is an application of FFT. Figures 6 and 7 show directional filtered images and enhanced image respectively. Apply this algorithm to smudges and corrupted fingerprints (see Fig. 1) and obtain the enhanced images (similar to Fig. 7). Filtered images for directions (a) $22.5°$ and (b) $90°$ are shown in Fig. 6. Obtain filtered images for other directions besides $22.5°$ and $90°$. Write a detailed project report based on your simulations. Review of the references listed at the end of this paper will be very helpful.

8.32.1 *Directional Bandpass Filters [LA15]*

Using polar coordinates (ρ, ϕ), express the filter as a separable function

$$H_\rho^F(\rho, \phi) = H_{\mathrm{radial}}^F(\rho) H_{\mathrm{angle}}^F(\phi) \tag{P8.14}$$

in order to allow independent manipulation of its directional and radial frequency responses.

Any good classical one-dimensional bandpass filter would be adequate for $H_{radial}^F(\rho)$; the *Butterworth filter* is chosen because its implementation is simpler than the Chebyshev or elliptic filter, especially if it is desired to vary the filter order n. This filter is defined as

$$\text{BPF, LPF} \qquad H_{radial}^F(\rho) = \frac{1}{\sqrt{1 + \left(\frac{\rho^2 - \rho_0^2}{\rho\,\rho_{bw}}\right)^{2n}}}, \qquad \rho = \sqrt{k_1^2 + k_2^2} \qquad (P8.15)$$

$$\text{HPF} \qquad H^F(\rho) = \frac{1}{\sqrt{1 + \left(\frac{\rho_{bw}}{\rho}\right)^{2n}}}, \qquad \rho = \sqrt{k_1^2 + k_2^2} \qquad (P8.16)$$

where ρ_{bw} and ρ_0 are desired bandwidth and center frequency respectively. The integer variable n is the order of the filter. A value of $n = 2$ works well for this project. The higher the order is, the more complex its implementation becomes. The 3-dB bandwidth of this filter is $W_2 - W_1 = 2\rho_{bw}$, where W_1 and W_2 are cutoff frequencies at which the square magnitude $|H_{radial}^F(\rho)|^2$ of the frequency response is $1/2$ of its maximum value ($10\log_{10}1/2 = -3$ dB). For $\rho_0 = 0$, this filter (P8.15) becomes a lowpass filter with bandwidth ρ_{bw}. High-frequency noise effects can be reduced by Fourier domain filtering with a zonal lowpass filter. The sharp cutoff characteristic of the zonal lowpass filter (Figs. 5.9 and 5.36) leads to *ringing* artifacts in a filtered image. This deleterious effect can be eliminated by the use of a smooth cutoff filter, such as the Butterworth lowpass filter [B41].

In designing $H_{angle}^F(\phi)$, one cannot be guided by analogy to one-dimensional filters because there is no meaningful concept of orientation. Thus the following function is used.

$$H_{angle}^F(\phi) = \begin{cases} \cos^2\left[\frac{\pi}{2}\frac{(\phi - \phi_c)}{\phi_{bw}}\right] & \text{if } |\phi - \phi_c| < \phi_{bw} \\ 0 & \text{otherwise} \end{cases} \qquad (P8.17)$$

where two times of ϕ_{bw} is the angular bandwidth of the filter, i.e. the range of angles for which $|H_{angle}^F(\phi)| \geq 1/2$, and ϕ_c is its orientation, i.e. the angle at which $|H_{angle}^F|$ is maximum.

If $(2\phi_{bw}) = \pi/8$, we can define $K = 8$ directional filters with equally spaced orientation $\phi_c = k\pi/8$, $k = 0, 1, \ldots, K-1$. These filters sum to unity everywhere; therefore they separate an image into K directional components which sum to the original image.

Appendix A
Performance Comparison of Various Discrete Transforms

This will not include the fast algorithms, separability, recursivity, orthogonality and fast algorithms (complexity of implementation). These topics are described elsewhere in detail (see Chapter 3). The focus is on their various properties. We will consider the random vector $\underline{x}$ is generated by I-order Markov process. When $\underline{x} = (x_0, x_1, \ldots, x_{N-1})^T$, $(x_0, x_1, \ldots, x_{N-1}$ are the N random variables) correlation matrix $[R_{xx}]$ is generated by the I-order Markov process,

$$\underbrace{[R_{xx}]}_{(N \times N)} = \underbrace{\left[\rho^{|j-k|} \right]}_{(N \times N)}, \quad \rho = \text{adjacent correlation coefficient}$$

$$j, k = 0, 1, \ldots, N-1 \tag{A.1}$$

The covariance matrix $[\Sigma]$ in the data domain is mapped into the transform domain as $[\tilde{\Sigma}]$ is

$$\underbrace{[\tilde{\Sigma}]}_{(N \times N)} = \underbrace{[\text{DOT}]}_{(N \times N)} \underbrace{[\Sigma]}_{(N \times N)} \underbrace{\left([\text{DOT}]^T \right)^*}_{(N \times N)} \tag{A.2}$$

DOT stands for *discrete orthogonal transform*. Superscripts T and $*$ denote transpose and complex conjugate respectively.

When DOT is KLT, $[\tilde{\Sigma}]$ is a diagonal matrix as all the transform coefficients in the KLT domain are uncorrelated. For all the other DOTs, residual correlation (correlation left undone in the DOT domain) is defined as [G6, G10]

$$r = \frac{1}{N} \left(||\Sigma||^2 - \sum_{n=0}^{N-1} |\tilde{\Sigma}_{nn}|^2 \right) = \frac{1}{N} \left(\sum_{m=0}^{N-1} \sum_{n=0}^{N-1} |\Sigma_{mn}|^2 - \sum_{n=0}^{N-1} |\tilde{\Sigma}_{nn}|^2 \right) \tag{A.3}$$

where $||\Sigma||^2$ is the Hilbert–Schmidt norm defined as $||\Sigma||^2 = \frac{1}{N} \sum_{m=0}^{N-1} \sum_{n=0}^{N-1} |\Sigma_{mn}|^2$.

Note that N is the size of discrete signal. Σ_{mn} is the element of $[\Sigma]$ in row m and column n ($m, n = 0, 1, \ldots, N-1$).

For a 2D-random signal such as an image assuming that row and column statistics are independent of each other, the variances of the $(N \times N)$ samples can be easily obtained. This concept is extended for computing the variances of the $(N \times N)$ transform coefficients.

A.1   Transform Coding Gain

$\underset{(N \times N)}{[\Sigma]}$ = Correlation or covariance matrix in data domain (see Section 5.6)

$\underset{(N \times N)}{[\tilde{\Sigma}]} = \underset{(N \times N)}{[A]} \; \underset{(N \times N)}{[\Sigma]} \; \underset{(N \times N)}{([A]^T)^*}$ (Note that $([A]^T)^* = [A]^{-1}$ for unitary transforms)

= Correlation or covariance matrix in transform domain

The transform coding gain, G_{TC} is defined as

$$G_{TC} = \frac{\dfrac{1}{N}\displaystyle\sum_{k=0}^{N-1}\tilde{\sigma}_{kk}^2}{\left(\displaystyle\prod_{k=0}^{N-1}\tilde{\sigma}_{kk}^2\right)^{1/N}} = \frac{\text{Arithmetic Mean}}{\text{Geometric Mean}}$$

where $\tilde{\sigma}_{kk}^2$ is the variance of the kth transform coefficient $(k = 0, 1, \ldots, N-1)$. As the sum of the variances in any orthogonal transform domain is invariant (total energy is preserved), G_{TC} can be maximized by minimizing the geometric mean [B23]. The lower bound on the gain is 1 (as seen in Fig. A.1), which is attained only if all the variances are equal.

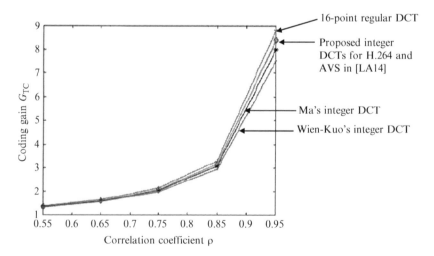

Fig. A.1 Comparing coding gain of orthogonal transforms [LA14] © 2009 IEEE

A.2 Variance Distribution in the Transform Domain

It is desirable to have few transform coefficients with large variances (this implies the remaining coefficients will have small variances, as the sum of the variances is invariant). This variance distribution can be described graphically or in a tabular form for $N = 8, 16, 32, \ldots$ and $\rho = 0.9, 0.95, \ldots$, etc.

The compaction of the energy in few transform coefficients can be represented by the *normalized basis restriction error* [B6] defined as

$$J_m(\rho) = \frac{\sum\limits_{k=m}^{N-1} \tilde{\sigma}_{kk}^2}{\sum\limits_{k=0}^{N-1} \tilde{\sigma}_{kk}^2}, \quad m = 0, 1, \ldots, N-1 \quad \text{Eq. (5.179) in [B6]} \qquad (5.72)$$

where $\tilde{\sigma}_{kk}^2$ have been arranged in decreasing order. See Table 5.2 and Fig. 5.33 about variance distribution of transform coefficients (See also Table 5.2 and Fig. 5.18 in [B6]).

A.3 Normalized MSE (see Figs. 5.35−5.39 and see also Figs. 5.21−5.23 in [B6])

$$J_s = \frac{\sum\limits_{k,l \in stopband} |v_{k,l}|^2}{\sum\limits_{k,l=0}^{N-1} |v_{k,l}|^2}$$

$J_s =$ Energy in stopband/total energy, $v_{k,l}$ $(k, l = 0, 1, \ldots, N-1)$ are the transform coefficients of an $(N \times N)$ image.

A.4 Rate versus Distortion (Rate-Distortion) [B6]

The rate distortion function R_D is the minimum average rate (bits/sample) for coding a signal at a specified distortion D (mean square error) [B6].

Let $x_0, x_1, \ldots, x_{N-1}$ be Gaussian random variables encoded independently and $\hat{x}_0, \hat{x}_1, \ldots, \hat{x}_{N-1}$ be their reproduced values. X_k and $\hat{X}_k$ $k = 0, 1, \ldots, N-1$ are the corresponding transform coefficients. Then the average mean square distortion is

$$D = \frac{1}{N} \sum_{n=0}^{N-1} E\left[(x_n - \hat{x}_n)^2\right] = \frac{1}{N} \sum_{k=0}^{N-1} E\left[(X_k - \hat{X}_k)^2\right]$$

For a fixed average distortion D, the rate distortion function R_D is

$$R_{D(0)} = \frac{1}{N} \sum_{k=0}^{N-1} \max\left(0, \frac{1}{2} \log_2 \frac{\tilde{\sigma}_{kk}^2}{\theta}\right) \qquad \text{Eq. (2.118) in [B6]}$$

where threshold θ is determined by solving

$$D(\theta) = \frac{1}{N} \sum_{k=0}^{N-1} \min\left(\theta, \ \tilde{\sigma}_{kk}^2\right), \quad \min_{kk} \{\tilde{\sigma}_{kk}^2\} \leq \theta \leq \max_{kk} \{\tilde{\sigma}_{kk}^2\} \quad \text{Eq. (2.119) in [B6]}$$

Select a value for
$$\theta$$

$$D(\theta) \quad \Leftrightarrow \quad R_{D(\theta)}$$

to get a point in the plot of R_D versus D.

Develop R_D versus D for various discrete transforms based on I-order Markov process given N and ρ adjacent correlation coefficient. Plot R_D vs. D, for $N = 8, 16, 32, \ldots$ and $\rho = 0.9, 0.95, \ldots$

For I-order Markov process (Eq. 2.68 in [B6]):

$$[\Sigma]_{jk} = \sigma_{jk}^2 = \rho^{|j-k|} \qquad j, k = 0, 1, \ldots, N-1$$

maximum achievable coding gain is

$$G_N(\rho) = \frac{(1/N)\mathrm{tr}[\Sigma]}{(\det[\Sigma])^{1/N}} = \left(1 - \rho^2\right)^{-(1-1/N)}$$

where tr = trace of the matrix, det = determinant of the matrix (see Appendix C in [B9]).

A.5   Residual Correlation [G1]

While the KLT completely decorrelates a random vector [B6], other discrete transforms fall short of this. An indication of the extent of decorrelation can be gauged by the correlation left undone by the discrete transform. This can be measured by the absolute sum of the cross covariance in the transform domain i.e.,

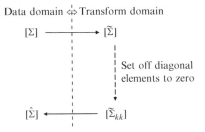

Fig. A.2 The relation between $[\Sigma]$ and $\left[\tilde{\Sigma}\right]$

$$\sum_{\substack{i=0 \\ i\neq j}}^{N-1} \sum_{j=0}^{N-1} |\tilde{\sigma}_{ij}^2| \tag{A.4}$$

for $N = 8, 16, 32, \ldots$ as a function of ρ (Fig. A.2).

$$\text{Given } \underset{(N\times N)}{\left[\tilde{\Sigma}\right]} = \underset{(N\times N)}{[A]} \underset{(N\times N)}{[\Sigma]} \underset{(N\times N)}{[A]^{T^*}} \tag{A.5}$$

$$\text{obtain } \underset{(N\times N)}{\left[\tilde{\Sigma}\right]} = \underset{(N\times N)}{[A]^{T^*}} \underset{(N\times N)}{\left[\tilde{\Sigma}_{kk}\right]} \underset{(N\times N)}{[A]} \tag{A.6}$$

where $\left[\tilde{\Sigma}_{kk}\right]$ is a diagonal matrix whose diagonal elements are the same as those of $\left[\tilde{\Sigma}\right]$, i.e.,

$$\left[\tilde{\Sigma}_{kk}\right] = \text{diag}\left(\tilde{\sigma}_{00}^2, \tilde{\sigma}_{11}^2, \ldots, \tilde{\sigma}_{(N-1)(N-1)}^2\right)$$

It should be recognized that the conjugate appears in (A.5) that is derived in (5.42a), whereas the 2-D discrete transform of $[\Sigma]$ is $[A][\Sigma][A]^T$ defined in (5.6a) and has no conjugate. Thus (A.5) can be regarded as a separable two-dimensional unitary transform of $[\Sigma]$ for purposes of computation. Plot residual correlation versus ρ for DCT, DFT, KLT and ST [B23].

Fractional correlation (correlation left undone by a transform – for KLT this is zero, as KLT diagonalizes a covariance or correlation matrix) is defined as

$$\frac{\|[\Sigma] - [\tilde{\Sigma}]\|^2}{\|[\Sigma] - [I]\|^2} \tag{A.7}$$

where $[I_N]$ is an $(N \times N)$ unit matrix and $\|[A]\|^2 = \sum_{j=0}^{N-1} \sum_{k=0}^{N-1} |[A]_{jk}|^2$. Note that the measures (A.3), (A.4) and (A.7) are zeros respectively for the KLT as $\left[\hat{\Sigma}\right] = \left[\tilde{\Sigma}_{kk}\right]$.

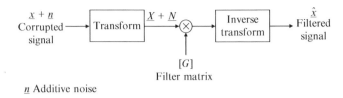

$\underline{x} + \underline{n}$
Corrupted signal ⟶ Transform $\xrightarrow{\underline{X} + \underline{N}}$ ⊗ ⟶ Inverse transform ⟶ Filtered signal $\hat{\underline{x}}$

$[G]$
Filter matrix

$\underline{n}$ Additive noise

Fig. A.3 Scalar Wiener filtering

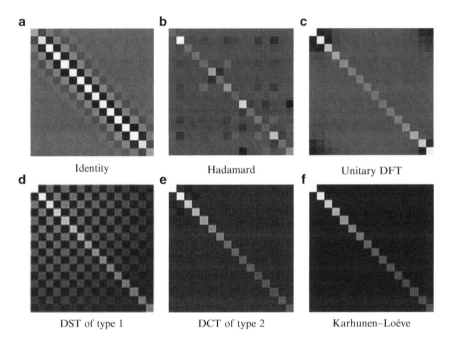

Fig. A.4 Magnitude displays of Wiener filter matrices $[G]$ for a vector length of 16 elements $(N = 16)$. *Dark pixels* represent zero values, *light pixels* represent one values, and *gray pixels* represent values in between. Signal-to-noise ratio is 0.3 and $\rho = 0.9$. Dynamic ranges of the Wiener filter magnitudes in the figure are compressed via the logarithmic transformation defined in (5.26)

A.6 Scalar Wiener Filtering

Filter matrix $[G]$ is optimized for a specific transform, such that the noise can be filtered (Fig. A.3) [G5]. Evaluate MSE $= E(||\underline{x} - \hat{\underline{x}}||^2)$ for the discrete transforms (appearing in Fig. A.4, plus Haar and slant transforms defined in [B6]) for $N = 4, 8, 16, 32,$ and $\rho = 0.9$ and 0.95.

Plot magnitude displays of various discrete transforms referring to Fig. A.4. Comparing the filter planes, the filter characteristic changes drastically for different unitary transforms. For the KLT the filtering operation is a scalar multiplication, while for the identity transform most elements of the filter matrix are of relatively large magnitude. The DFT filter matrix contains large magnitude terms along the diagonal and terms of decreasing magnitude away from the diagonal [LA13].

A.7   Geometrical Zonal Sampling (GZS)

Geometrical zonal filter can be 2:1, 4:1, 8:1, or 16:1 (sample reduction) (Fig. A.5). See Fig. A.6 for 2:1 and 4:1 sample reduction in the 2D-DCT domain.

Note that for 2D-DFT, the low frequency zones need to be appropriately identified (see Figs. 5.8 and 5.9).

The reconstructed images for various sample reductions can be obtained and a plot of the normalized MSE vs. various sample reduction ratios for all the DOTs can be implemented.

$$\text{Normalized MSE} = \frac{\sum_{m=0}^{N-1}\sum_{n=0}^{N-1} E(|x(m,n) - \hat{x}(m,n)|^2)}{\sum_{m=0}^{N-1}\sum_{n=0}^{N-1} E(|x(m,n)|^2)} \qquad (A.8)$$

```
Original Image                          Retain only                        Reconstructed
   (N × N)                             a fraction of the low      2-D          Image
                     2-D               frequency transform      Inverse
      [x]            DOT               coefficients with the      DOT           [x̂]
   (N × N)                                rest set to zero                    (N × N)
```

Fig. A.5 Geometrical zonal sampling (DOT: Discrete orthogonal transform)

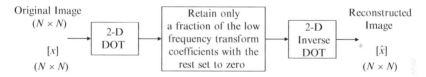

Fig. A.6 Sample reduction in the 2D-DCT domain

(Passband) Retain
(Stopband) Set to zero
2:1 Sample reduction

(Passband) Retain
(Stopband) Set to zero
4:1 Sample reduction

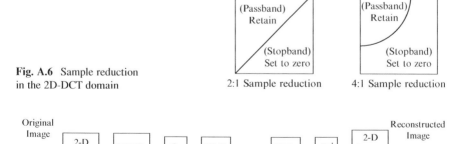

```
Original                                                                    Reconstructed
  Image                                                                        Image
           2-D                                                      2-D
    [x]    DOT   MVZS    Q    VLC  ········  VLD   Q⁻¹   Inverse               [x̂]
 (N × N)                                                   DOT               (N × N)
```

Fig. A.7 Maximum variance zonal sampling

A.8 Maximum Variance Zonal Sampling (MVZS)

In MVZS, transform coefficients with large variances can be selected for quantization and coding with the remainder (transform coefficients with small variances) set to zero. At the receiver side inverse operations are carried out resulting in the reconstructed signal or image (Fig. A.7).

Appendix B
Spectral Distance Measures of Image Quality

The following material is based on [IP36]. The same notation as in [IP36] is used in this appendix. This appendix discusses various image quality measures using the distortion metrics evaluated in 2D-DFT domain. This exemplifies another application of 2D-DFT. 2D-DFT can be used as a measure of image quality in terms of the reconstructed image [IP36].

$$C_k(n_1, n_2) \longleftarrow (n_1, n_2)\text{th pel of the } k\text{th band of } \underline{C}(n_1, n_2),$$
$$k = 1, 2, \ldots, K, \# \text{ of bands} = K$$
$$\text{or } k\text{th spectral component at location } (n_1, n_2).$$
$$\text{Each band is of size } (N \times N).$$

e.g., Color images in (RGB), or YIQ or Y C_R C_B

$$\underbrace{\underline{C}(n_1, n_2)}_{(3 \times 1)} = \begin{bmatrix} R(n_1, n_2) \\ G(n_1, n_2) \\ B(n_1, n_2) \end{bmatrix}, \quad (n_1, n_2)\text{th multispectral (with } K = 3 \text{ bands) pel vector}$$

$\underline{C}$ Multispectral image
C_k kth band of multispectral image $\underline{C}$

$$\underbrace{\hat{\underline{C}}(n_1, n_2)}_{(3 \times 1)} = \begin{bmatrix} \hat{R}(n_1, n_2) \\ \hat{G}(n_1, n_2) \\ \hat{B}(n_1, n_2) \end{bmatrix} \quad \begin{array}{l} \text{processed or reconstructed multispectral image} \\ \text{at location } (n_1, n_2) \end{array}$$

$$\epsilon_k = C_k - \hat{C}_k, \quad \text{error over all the pels in the } k\text{th band of } \underline{C}$$

$$\text{Power in } k\text{th band} \quad \sigma_k^2 = \sum_{n_1=0}^{N-1} \sum_{n_2=0}^{N-1} C_k^2(n_1, n_2)$$

$\hat{C}_k(n_1, n_2)$, $\hat{\underline{C}}$, $\hat{\underline{C}}(n_1, n_2)$ refer to processed or reconstructed (distorted) images.

Original
image $\quad\longrightarrow\quad$ [Image processing / e.g., Codec / (encoder / decoder)] $\quad\longrightarrow\quad$ Reconstructed
image

$\underline{C}(n_1, n_2)$ $\qquad\qquad\qquad\qquad\qquad\qquad\qquad$ $\hat{\underline{C}}(n_1, n_2)$

Note that $\displaystyle\sum_{n_1=0}^{N-1}\sum_{n_2=0}^{N-1} = \sum_{n_1,n_2=0}^{N-1}$.

Sum of the errors in all K bands at pel (n_1, n_2)

$$||\underline{C}(n_1, n_2) - \hat{\underline{C}}(n_1, n_2)||^2 = \sum_{k=1}^{K}\left[C_k(n_1, n_2) - \hat{C}_k(n_1, n_2)\right]^2$$

(Square of the error in the kth band of pel (n_1, n_2))
(K = # of bands, $k = 1, 2, \ldots, K$)

$$\epsilon_k^2 = \sum_{n_1=0}^{N-1}\sum_{n_2=0}^{N-1}\left[C_k(n_1, n_2) - \hat{C}_k(n_1, n_2)\right]^2$$

$N \times N$ image ($N \times N$)

Define

$$W_N = \exp\left(\frac{-j2\pi}{N}\right) = N\text{th root of unity}$$

$$\Gamma_k(k_1, k_2) = \sum_{n_1,n_2=0}^{N-1}C_k(n_1, n_2)W_N^{n_1 k_1}W_N^{n_2 k_2}, \quad k = 1, 2, \ldots, K\ (k_1, k_2 = 0, 1, \ldots, N-1)$$

$$= \text{2D-DFT of } C_k(n_1, n_2)$$

$$\hat{\Gamma}_k(k_1, k_2) = \sum_{n_1,n_2=0}^{N-1}\hat{C}_k(n_1, n_2)W_N^{n_1 k_1}W_N^{n_2 k_2}, \quad k = 1, 2, \ldots, K\ (k_1, k_2 = 0, 1, \ldots, N-1)$$

$$= \text{2D-DFT of } \hat{C}_k(n_1, n_2)$$

Phase spectra

$$\phi(k_1, k_2) = \arctan[\Gamma(k_1, k_2)], \quad \hat{\phi}(k_1, k_2) = \arctan\left[\hat{\Gamma}(k_1, k_2)\right]$$

Magnitude spectra

$$M(k_1, k_2) = |\Gamma(k_1, k_2)|, \ \hat{M}(k_1, k_2) = |\hat{\Gamma}(k_1, k_2)|$$

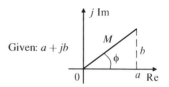

Given: $a + jb$

Spectral magnitude distortion

$$S = \frac{1}{N^2} \sum_{k_1, k_2 = 0}^{N-1} |M(k_1, k_2) - \hat{M}(k_1, k_2)|^2$$

Spectral phase distortion

$$S1 = \frac{1}{N^2} \sum_{k_1, k_2 = 0}^{N-1} |\phi(k_1, k_2) - \hat{\phi}(k_1, k_2)|^2$$

Magnitude

$$M = \sqrt{a^2 + b^2} = |a + jb|$$

Phase

$$\phi = \tan^{-1}\left(\frac{b}{a}\right)$$

Weighted spectral distortion

$$S2 = \frac{1}{N^2} \left[\lambda \left(\sum_{k_1, k_2 = 0}^{N-1} |\phi(k_1, k_2) - \hat{\phi}(k_1, k_2)|^2 \right) \right.$$
$$\left. + (1 - \lambda) \left(\sum_{k_1, k_2 = 0}^{N-1} |M(k_1, k_2) - \hat{M}(k_1, k_2)|^2 \right) \right]$$

λ is chosen to match commensurate weights to the phase and magnitude terms. $0 \leq \lambda \leq 1$. Note that for $\lambda = 1$, $S2$ reduces to $S1$ while for $\lambda = 0$, $S2$ reduces to S.

Let 2D-DFT of lth block of kth band image $C_k^l(n_1, n_2)$ be

Each of size $(b \times b)$

$$\Gamma_k^l(k_1, k_2) = \sum_{n_1, n_2 = 0}^{b-1} C_k^l(n_1, n_2)\, W_b^{n_1 k_1} W_b^{n_2 k_2}$$

$$W_b = \exp\left(\frac{-j2\pi}{b}\right)$$

$$\Gamma_k^l(k_1, k_2) = |\Gamma_k^l(k_1, k_2)| e^{j\phi_k^l(k_1, k_2)}$$
$$= m_k^l(k_1, k_2) e^{j\phi_k^l(k_1, k_2)}$$

$$k_1, k_2 = 0, 1, \ldots, N - 1 \qquad l = 1, 2, \ldots, L$$

L is number of $(b \times b)$ size overlapping or non-overlapping blocks.

$$J_M^l = \frac{1}{K} \sum_{k=1}^{K} \left(\sum_{k_1,k_2=0}^{b-1} \left[|\Gamma_k^l(k_1,k_2)| - |\hat{\Gamma}_k^l(k_1,k_2)| \right]^\gamma \right)^{1/\gamma}$$

$$J_\phi^l = \frac{1}{K} \sum_{k=1}^{K} \left(\sum_{k_1,k_2=0}^{b-1} \left[|\phi_k^l(k_1,k_2)| - |\hat{\phi}_k^l(k_1,k_2)| \right]^\gamma \right)^{1/\gamma}$$

$$J^l = \lambda J_M^l + (1 - \lambda) J_\phi^l$$

λ is the relative weighting factor of the magnitude and phase spectra.

Various rank order operations of the block spectral difference J_M and/or J_ϕ can be useful. Let $J^{(1)}$, $J^{(2)}$, ..., $J^{(L)}$ be the rank ordered block distortions such that, $J^{(L)} = \max_l \left(J^{(l)} \right)$.

Consider rank order averages:

Median block distortion $\frac{1}{2} \left[J^{L/2} + J^{(L/2)+1} \right]$ when L is an even number

Maximum block distortion $J^{(L)}$

Average block distortion $\frac{1}{L} \left(\sum_{i=1}^{L} J^{(i)} \right)$

The median of the block distortions is the most effective average of rank ordered block spectral distortion.

$$S3 = \underset{l}{\text{median}}\, J_m^l$$

$$S4 = \underset{l}{\text{median}}\, J_\phi^l$$

$$S5 = \underset{l}{\text{median}}\, J^l$$

(use $\gamma = 2$), block sizes of (32×32) and (64×64) yield better results than lower or higher range.

Project B

This project is related to JPEG baseline system at 0.25, 0.5, 0.75 and 1 bpp. Please refer to Appendix B and [IP36]. JPEG software is on the web in [IP22].

$$C(n_1, n_2) \longrightarrow \boxed{\begin{array}{c} \text{JPEG} \\ \text{Encoder} / \\ \text{Decoder} \end{array}} \longrightarrow \hat{C}(n_1, n_2)$$

$$\begin{pmatrix} 0.25,\ 0.5,\ 0.75 \\ \text{and 1 bpp} \end{pmatrix}$$

Fig. B.1 The DFTs of the
original and reconstructed
images

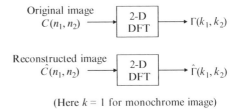

(Here $k = 1$ for monochrome image)

The objective of this project is how to correlate MSE, SNR, peak-to-peak SNR in dB with spectral distance measures described in [IP36] for Lena image for the above bit rates (Fig. B.1).

Compute spectral magnitude distortion S, spectral phase distortion $S1$, weighted spectral distortion $S2$, and median block distortion $S5$. Referring to Fig. B.2 and Table B.1, plot (also show in tabular form):

(a) Bit rate vs. S
(b) Bit rate vs. $S1$
(c) Bit rate vs. $S2$ (use $\lambda = 0.5$)
(d) Bit rate vs. $S5$ (use $\lambda = 0.5$ and block size $b = 32$)

What conclusions can you draw from these plots (tables)?

For JPEG, bit rate can be controlled by the quality factor [IP28]. Quantized DCT coefficients $S_q(u, v)$ are defined as

$$S_q(u, v) = \text{Nearest integer of} \left[\frac{S(u, v)}{Q(u, v)} \right]$$

where $S(u, v)$ is the DCT coefficient, $Q(u, v)$ is the quantization matrix element and (u, v) indexes the DCT frequency. Inverse quantization removes the normalized factor.

$$\tilde{S}(u, v) = S_q(u, v) \, Q(u, v)$$

where $\tilde{S}(u, v)$ is a dequantized DCT coefficient to be fed into the 2D (8×8) IDCT operation.

A quality factor q_JPEG, an integer in a range of 1–100, is used to control the elements of quantization matrix, $Q(u, v)$ [IP27], [IP28] and thus the quality factor can adjust the bit rate. $Q(u, v)$ is multiplied by the compression factor α, defined as

$$\alpha = \frac{50}{q\_JPEG} \qquad 1 \leq q\_JPEG < 50$$

$$\alpha = 2 - \frac{2 \times q\_JPEG}{100} \qquad 50 \leq q\_JPEG \leq 99$$

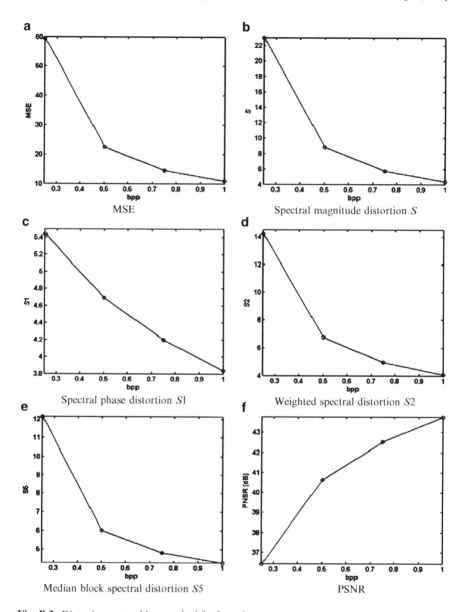

Fig. B.2 Distortion versus bits per pixel for Lena image

Table B.1 Various measures of JPEG [IP27] [IP28] reconstructed images

Image	bpp	Quality factor	CR	MSE	S	S1	S2	S5	PSNR (dB)
Lena10	0.25	10	32:1	59.481	22.958	5.436	14.197	12.144	36.442
Lena34	0.5	34	16:1	22.601	8.881	4.693	6.787	6.027	40.644
Lena61	0.75	61	32:3	14.678	5.738	4.199	4.968	4.823	42.519
Lena75	1	75	8:1	10.972	4.348	3.835	4.092	4.271	43.783

subject to the condition that the minimum value of the modified quantization matrix elements, $\alpha Q(u, v)$, is 1. For $q\_JPEG = 100$, all the elements of $\alpha Q(u, v)$ are set to 1. For example $\alpha = 1$ for $q\_JPEG = 50$, $\alpha = 2$ for $q\_JPEG = 25$ and $\alpha = 1/2$ for $q\_JPEG = 75$.

Appendix C
Integer Discrete Cosine Transform (INTDCT)

Several versions of the integer DCT have been developed. The objective is to implement the INTDCT with integer additions and integer multiplications (bit adds and shifts) and also implement the inverse INTDCT similarly. This also ensures that there is no mismatch nor error accumulation that is inherent in going through DCT and IDCT in the hybrid MC transform DPCM loop standardized in H.261, MPEG-1, MPEG-2, MPEG-4 visual [D37], AVS China [I-21] and WMV-9 [I-23]. As an example several (8 × 8) INTDCTs are listed below.

1. Cham [I-13]
2. H.264 [I-22, LA19]
3. WMV-9 [I-23]
4. AVS-China [I-21]

For the theory behind developing the INTDCTs the reader is referred to [I-13].

C.1   Integer DCT Via Lifting [B28]

The DCT and its inverse are defined as

$$X^{CII}(k) = \sqrt{\frac{2}{N}}\epsilon(k) \sum_{n=0}^{N-1} x(n) \cos\left[\frac{\pi k\left(n+\frac{1}{2}\right)}{N}\right] \qquad k = 0, 1, \ldots, N-1 \qquad \text{(C.1a)}$$

$$x(n) = \sqrt{\frac{2}{N}} \sum_{k=0}^{N-1} \epsilon(k)X^{CII}(k) \cos\left[\frac{\pi k\left(n+\frac{1}{2}\right)}{N}\right] \qquad n = 0, 1, \ldots, N-1 \qquad \text{(C.1b)}$$

where

$$\epsilon(k) = \begin{cases} 1/\sqrt{2} & \text{when } k = 0 \\ 1 & \text{when } k \neq 0 \end{cases}$$

In the same way as with Integer FFT (see Chapter 4), Integer DCT can be developed. To construct Integer DCT, the DCT kernel is decomposed into Givens rotations via Walsh–Hadamard transform [I-15].

$$[R_0] = \begin{bmatrix} \cos\theta & -\sin\theta \\ \sin\theta & \cos\theta \end{bmatrix} \Leftrightarrow [R_0]^{-1} = \begin{bmatrix} \cos\theta & \sin\theta \\ -\sin\theta & \cos\theta \end{bmatrix} = [R_{-\theta}] = [R_0]^T$$

(C.2)

Then a Givens rotation can be lifted as follows.

$$[R_0] = \begin{bmatrix} 1 & \frac{\cos 0 - 1}{\sin 0} \\ 0 & 1 \end{bmatrix} \begin{bmatrix} 1 & 0 \\ \sin\theta & 1 \end{bmatrix} \begin{bmatrix} 1 & \frac{\cos 0 - 1}{\sin 0} \\ 0 & 1 \end{bmatrix}$$

$$\Leftrightarrow [R_0]^{-1} = \begin{bmatrix} 1 & \frac{\cos 0 - 1}{\sin 0} \\ 0 & 1 \end{bmatrix} \begin{bmatrix} 1 & 0 \\ \sin\theta & 1 \end{bmatrix} \begin{bmatrix} 1 & \frac{\cos 0 - 1}{\sin 0} \\ 0 & 1 \end{bmatrix}$$

(C.3)

Once a butterfly structure of (C.2) changes into a lattice structure of (C.3), non-linear quantizing operation Q such as "flooring" or "rounding" can be included in the lifting steps to construct integer DCT.

$$[R_g] = \begin{bmatrix} 1 & a \\ 0 & 1 \end{bmatrix} \begin{bmatrix} 1 & 0 \\ b & 1 \end{bmatrix} \begin{bmatrix} 1 & c \\ 0 & 1 \end{bmatrix} \Leftrightarrow [R_g]^{-1} = \begin{bmatrix} 1 & -c \\ 0 & 1 \end{bmatrix} \begin{bmatrix} 1 & 0 \\ -b & 1 \end{bmatrix} \begin{bmatrix} 1 & -a \\ 0 & 1 \end{bmatrix}$$

(C.4)

where a, b and c are any real numbers. Note that the quantization of the lifting coefficients in the forward structure is cancelled out in the inverse structure. Thus the lattice structure is reversible or invertible. Also, the lattice structure of (C.4) has the property:

$$[R_g] = \begin{bmatrix} d & e \\ f & g \end{bmatrix} \Leftrightarrow [R_g]^{-1} = \begin{bmatrix} g & -e \\ -f & d \end{bmatrix}$$

(C.5)

C.1.1 Decomposition of DCT Via Walsh–Hadamard Transform

The Walsh–Hadamard transform (WHT) can be used to develop Integer DCT [I-15]. The WHT requires only additions (subtractions) as all of its elements are ± 1. The eight-point Walsh-ordered WHT matrix is given by

$$[H_w] = \begin{bmatrix} 1 & 1 & 1 & 1 & 1 & 1 & 1 & 1 \\ 1 & 1 & 1 & 1 & -1 & -1 & -1 & -1 \\ 1 & 1 & -1 & -1 & -1 & -1 & 1 & 1 \\ 1 & 1 & -1 & -1 & 1 & 1 & -1 & -1 \\ 1 & -1 & -1 & 1 & 1 & -1 & -1 & 1 \\ 1 & -1 & -1 & 1 & -1 & 1 & 1 & -1 \\ 1 & -1 & 1 & -1 & -1 & 1 & -1 & 1 \\ 1 & -1 & 1 & -1 & 1 & -1 & 1 & -1 \end{bmatrix}$$

(C.6)

A bit-reversal matrix $[B]$ is one which rearranges the input sequence into the bit-reversal order. Let $\underline{X} = \{x(0), x(1), x(2), x(3), x(4), x(5), x(6), x(7)\}^T$, then the product

$$[B]\underline{X} = \{x(0), x(4), x(2), x(6), x(1), x(5), x(3), x(7)\}^T$$

has its entries in bit-reversed order.

The DCT of input sequence $\underline{x}$ is $\underline{X} = [C^{II}]\underline{x}$, where $[C^{II}]$ is the type-2 DCT kernel (see (C.1a)). Now let us normalize $[H_w]$ such that $[\hat{H}_w] = \sqrt{1/N}\,[H_w]$ is an orthonormal matrix. Hence $[\hat{H}_w]^{-1} = [\hat{H}_w]^T$. Then one can prove that

$$\underline{X}_{BRO} \equiv [B]\underline{X} = \sqrt{1/N}\,[T_{BRO}][B][H_w]\,\underline{x}$$

where $[T_{BRO}] = [B]\left(\underline{X}[\hat{H}_w]\right)[B]^T$ is block-diagonal. In terms of the conversion matrix $[T_{BRO}]$, the DCT can be decomposed as

$$\underline{X} = [B]^T \underline{X}_{BRO} = [B]^T [T_{BRO}][B][\hat{H}_w]\,\underline{x} \tag{C.7}$$

For eight-point input sequence, $[T_{BRO}]$ is given explicitly by

$$[T_{BRO}] = \begin{bmatrix} \begin{matrix} 1 & 0 \\ 0 & 1 \end{matrix} & & & \\ & \begin{bmatrix} 0.924 & 0.383 \\ -0.383 & 0.924 \end{bmatrix} & & O \\ & & O & \\ & & \begin{bmatrix} 0.906 & -0.075 & 0.375 & 0.180 \\ 0.213 & 0.768 & -0.513 & 0.318 \\ -0.318 & 0.513 & 0.768 & 0.213 \\ -0.180 & -0.375 & -0.075 & 0.906 \end{bmatrix} \end{bmatrix} \tag{C.8}$$

$$= \begin{bmatrix} \begin{matrix} 1 & 0 \\ 0 & 1 \end{matrix} & & \\ & [U_{-\pi/8}] & O \\ & O & [U_4] \end{bmatrix} \tag{C.9}$$

$$[U_4] = [B_4] \begin{bmatrix} 0.981 & 0 & 0 & 0.195 \\ 0 & & & 0 \\ 0 & [U_{-3\pi/16}] & & 0 \\ -0.195 & 0 & 0 & 0.981 \end{bmatrix} \begin{bmatrix} [U_{-\pi/8}] & O \\ O & [U_{-\pi/8}] \end{bmatrix} [B_4] \tag{C.10}$$

where

$$[B_4] = \begin{bmatrix} 1 & 0 & 0 & 0 \\ 0 & 0 & 1 & 0 \\ 0 & 1 & 0 & 0 \\ 0 & 0 & 0 & 1 \end{bmatrix} \qquad [U_{-3\pi/16}] = \begin{bmatrix} 0.832 & 0.557 \\ -0.557 & 0.832 \end{bmatrix}$$

$$[U_{\pi/16}] = \begin{bmatrix} 0.981 & 0.195 \\ -0.195 & 0.981 \end{bmatrix} \qquad [U_{-\pi/8}] = \begin{bmatrix} 0.924 & 0.383 \\ -0.383 & 0.924 \end{bmatrix}$$

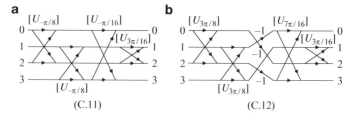

Fig. C.1 Flowgraphs of $[U_4]$. [I-17] © 2000 IEEE

Flowgraph of $[U_4]$ is shown in Fig. C.1. When we draw the flowgraph of (C.10), the bit-reversed ordering matrices $[B_4]$ can be absorbed. Thus $[U_{-3\pi/16}]$ is changed to $[U_{3\pi/16}]$. Figure C.1a can be represented in matrix form as

$$[U_4] = \begin{bmatrix} 0.981 & 0 & 0 & 0.195 \\ 0 & & & 0 \\ 0 & [U_{3\pi/16}] & & 0 \\ -0.195 & 0 & 0 & 0.981 \end{bmatrix}$$

$$\times \begin{bmatrix} 0.924 & 0 & 0.383 & 0 \\ 0 & 0.924 & 0 & 0.383 \\ -0.383 & 0 & 0.924 & 0 \\ 0 & -0.383 & 0 & 0.924 \end{bmatrix} \quad \text{(C.11)}$$

Equation (C.11) can be factored as

$$[U_4] = \begin{bmatrix} 0.195 & 0 & 0 & -0.981 \\ 0 & & & 0 \\ 0 & [U_{3\pi/16}] & & 0 \\ 0.981 & 0 & 0 & 0.195 \end{bmatrix} \begin{bmatrix} 0 & 0 & 0 & 1 \\ 0 & 1 & 0 & 0 \\ 0 & 0 & 1 & 0 \\ -1 & 0 & 0 & 0 \end{bmatrix}$$

$$\begin{bmatrix} 0 & 0 & 1 & 0 \\ 0 & 0 & 0 & 1 \\ -1 & 0 & 0 & 0 \\ 0 & -1 & 0 & 0 \end{bmatrix} \begin{bmatrix} 0.383 & 0 & -0.924 & 0 \\ 0 & 0.383 & 0 & -0.924 \\ 0.924 & 0 & 0.383 & 0 \\ 0 & 0.924 & 0 & 0.383 \end{bmatrix}$$

$$= \begin{bmatrix} 0.195 & 0 & 0 & -0.981 \\ 0 & & & 0 \\ 0 & [U_{3\pi/16}] & & 0 \\ 0.981 & 0 & 0 & 0.195 \end{bmatrix} \begin{bmatrix} 0 & -1 & 0 & 0 \\ 0 & 0 & 0 & 1 \\ -1 & 0 & 0 & 0 \\ 0 & 0 & -1 & 0 \end{bmatrix} \begin{bmatrix} 0.383 & 0 & -0.924 & 0 \\ 0 & 0.383 & 0 & -0.924 \\ 0.924 & 0 & 0.383 & 0 \\ 0 & 0.924 & 0 & 0.383 \end{bmatrix}$$

$$\text{(C.12)}$$

$$[U_{3\pi/16}] = \begin{bmatrix} 0.832 & -0.557 \\ 0.557 & 0.832 \end{bmatrix} = [U_{-3\pi/16}]^T \qquad \text{See (C.2)}$$

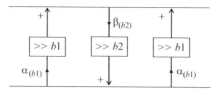

Fig. C.2 The three integer lifting steps approximately implementing a Givens rotation angle. α_b and β_b are the at-most-b-bit lifting multipliers used, which are tabulated in Table C.1. [I-17] © 2000 IEEE

$$\left[U_{7\pi/16}\right] = \begin{bmatrix} 0.195 & -0.981 \\ 0.981 & 0.195 \end{bmatrix} \qquad \left[U_{3\pi/8}\right] = \begin{bmatrix} 0.383 & -0.924 \\ 0.924 & 0.383 \end{bmatrix}$$

The signal flowgraph of (C.12) is shown in Fig. C.1b. From the WHT decomposition of DCT-II, one may then apply integer lifting to obtain an integer-to integer mapping which is reversible and mimics the behavior of the floating-point DCT.

C.1.2 Implementation of Integer DCT

Since Givens rotations can be lifted as shown in (C.2) through (C.4), butterflies in (C.9) can be changed into lattice structures. Then floating-point multipliers β in lattice structures are quantized to take the form $\beta_Q = \pm\, \beta_Q/2^b$ so that they can be implemented only with shifts and adds where $1/2^b$ denotes right shifts by b bits.

Figure C.2 shows three integer lifting steps approximately implementing a Givens rotation. Two integer multipliers α_b and β_b are right shifted by b bits. Table C.1 lists α_b and β_b for four Givens rotation angles with the number of bits used to represent an integer multiplier $b = 1, 2, \ldots, 8$. Simulations show $b = 8$ has comparable performance to the floating-point DCT even when all the internal nodes are limited to 16 bits [I-15]. A general-purpose eight-point IntDCT which requires 45 adds and 18 shifts is shown in Fig. C.3.

C.2 Integer DCT by the Principle of Dyadic Symmetry [I-13]

Another integer DCT can be developed by replacing transform kernel components of the order-eight DCT by a new set of numbers. For the order-eight DCT, many transforms of this kind can be obtained. Boundary conditions are applied to ensure

Table C.1 The multipliers α_b and β_b for $b = 1, 2, \ldots , 8$ and various rotation angles [I-17] © 2000 IEEE

Angle $-\pi/8$								
b bits	1	2	3	4	5	6	7	9
α_b	0	0	1	3	6	12	25	50
β_b	0	−1	−3	−6	−12	−24	−48	−97
Angle $3\pi/8$								
b bits	1	2	3	4	5	6	7	9
α_b	−1	−2	−5	−10	−21	−42	−85	−171
β_b	1	3	7	14	29	59	118	236
Angle $7\pi/16$								
b bits	1	2	3	4	5	6	7	9
α_b	−1	−3	−6	−13	−26	−52	−105	−210
β_b	1	3	7	15	31	62	125	251
Angle $3\pi/16$								
b bits	1	2	3	4	5	6	7	9
α_b	0	1	2	4	9	19	38	77
β_b	−1	−2	−4	−8	−17	−35	−71	−142

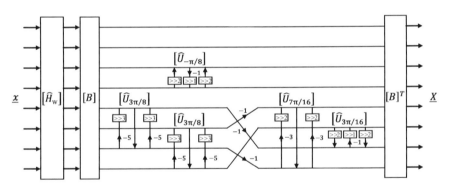

Fig. C.3 The flowgraph of an example of the low-cost, multiplierless integer DCT. $[\hat{U}_{-\pi/8}]$ is an integer approximation of $[U_{-\pi/8}]$ and is chosen from Table C.1. [I-17] © 2000 IEEE

that the new transforms contain only integers, while they still approximate the DCT. As the magnitude of these integers can be very small, the transforms can be simple to implement.

C.2.1 Generation of the Order-Eight Integer DCT

Let the matrix $[TR]$ represents the kernel of the eight-point DCT. The following describes the steps to convert the order-eight DCT into integer DCT kernels.

If the basis vectors $\underline{J}_m$, $m = 0, 1, \ldots , 7$ of $[TR]$ are scaled by $[K]$, all the basis vectors can be expressed as the variables of a, b, c, d, e, f and g as follows.

$$[TR] = [K] \begin{bmatrix} g & g & g & g & g & g & g & g \\ a & b & c & d & -d & -c & -b & -a \\ e & f & -f & -e & -e & -f & f & e \\ b & -d & -a & -c & c & a & d & -b \\ g & -g & -g & g & g & -g & -g & g \\ c & a & d & b & b & d & a & c \\ f & -e & e & -f & -f & e & -e & f \\ d & -c & b & -a & a & -b & c & -d \end{bmatrix} \tag{C.13}$$

$$= \left(k_0\underline{J}_0, k_1\underline{J}_1, k_2\underline{J}_2, k_3\underline{J}_3, k_4\underline{J}_4, k_5\underline{J}_5, k_6\underline{J}_6, k_7\underline{J}_7 \right)^T = [K][J]$$

where $[K] = \mathrm{diag}(k_0, k_1, \ldots , k_7)$ and $\|k_m\underline{J}_m\|^2 = 1$ for $m = 0, 1, \ldots , 7$.

From Table C.2, the only condition to make the transform $[TR]$ be orthogonal is

$$ab = ac + bd + cd \tag{C.14}$$

As there are four variables in (C.14), there are an infinite number of solutions. Thus an infinite number of new orthogonal transforms can be generated.

If we choose k_m, $m = 1, 3, 5, 7$ such that d is unity, then $a = 5.027$, $b = 4.2620$ and $c = 2.8478$. If we choose k_m, $m = 2, 6$ such that f is unity, then $e = 2.4142$. From these numbers we can set up boundary conditions as follows expecting that the new transform approximates the DCT and has as good performance as the DCT.

$$a \geq b \geq c \geq d \text{ and } e \geq f \tag{C.15}$$

Furthermore to eliminate truncation error due to operations of irrational numbers the following condition has to be satisfied.

$$a, b, c, d, e \text{ and } f \text{ are integers.} \tag{C.16}$$

Table C.2 Conditions under which basis row vectors are orthogonal

	$\underline{J}_1$	$\underline{J}_2$	$\underline{J}_3$	$\underline{J}_4$	$\underline{J}_5$	$\underline{J}_6$	$\underline{J}_7$
$\underline{J}_0$	1	1	1	1	1	1	1
$\underline{J}_1$		1	2	1	2	1	1
$\underline{J}_2$			1	1	1	1	1
$\underline{J}_3$				1	1	1	2
$\underline{J}_4$					1	1	1
$\underline{J}_5$						1	2
$\underline{J}_6$							1

1. Orthogonal as their inner product equals zero
2. Orthogonal if $ab = ac + bd + cd$

Data vector   Transform vector

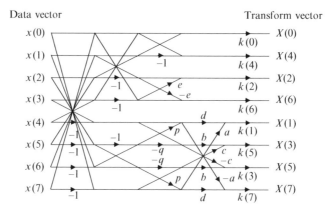

Fig. C.4 Fast algorithm for eight-point forward integer DCT where $p = (b + c)/2a$ and $q = (a - d)/2c$ [I-13]

Transforms that satisfy the conditions of (C.14), (C.15) and (C.16) are referred to as eight-point integer DCTs. Once the variables of $[J]$ are fixed, we can get the scaling factors that make the new transform orthonormal. From (C.13) we can get

$$k_0 = k_4 = 1/\left(2\sqrt{2}\,g\right) \tag{C.17a}$$

$$k_1 = k_3 = k_5 = k_7 = 1/\left(\sqrt{2}\sqrt{a^2 + b^2 + c^2 + d^2}\right) \tag{C.17b}$$

$$k_2 = k_6 = 1/\left(2\sqrt{e^2 + f^2}\right) \tag{C.17c}$$

A fast algorithm for this eight-point integer DCT is shown in Fig. C.4.

C.2.2   Integer DCTs in Video Coding Standards

H.264

The integer DCT is developed for the Fidelity Range Extension (FRExt) of the H.264 / MPEG-4 Advanced Video Coding (AVC) standard [I-22]. We refer to the eight-point integer DCT with $g = 8$, $a = 12$, $b = 10$, $c = 6$, $d = 3$, $e = 8$, $f = 4$ as $DCT(8, 12, 10, 6, 3, 8, 4)$. Substitute these variables in (C.13) and compute the scaling factors that makes $[TR][TR]^T = [I]$ where $[I]$ is the identity matrix. Then the scaling factors are as follows.

$$k_0 = \frac{1}{16\sqrt{2}}, \quad k_1 = \frac{1}{17\sqrt{2}}, \quad k_2 = \frac{1}{8\sqrt{5}} \tag{C.18}$$

Once $[TR][TR]^T = [I]$, then $[TR]^T[TR] = [I]$. Thus $[TR_{h,8}] = [K_{h,8}][J_{h,8}]$ is an orthogonal matrix/transform where

$$[K_{h,8}] = \text{diag}\left(\frac{1}{16\sqrt{2}}, \frac{1}{17\sqrt{2}}, \frac{1}{8\sqrt{5}}, \frac{1}{17\sqrt{2}}, \frac{1}{16\sqrt{2}}, \frac{1}{17\sqrt{2}}, \frac{1}{8\sqrt{5}}, \frac{1}{17\sqrt{2}}\right) \quad \text{(C.19)}$$

$$[J_{h,8}] = \begin{bmatrix} 8 & 8 & 8 & 8 & 8 & 8 & 8 & 8 \\ 12 & 10 & 6 & 3 & -3 & -6 & -10 & -12 \\ 8 & 4 & -4 & -8 & -8 & -4 & 4 & 8 \\ 10 & -3 & -12 & -6 & 6 & 12 & 3 & -10 \\ 8 & -8 & -8 & 8 & 8 & -8 & -8 & 8 \\ 6 & -12 & 3 & 10 & -10 & -3 & 12 & -6 \\ 4 & -8 & 8 & -4 & -4 & 8 & -8 & 4 \\ 3 & -6 & 10 & -12 & 12 & -10 & 6 & -3 \end{bmatrix} \quad \text{(C.20)}$$

WMV-9

$DCT(12, 16, 15, 9, 4, 16, 6)$ is developed for Microsoft® Windows Media Video 9 (WMV-9, SMPTE 421M VC-1) [I-23]. The transform's design lets decoders implement the transform with operations on 16-bit registers. Its scaling factors are as follows.

$$k_0 = \frac{1}{24\sqrt{2}}, \quad k_1 = \frac{1}{34}, \quad k_2 \frac{1}{4\sqrt{73}} \quad \text{(C.21)}$$

$$[K_{w,8}] = \text{diag}\left(\frac{1}{24\sqrt{2}}, \frac{1}{34}, \frac{1}{4\sqrt{73}}, \frac{1}{24\sqrt{2}}, \frac{1}{34}, \frac{1}{4\sqrt{73}}, \frac{1}{34}\right) \quad \text{(C.22)}$$

$$[J_{w,8}] = \begin{bmatrix} 12 & 12 & 12 & 12 & 12 & 12 & 12 & 12 \\ 16 & 15 & 9 & 4 & -4 & -9 & -15 & -16 \\ 16 & 6 & -6 & -16 & -16 & -6 & 6 & 16 \\ 15 & -4 & -16 & -9 & 9 & 16 & 4 & -15 \\ 12 & -12 & -12 & 12 & 12 & -12 & -12 & 12 \\ 9 & -16 & 4 & 15 & -15 & -4 & 16 & -9 \\ 6 & -16 & 16 & -6 & -6 & 16 & -16 & 6 \\ 4 & -9 & 15 & -16 & 16 & -15 & 9 & -4 \end{bmatrix} \quad \text{(C.23)}$$

$$[J_{w,8}][J_{w,8}]^T = \text{diag}(1152, 1156, 1168, 1156, 1152, 1156, 1168, 1156) \quad \text{(C.24)}$$

$$[I] = [J_{w,8}]^T \left([J_{w,8}]^T\right)^{-1}$$

$$\left([J_{w,8}]^T\right)^{-1} = \text{diag}\left(\frac{1}{1152}, \frac{1}{1156}, \frac{1}{1168}, \frac{1}{1156}, \frac{1}{1152}, \frac{1}{1156}, \frac{1}{1168}, \frac{1}{1156}\right)[J_{w,8}] \quad \text{(C.25)}$$

$$= \frac{1}{32}[J_{w,8}]^T \text{diag}\left(\frac{8}{288}, \frac{8}{289}, \frac{8}{292}, \frac{8}{289}, \frac{8}{288}, \frac{8}{289}, \frac{8}{292}\right)[J_{w,8}] \quad \text{(C.26)}$$

Thus $\frac{1}{32}\left[J_{w,8}\right]^{T}$ is used for the VC1 decoder as the inverse transform (Annex A in [LA22]).

AVS China

$DCT(8, 10, 9, 6, 2, 10, 4)$ is developed for the Chinese next-generation Audio Video coding Standard (AVS 1.0) [I-21]. Its scaling factors are as follows.

$$k_0 = \frac{1}{\sqrt{512}} = \frac{1}{16\sqrt{2}}, \quad k_1 = \frac{1}{\sqrt{442}}, \quad k_2 = \frac{1}{\sqrt{464}} = \frac{1}{4\sqrt{29}} \tag{C.27}$$

$$\left[K_{c,8}\right] = \operatorname{diag}\left(\frac{1}{16\sqrt{2}}, \frac{1}{\sqrt{442}}, \frac{1}{4\sqrt{29}}, \frac{1}{\sqrt{442}}, \frac{1}{16\sqrt{2}}, \frac{1}{\sqrt{442}}, \frac{1}{4\sqrt{29}}, \frac{1}{\sqrt{442}}\right) \tag{C.28}$$

$$\left[J_{c,8}\right] = \begin{bmatrix} 8 & 8 & 8 & 8 & 8 & 8 & 8 & 8 \\ 10 & 9 & 6 & 2 & -2 & -6 & -9 & -10 \\ 10 & 4 & -4 & -10 & -10 & -4 & 4 & 10 \\ 9 & -2 & -10 & -6 & 6 & 10 & 2 & -9 \\ 8 & -8 & -8 & 8 & 8 & -8 & -8 & 8 \\ 6 & -10 & 2 & 9 & -9 & -2 & 10 & -6 \\ 4 & -10 & 10 & -4 & -4 & 10 & -10 & 4 \\ 2 & -6 & 9 & -10 & 10 & -9 & 6 & -2 \end{bmatrix} \tag{C.29}$$

Note that eight-point integer DCT used in AVS China is actually the transpose of the matrix $[J_{c,8}]$. DCT kernel of four-point integer DCT for H.264 [I-22] is denoted as $[J_{h,4}]$:

$$\left[J_{h,4}\right] = \begin{bmatrix} 1 & 1 & 1 & 1 \\ 2 & 1 & -1 & -2 \\ 1 & -1 & -1 & 1 \\ 1 & -2 & 2 & -1 \end{bmatrix} \tag{C.30}$$

We want to represent $[J_{h,4}]^{-1}$ in terms of $[J_{h,4}]$. The rows of a transform matrix are often referred to as the basis vectors for the transform because they form an orthonormal basis set. As only rows of $[J_{h,4}]$ are orthogonal

$$\left[J_{h,4}\right]\left[J_{h,4}\right]^{T} = \operatorname{diag}(4, 10, 4, 10) \tag{C.31}$$

$$\left[J_{h,4}\right] = \operatorname{diag}(4, 10, 4, 10)\left(\left[J_{h,4}\right]^{T}\right)^{-1} \tag{C.32}$$

$$[I] = \left[J_{h,4}\right]^{T}\left(\left[J_{h,4}\right]^{T}\right)^{-1} = \left[J_{h,4}\right]^{T}\operatorname{diag}\left(\frac{1}{4}, \frac{1}{10}, \frac{1}{4}, \frac{1}{10}\right)\left[J_{h,4}\right] \tag{C.33}$$

Distribute the two fractions of each element of the diagonal matrix properly to the forward and inverse transforms respectively.

$$[I] = [J_{h,4}]^T \left([J_{h,4}]^T\right)^{-1} = [J_{h,4}]^T \text{diag}\left(\frac{1}{4}, \frac{1}{10}, \frac{1}{4}, \frac{1}{10}\right)[J_{h,4}]$$

$$= \left\{\text{diag}\left(1, \frac{1}{2}, 1, \frac{1}{2}\right)[J_{h,4}]\right\}^T \text{diag}\left(\frac{1}{4}, \frac{1}{5}, \frac{1}{4}, \frac{1}{5}\right)[J_{h,4}]$$

$$= \left[J_{h,4}^{\text{inv}}\right]\text{diag}\left(\frac{1}{4}, \frac{1}{5}, \frac{1}{4}, \frac{1}{5}\right)[J_{h,4}] \tag{C.34}$$

where $\left[J_{h,4}^{\text{inv}}\right]$ denotes the inverse transform for H.264 [LA2] and is defined as

$$\left[J_{h,4}^{\text{inv}}\right] = [J_{h,4}]^T \text{diag}\left(1, \frac{1}{2}, 1, \frac{1}{2}\right) = \begin{bmatrix} 1 & 1 & 1 & 1/2 \\ 1 & 1/2 & -1 & -1 \\ 1 & -1/2 & -1 & 1 \\ 1 & -1 & 1 & -1/2 \end{bmatrix} \tag{C.35}$$

Using (C.34), (C.35) and Appendix F.4, this can be extended to get the integer DCT coefficients $[X]$ of 2-D input data $[x]$.

$$[X] = \left[\left(\frac{1}{4}, \frac{1}{5}, \frac{1}{4}, \frac{1}{5}\right)^T \left(\frac{1}{4}, \frac{1}{5}, \frac{1}{4}, \frac{1}{5}\right)\right] \circ \left([J_{h,4}][x][J_{h,4}]^T\right) \tag{C.36}$$

where $\circ$ denotes element-by-element multiplication of the two matrices. The 2-D 4×4 transform is computed in a separable way as a 1-D transform is applied to row vectors of an input matrix/block and then to column vectors. The 2-D inverse transform is computed similarly.

$$[x] = [J_{h,4}]^T \left\{\left[\left(1, \frac{1}{2}, 1, \frac{1}{2}\right)^T \left(1, \frac{1}{2}, 1, \frac{1}{2}\right)\right] \circ [X]\right\}[J_{h,4}] = \left[J_{h,4}^{\text{inv}}\right][X]\left[J_{h,4}^{\text{inv}}\right]^T \tag{C.37}$$

Equation (C.34) can be also factored to get orthogonal matrices for the encoder (C.38a) and decoder (C.38b) as [LA19]

$$[I] = \left[J_{h,4}^{\text{inv}}\right]\text{diag}\left(\frac{1}{2}, \sqrt{\frac{2}{5}}, \frac{1}{2}, \sqrt{\frac{2}{5}}\right)\text{diag}\left(\frac{1}{2}, \frac{1}{\sqrt{10}}, \frac{1}{2}, \frac{1}{\sqrt{10}}\right)[J_{h,4}]$$

$$= \left\{\text{diag}\left(\frac{1}{2}, \frac{1}{\sqrt{10}}, \frac{1}{2}, \frac{1}{\sqrt{10}}\right)[J_{h,4}]\right\}^T \text{diag}\left(\frac{1}{2}, \frac{1}{\sqrt{10}}, \frac{1}{2}, \frac{1}{\sqrt{10}}\right)[J_{h,4}] \tag{C.38a}$$

$$= \left[J_{h,4}^{\text{inv}}\right]\text{diag}\left(\frac{1}{2}, \sqrt{\frac{2}{5}}, \frac{1}{2}, \sqrt{\frac{2}{5}}\right)\left\{\left[J_{h,4}^{\text{inv}}\right]\text{diag}\left(\frac{1}{2}, \sqrt{\frac{2}{5}}, \frac{1}{2}, \sqrt{\frac{2}{5}}\right)\right\}^T \tag{C.38b}$$

DCT kernel $[J_{w,4}]$ of four-point integer DCT for WMV-9 has the form:

$$[J_{w,4}] = \begin{bmatrix} 17 & 17 & 17 & 17 \\ 22 & 10 & -10 & -22 \\ 17 & -17 & -17 & 17 \\ 10 & -22 & 22 & -10 \end{bmatrix} \tag{C.39}$$

$$[J_{w,4}]^T \text{diag}\left(\frac{1}{1156}, \frac{1}{1168}, \frac{1}{1156}, \frac{1}{1168}\right)[J_{w,4}]$$

$$= [J_{w,4}]^T \text{diag}\left(\frac{1}{32}\frac{8}{289}, \frac{1}{32}\frac{8}{292}, \frac{1}{32}\frac{8}{289}, \frac{1}{32}\frac{8}{292}\right)[J_{w,4}]$$

$$= \frac{1}{32}[J_{w,4}]^T \text{diag}\left(\frac{8}{289}, \frac{8}{292}, \frac{8}{289}, \frac{8}{292}\right)[J_{w,4}] = [I] \tag{C.40}$$

Thus $\frac{1}{32}[J_{w,4}]^T$ is used for the VC1 decoder as the inverse transform (Annex A in [LA22]). DCT kernel $[J_{c,4}]$ of four-point integer DCTs for AVS China has the form:

$$[J_{c,4}] = \begin{bmatrix} 2 & 2 & 2 & 2 \\ 3 & 1 & -1 & -3 \\ 2 & -2 & -2 & 2 \\ 1 & -3 & 3 & -1 \end{bmatrix} \tag{C.41}$$

Property C.1. Let an $N \times N$ matrix $[A]$ be nonsingular or invertible. Let $\det [A] = a$. Divide the m rows of $[A]$ by the m ($\leq N$) constants $a_0, a_1, \ldots, a_{m-1}$ which satisfies [I-18]

$$\prod_{k=0}^{m-1} a_k = a = \det [A] \tag{C.42}$$

and denote it as $[B]$. Then $\det [B] = 1$. $\det [J_{h,8}] = 1/ (\det [K_{h,8}])$

For example, let $[A] = \begin{bmatrix} 2 & 3 \\ 1 & 5 \end{bmatrix}$. Then $\det [A] = 7$. Thus $[B] = \begin{bmatrix} 2/7 & 3/7 \\ 1 & 5 \end{bmatrix}$.

Property C.2.   For an orthogonal matrix $[TR]$, det $[TR] = 1$.

From those two properties, det $[J_{h,8}] = 1/(\det [K_{h,8}])$ where $[K_{h,8}]$ and $[J_{h,8}]$ are defined in (C.19) and (C.20).

C.2.3   Performance of Eight-Point Integer DCTs

Dong et al. [LA14] have developed two types of order-16 integer DCTs and then integrated them into the Chinese audio and video coding standard (AVS China) enhanced profile (EP) [LA18] and the H.264 high profile (HP) [I-22], respectively, and used them adaptively as an alternative to the order-eight integer DCT according to local activities. The fidelity range extensions (FRExt) as the first amendment to the initial H.264 standard includes a suite of four high profiles.

Simulation results show that order-16 integer DCTs provide significant performance improvement for both AVS Enhanced Profile and H.264 High Profile, which means they can be efficient coding tools especially for HD video coding (Fig. A.1) [LA14].

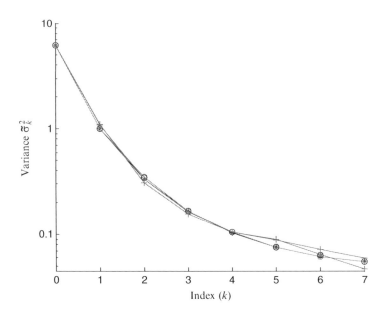

Fig. C.5 Variance distribution of the transform coefficients of a stationary Markov sequence with $\rho = 0.9$ and $N = 8$ (see Table C.3). * denotes KLT and DCT^{II}. O denotes Cham's Int DCT, H.264, AVC, FRExt, WMV-9 and AVS. + denotes Chen-Oraintara-Nguyen's integer DCT C1 and C4 [I-20]

Table C.3 Variances $\tilde{\sigma}_k^2$ of transform coefficients of a stationary Markov sequence with $\rho = 0.9$ and $N = 8$

Transform ↓ k	KLT	DCT^{II}	Int DCT	H.264	WMV-9	AVS	C1	C4
0	6.203	6.186	6.186	6.186	6.186	6.186	6.186	6.186
1	1.007	1.006	1.007	1.001	1.005	1.007	1.084	1.100
2	0.330	0.346	0.345	0.345	0.346	0.346	0.330	0.305
3	0.165	0.166	0.165	0.167	0.165	0.165	0.164	0.155
4	0.104	0.105	0.105	0.105	0.105	0.105	0.105	0.105
5	0.076	0.076	0.076	0.077	0.076	0.076	0.089	0.089
6	0.062	0.062	0.063	0.063	0.062	0.062	0.065	0.072
7	0.055	0.055	0.055	0.057	0.057	0.055	0.047	0.060

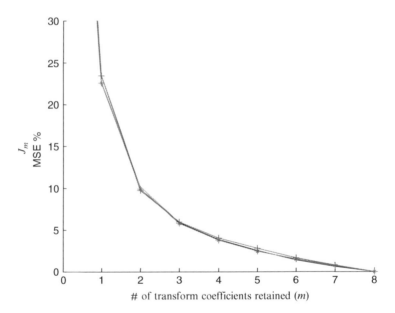

Fig. C.6 Performance of different transforms with respect to basis restriction errors (J_m) versus the number of basis (m) for a stationary Markov sequence with $\rho = 0.9$ and $N = 8$ (see Table C.4). * denotes KLT and DCT^{II}, Cham's Int DCT, H.264 AVC FRExt, WMV-9 and AVS. + denotes Chen-Oraintara-Nguyen's integer DCT C1 and C4 [I-20]

Problems

C.1 Develop scaling factors similar to (C.36) and (C.37) for the four-point integer DCT kernels defined in (C.39) and (C.41).

C.2 Compute det $[K_{h,8}]$ and det $[J_{h,8}]$ defined in (C.19) and (C.20) to verify Property C.2.

Table C.4 Performance of different transforms with respect to basis restriction errors (J_m) versus the number of basis (m) for a stationary Markov sequence with $\rho = 0.9$ and $N = 8$. Int DCT [I-13], H.264 [I-22], WMV-9 [I-23], AVS [I-21], C1 and C4 [I-20]

Transform $\downarrow m$	KLT	DCT^{II}	Int DCT	H.264	WMV-9	AVS	C1	C4
0	100	100	100	100	100	100	100	100
1	22.5	22.7	22.7	22.7	22.7	22.7	23.3	23.4
2	9.9	10.1	10.1	10.2	10.1	10.1	9.9	9.7
3	5.8	5.8	5.8	5.9	5.8	5.8	5.8	5.9
4	3.7	3.7	3.7	3.8	3.7	3.7	3.8	4.0
5	2.4	2.4	2.4	2.5	2.4	2.4	2.5	2.7
6	1.5	1.5	1.5	1.5	1.5	1.5	1.4	1.6
7	0.7	0.7	0.7	0.7	0.7	0.7	0.6	0.7

Projects

C.1 Compare the performance of the (8×8) DCTs:
   (I) (8×8) DCT^{II} defined in (C.1a)
   (II) (8×8) Integer DCT adopted in H.264 (defined in (C.19) and (C.20))
   (III) (8×8) Integer DCT adopted in WMV-9 (VC-1) of Microsoft (defined in (C.22) and (C.23))
   (IV) (8×8) Integer DCT adopted in AVS China (defined in (C.25) and (C.26))

Repeat the following problems for (I), (II), (III) and (IV).

(a) Variance distribution for I order Markov process, $\rho = 0.9$ (Plot and Tabulate). See Fig. C.5 and Table C.3.
(b) Normalized basis restriction error vs. number of basis function (Plot and Tabulate). See Fig. C.6 and Table C.4.
(c) Obtain transform coding gains.
(d) Plot fractional correlation ($0 < \rho < 1$) defined in (A.7).

Appendix D
DCT and DST

D.1 Kernels for DCT and DST [I-10]

Kernels for DCT and DST (types I–IV) are listed below [B23].

DCT of types I–IV

$$\left[C_{N+1}^{\mathrm{I}}\right]_{kn} = \sqrt{\frac{2}{N}}\, \epsilon_k \epsilon_n \cos\left[\frac{\pi k n}{N}\right] \qquad k, n = 0, 1, \ldots, N$$

$$\left[C_N^{\mathrm{II}}\right]_{kn} = \sqrt{\frac{2}{N}}\, \epsilon_k \cos\left[\frac{\pi k \left(n + \frac{1}{2}\right)}{N}\right] \qquad k, n = 0, 1, \ldots, N-1$$

$$\left[C_N^{\mathrm{III}}\right]_{kn} = \sqrt{\frac{2}{N}}\, \epsilon_n \cos\left[\frac{\pi \left(k + \frac{1}{2}\right) n}{N}\right] \qquad k, n = 0, 1, \ldots, N-1$$

$$\left[C_N^{\mathrm{IV}}\right]_{kn} = \sqrt{\frac{2}{N}} \cos\left[\frac{\pi \left(k + \frac{1}{2}\right)\left(n + \frac{1}{2}\right)}{N}\right] \qquad k, n = 0, 1, \ldots, N-1$$

The matrices $\left[C_{N+1}^{\mathrm{I}}\right]$ and $\left[C_N^{\mathrm{IV}}\right]$ are symmetric, and $\left[C_N^{\mathrm{III}}\right]$ is the inverse of $\left[C_N^{\mathrm{II}}\right]$ and vice versa.

$$\left[C_{N+1}^{\mathrm{I}}\right]^{-1} = \left[C_{N+1}^{\mathrm{I}}\right]^{T} = \left[C_{N+1}^{\mathrm{I}}\right] \qquad\qquad \left[C_N^{\mathrm{II}}\right]^{-1} = \left[C_N^{\mathrm{II}}\right]^{T} = \left[C_N^{\mathrm{III}}\right]$$

$$\left[C_N^{\mathrm{III}}\right]^{-1} = \left[C_N^{\mathrm{III}}\right]^{T} = \left[C_N^{\mathrm{II}}\right] \qquad\qquad \left[C_N^{\mathrm{IV}}\right]^{-1} = \left[C_N^{\mathrm{IV}}\right]^{T} = \left[C_N^{\mathrm{IV}}\right]$$

DST of types I–IV

$$\left[S_{N-1}^{\mathrm{I}}\right]_{kn} = \sqrt{\frac{2}{N}} \sin\left[\frac{\pi k n}{N}\right] \qquad k, n = 1, 2, \ldots, N-1$$

$$\left[S_N^{\mathrm{II}}\right]_{kn} = \sqrt{\frac{2}{N}}\, \epsilon_k \sin\left[\frac{\pi \left(n - \frac{1}{2}\right)}{N}\right] \qquad k, n = 1, 2, \ldots, N$$

$$\left[S_N^{\mathrm{III}}\right]_{kn} = \sqrt{\frac{2}{N}}\, \epsilon_n \sin\left[\frac{\pi \left(k - \frac{1}{2}\right) n}{N}\right] \qquad k, n = 1, 2, \ldots, N$$

$$\left[S_N^{\mathrm{IV}}\right]_{kn} = \sqrt{\frac{2}{N}} \sin\left[\frac{\pi \left(k + \frac{1}{2}\right)\left(n + \frac{1}{2}\right)}{N}\right] \qquad k, n = 0, 1, \ldots, N-1$$

$$[S_{N-1}^{\text{I}}]^{-1} = [S_{N-1}^{\text{I}}]^{T} = [S_{N-1}^{\text{I}}] \qquad\qquad [S_N^{\text{II}}]^{-1} = [S_N^{\text{II}}]^{T} = [S_N^{\text{III}}]$$

$$[S_N^{\text{III}}]^{-1} = [S_N^{\text{III}}]^{T} = [S_N^{\text{II}}] \qquad\qquad [S_N^{\text{IV}}]^{-1} = [S_N^{\text{IV}}]^{T} = [S_N^{\text{IV}}]$$

$$[C_N^{\text{II}}] = [J][S_N^{\text{II}}][D] \qquad [C_N^{\text{III}}] = [D][S_N^{\text{III}}][J] \qquad [C_N^{\text{IV}}] = [D][S_N^{\text{IV}}][J] \quad \text{(D.1)}$$

where $[J]$ is the reverse identity matrix and $[D]$ is diagonal with $[D]_{kk} = (-1)^{k}$, $k = 0, 1, \ldots, N-1$ (p. 76 in [B28]).

Kernels for DCT and DST (types V–VIII) are listed below.

DCT of types V–VIII

$$[C_N^{\text{V}}]_{kn} = \frac{2}{\sqrt{2N-1}} \epsilon_k \epsilon_n \cos\left[\frac{2\pi kn}{2N-1}\right] \qquad\qquad k, n = 0, 1, \ldots, N-1$$

$$[C_N^{\text{VI}}]_{kn} = \frac{2}{\sqrt{2N-1}} \epsilon_k l_n \cos\left[\frac{2\pi k\left(n+\frac{1}{2}\right)}{2N-1}\right] \qquad\qquad k, n = 0, 1, \ldots, N-1$$

$$[C_N^{\text{VII}}]_{kn} = \frac{2}{\sqrt{2N-1}} l_k \epsilon_n \cos\left[\frac{2\pi\left(k+\frac{1}{2}\right)n}{2N-1}\right] \qquad\qquad k, n = 0, 1, \ldots, N-1$$

$$[C_{N-1}^{\text{VIII}}]_{kn} = \frac{2}{\sqrt{2N-1}} \cos\left[\frac{2\pi\left(k+\frac{1}{2}\right)\left(n+\frac{1}{2}\right)}{2N-1}\right] \qquad\qquad k, n = 0, 1, \ldots, N-2$$

$$[C_N^{\text{V}}]^{-1} = [C_N^{\text{V}}]^{T} = [C_N^{\text{V}}] \qquad\qquad [C_N^{\text{VI}}]^{-1} = [C_N^{\text{VI}}]^{T} = [C_N^{\text{VII}}]$$

$$[C_N^{\text{VII}}]^{-1} = [C_N^{\text{VII}}]^{T} = [C_N^{\text{VI}}] \qquad\qquad [C_{N-1}^{\text{VIII}}]^{-1} = [C_{N-1}^{\text{VIII}}]^{T} = [C_{N-1}^{\text{VIII}}]$$

DST of types V–VIII

$$[S_{N-1}^{\text{V}}]_{kn} = \frac{2}{\sqrt{2N-1}} \sin\left[\frac{2\pi kn}{2N-1}\right] \qquad\qquad k, n = 1, 2, \ldots, N-1$$

$$[S_{N-1}^{\text{VI}}]_{kn} = \frac{2}{\sqrt{2N-1}} \sin\left[\frac{2\pi k\left(n-\frac{1}{2}\right)}{2N-1}\right] \qquad\qquad k, n = 1, 2, \ldots, N-1$$

$$[S_{N-1}^{\text{VII}}]_{kn} = \frac{2}{\sqrt{2N-1}} \sin\left[\frac{2\pi\left(k-\frac{1}{2}\right)n}{2N-1}\right] \qquad\qquad k, n = 1, 2, \ldots, N-1$$

$$[S_N^{\text{VIII}}]_{kn} = \frac{2}{\sqrt{2N-1}} l_k l_n \sin\left[\frac{2\pi\left(k+\frac{1}{2}\right)\left(n+\frac{1}{2}\right)}{2N-1}\right] \qquad\qquad k, n = 0, 1, \ldots, N-1$$

$$[S_{N-1}^{\text{V}}]^{-1} = [S_{N-1}^{\text{V}}]^{T} = [S_{N-1}^{\text{V}}] \qquad\qquad [S_{N-1}^{\text{VI}}]^{-1} = [S_{N-1}^{\text{VI}}]^{T} = [S_{N-1}^{\text{VII}}]$$

$$[S_{N-1}^{\text{VII}}]^{-1} = [S_{N-1}^{\text{VII}}]^{T} = [S_{N-1}^{\text{VI}}] \qquad\qquad [S_N^{\text{VIII}}]^{-1} = [S_N^{\text{VIII}}]^{T} = [S_N^{\text{VIII}}]$$

where

$$\epsilon_p = \begin{cases} \frac{1}{\sqrt{2}} & \text{when } p = 0 \text{ or } N \\ 1 & \text{when } p \neq 0 \text{ and } N \end{cases}$$

$$l_p = \begin{cases} \frac{1}{\sqrt{2}} & \text{when } p = N - 1 \\ 1 & \text{when } p \neq N - 1 \end{cases}$$

The DCTs and DSTs are related with the M-point generalized DFT (GDFT). Let N be a positive integer. For types I–IV, $M = 2N$; for types V–IIIV, $M = 2N - 1$. Thus M is even for types I–IV (even length GDFT) and M is odd for types V–VIII (odd length GDFT).

MATLAB code for the GDFT and all 16 discrete trigonometric transforms, DCTs and DSTs of types I–VIII, in both orthogonal and non-orthogonal forms is provided in [LA3]. To compute the GDFT, FFT is used in gdft.m of [LA3] as follows.

$$[G_{a,b}] = X^F(k) = \frac{1}{\sqrt{N}} \sum_{n=0}^{N-1} x(n) W_N^{(k+a)(n+b)} \qquad k = 0, 1, \ldots, N - 1$$

$$= \frac{1}{\sqrt{N}} \sum_{n=0}^{N-1} x(n) W_N^{(k+a)b + kn + an} = \frac{1}{\sqrt{N}} W_N^{(k+a)b} \sum_{n=0}^{N-1} \left[x(n) W_N^{an} \right] W_N^{kn}$$

$$= \frac{1}{\sqrt{N}} W_N^{(k+a)b} \text{FFT} \left[x(n) W_N^{an} \right] \tag{D.2}$$

$a = 0, b = 0$: normal DFT
$a = \frac{1}{2}, b = 0$: odd frequency DFT
$a = 0, b = \frac{1}{2}$: odd time DFT
$a = \frac{1}{2}, b = \frac{1}{2}$: odd-time odd-frequency DFT

D.2 Derivation of Unitary DCTs and DSTs

In this section, unitary discrete trigonometric transforms (DTT) are derived in three steps.

1. Choose the same/corresponding type of DTTs for a special form of the generalized DFT (GDFT) as shown in Table III of [I-12].
2. Apply symmetrically extended input to the GDFT to get the DTT (see Table III and Fig. 2 in [I-12]).
3. Make the DTT orthogonal.

This derivation will give us insight into new DTT properties. The same notation as in [I-12] is used in this section.

DCT-II

Let $\underline{X}$ be the type-2 DCT of an input vector $\underline{x}$ of size $(N \times 1)$.

$$\underline{X} = \left[G_{0,1/2}\right]\left[E_{\text{HSHS}}\right]\underline{x} = \left[C_{2\text{e}}\right]\underline{x} \tag{D.3}$$

$$\left[G_{0,1/2}\right]_{kn} = \exp\left(-j\frac{2\pi k\left(n+\frac{1}{2}\right)}{2N}\right) = \exp\left(-j\frac{\pi k\left(n+\frac{1}{2}\right)}{N}\right) \tag{D.4}$$

For example, let DCT size $N = 4$. M is the size of GDFT input sequence.

$$L \in \{N-1, N, N+1\}$$

$$(L \times M) = (N \times 2N)$$

$$\left[G_{0,1/2}\right]\left[E_{\text{HSHS}}\right] = \begin{bmatrix} e^{-j\frac{0\pi(0+1/2)}{N}} & e^{-j\frac{0\pi(1+1/2)}{N}} & \cdots & e^{-j\frac{0\pi(6+1/2)}{N}} & e^{-j\frac{0\pi(7+1/2)}{N}} \\ e^{-j\frac{\pi(0+1/2)}{N}} & e^{-j\frac{\pi(1+1/2)}{N}} & \cdots & e^{-j\frac{\pi(6+1/2)}{N}} & e^{-j\frac{\pi(7+1/2)}{N}} \\ e^{-j\frac{2\pi(0+1/2)}{N}} & e^{-j\frac{2\pi(1+1/2)}{N}} & \cdots & e^{-j\frac{2\pi(6+1/2)}{N}} & e^{-j\frac{2\pi(7+1/2)}{N}} \\ e^{-j\frac{3\pi(0+1/2)}{N}} & e^{-j\frac{3\pi(1+1/2)}{N}} & \cdots & e^{-j\frac{3\pi(6+1/2)}{N}} & e^{-j\frac{3\pi(7+1/2)}{N}} \end{bmatrix}\left[E_{\text{HSHS}}\right] \tag{D.5}$$

$$(M = 2N) \qquad\qquad (N)$$

$$= \begin{bmatrix} e^{-j\frac{0\pi(0+1/2)}{N}} & e^{-j\frac{0\pi(1+1/2)}{N}} & \cdots & e^{j\frac{0\pi(1+1/2)}{N}} & e^{j\frac{0\pi(1/2)}{N}} \\ e^{-j\frac{\pi(0+1/2)}{N}} & e^{-j\frac{\pi(1+1/2)}{N}} & \cdots & e^{j\frac{\pi(1+1/2)}{N}} & e^{j\frac{\pi(1/2)}{N}} \\ e^{-j\frac{2\pi(0+1/2)}{N}} & e^{-j\frac{2\pi(1+1/2)}{N}} & \cdots & e^{j\frac{2\pi(1+1/2)}{N}} & e^{j\frac{2\pi(1/2)}{N}} \\ e^{-j\frac{3\pi(0+1/2)}{N}} & e^{-j\frac{3\pi(1+1/2)}{N}} & \cdots & e^{j\frac{3\pi(1+1/2)}{N}} & e^{j\frac{3\pi(1/2)}{N}} \end{bmatrix}\begin{bmatrix} 1 & & & & & & & \\ & 1 & & & & & & \\ & & 1 & & & & & \\ & & & 1 & & & & \\ & & & & 1 & & & \\ & & & & & 1 & & \\ & & & & & & 1 & \\ & & & & & & & 1 \end{bmatrix} \tag{D.6}$$

$$= 2\begin{bmatrix} \cos\left(\frac{0\pi(0+1/2)}{N}\right) & \cos\left(\frac{0\pi(1+1/2)}{N}\right) & \cos\left(\frac{0\pi(2+1/2)}{N}\right) & \cos\left(\frac{0\pi(3+1/2)}{N}\right) \\ \cos\left(\frac{1\pi(0+1/2)}{N}\right) & \cos\left(\frac{1\pi(1+1/2)}{N}\right) & \cos\left(\frac{1\pi(2+1/2)}{N}\right) & \cos\left(\frac{1\pi(3+1/2)}{N}\right) \\ \cos\left(\frac{2\pi(0+1/2)}{N}\right) & \cos\left(\frac{2\pi(1+1/2)}{N}\right) & \cos\left(\frac{2\pi(2+1/2)}{N}\right) & \cos\left(\frac{2\pi(3+1/2)}{N}\right) \\ \cos\left(\frac{3\pi(0+1/2)}{N}\right) & \cos\left(\frac{3\pi(1+1/2)}{N}\right) & \cos\left(\frac{3\pi(2+1/2)}{N}\right) & \cos\left(\frac{3\pi(3+1/2)}{N}\right) \end{bmatrix} \tag{D.7}$$

$$= 2\left[\cos\left(\frac{\pi k\left(n+\frac{1}{2}\right)}{N}\right)\right] = 2[\cos] = [C_{2\text{e}}] \qquad \text{(biorthogonal)}$$

Make this matrix self-orthogonal.

$$\left(\frac{1}{M}[C_{3e}]\right)[C_{2e}] = [I]$$

$$\left(\frac{1}{M}2[\cos]^T\left[\epsilon_k^2\right]\right)2[\cos] = [I], \qquad\qquad \left[\epsilon_k^2\right] = \mathrm{diag}\left(\frac{1}{2}, 1, 1, 1\right)$$

$$\left(\frac{1}{N}2[\cos]^T\left[\epsilon_k^2\right]\right)[\cos] = [I]$$

$$\left(\sqrt{2/N}[\cos]^T[\epsilon_k]\right)\left(\sqrt{2/N}[\epsilon_k][\cos]\right) = [I], \qquad [\epsilon_k] = \mathrm{diag}\left(\frac{1}{\sqrt{2}}, 1, 1, 1\right)$$

$$\left(\sqrt{2/N}[\epsilon_k][\cos]\right)^T\left(\sqrt{2/N}[\epsilon_k][\cos]\right) = [I]$$

$$[C_{\mathrm{IIE}}]^T[C_{\mathrm{IIE}}] = [I] \qquad\qquad \text{(self-orthogonal)} \qquad (D.8)$$

$$[C_{2e}] = \sqrt{2N}\,[\epsilon_k]^{-1}[C_{\mathrm{IIE}}] \qquad\qquad [\epsilon_k]^{-1} = \mathrm{diag}\left(\sqrt{2}, 1, 1, 1\right) \quad (D.9)$$

DCT-I
Let $\underline{X}$ be the type-1 DCT of an input vector $\underline{x}$ of size $(N+1) \times 1$.

$$\underline{X} = [G_{0,0}][E_{\mathrm{WSWS}}]\underline{x} = [C_{1e}]\underline{x} \qquad (D.10)$$

$$[G_{0,0}]_{kn} = \exp\left(-j\frac{2\pi kn}{2N}\right) = \exp\left(-j\frac{\pi kn}{N}\right) \qquad (D.11)$$

For example, let $N = 4$.

$$\begin{bmatrix} e^{-j\frac{0\pi(0)}{N}} & e^{-j\frac{0\pi(1)}{N}} & e^{-j\frac{0\pi(2)}{N}} & e^{-j\frac{0\pi(3)}{N}} & e^{-j\frac{0\pi(4)}{N}} & e^{-j\frac{0\pi(5)}{N}} & e^{-j\frac{0\pi(6)}{N}} & e^{-j\frac{0\pi(7)}{N}} \\ e^{-j\frac{\pi(0)}{N}} & e^{-j\frac{\pi(1)}{N}} & \cdots & & & & e^{-j\frac{\pi(6)}{N}} & e^{-j\frac{\pi(7)}{N}} \\ e^{-j\frac{2\pi(0)}{N}} & e^{-j\frac{2\pi(1)}{N}} & \cdots & & & & e^{-j\frac{2\pi(6)}{N}} & e^{-j\frac{2\pi(7)}{N}} \\ e^{-j\frac{3\pi(0)}{N}} & e^{-j\frac{3\pi(1)}{N}} & \cdots & & & & e^{-j\frac{3\pi(6)}{N}} & e^{-j\frac{3\pi(7)}{N}} \\ e^{-j\frac{4\pi(0)}{N}} & e^{-j\frac{4\pi(1)}{N}} & \cdots & & & & e^{-j\frac{4\pi(6)}{N}} & e^{-j\frac{4\pi(7)}{N}} \end{bmatrix} [E_{\mathrm{WSWS}}]$$

$$(L \times M) \qquad\qquad L = N+1, \ M = 2N$$

$$(D.12)$$

$$(M = 2N) \hspace{3cm} (N+1)$$

$$= \begin{bmatrix} e^{-j\frac{0\pi(0)}{N}} & e^{-j\frac{0\pi(1)}{N}} & \cdots & e^{j\frac{0\pi(1)}{N}} \\ e^{-j\frac{\pi(0)}{N}} & e^{-j\frac{\pi(1)}{N}} & \cdots & e^{j\frac{\pi(1)}{N}} \\ e^{-j\frac{2\pi(0)}{N}} & e^{-j\frac{2\pi(1)}{N}} & \cdots & e^{j\frac{2\pi(1)}{N}} \\ e^{-j\frac{3\pi(0)}{N}} & e^{-j\frac{3\pi(1)}{N}} & \cdots & e^{j\frac{3\pi(1)}{N}} \\ e^{-j\frac{4\pi(0)}{N}} & e^{-j\frac{4\pi(1)}{N}} & \cdots & e^{j\frac{4\pi(1)}{N}} \end{bmatrix} \begin{bmatrix} 1 & & & & & \\ & 1 & & & & \\ & & 1 & & & \\ & & & 1 & & \\ & & & & 1 & \\ & & & 1 & \\ & & 1 & \\ & 1 \end{bmatrix} \quad \text{(D.13)}$$

$$= \begin{bmatrix} \cos\left(\frac{0\pi(0)}{N}\right) & 2\cos\left(\frac{0\pi(1)}{N}\right) & 2\cos\left(\frac{0\pi(2)}{N}\right) & 2\cos\left(\frac{0\pi(3)}{N}\right) & \cos\left(\frac{0\pi(4)}{N}\right) \\ \cos\left(\frac{1\pi(0)}{N}\right) & 2\cos\left(\frac{1\pi(1)}{N}\right) & 2\cos\left(\frac{1\pi(2)}{N}\right) & 2\cos\left(\frac{1\pi(3)}{N}\right) & \cos\left(\frac{1\pi(4)}{N}\right) \\ \cos\left(\frac{2\pi(0)}{N}\right) & 2\cos\left(\frac{2\pi(1)}{N}\right) & 2\cos\left(\frac{2\pi(2)}{N}\right) & 2\cos\left(\frac{2\pi(3)}{N}\right) & \cos\left(\frac{2\pi(4)}{N}\right) \\ \cos\left(\frac{3\pi(0)}{N}\right) & 2\cos\left(\frac{3\pi(1)}{N}\right) & 2\cos\left(\frac{3\pi(2)}{N}\right) & 2\cos\left(\frac{3\pi(3)}{N}\right) & \cos\left(\frac{3\pi(4)}{N}\right) \\ \cos\left(\frac{4\pi(0)}{N}\right) & 2\cos\left(\frac{4\pi(1)}{N}\right) & 2\cos\left(\frac{4\pi(2)}{N}\right) & 2\cos\left(\frac{4\pi(3)}{N}\right) & \cos\left(\frac{4\pi(4)}{N}\right) \end{bmatrix}$$

$$\text{(D.14)}$$

$$= 2\left[\cos\left(\frac{\pi kn}{N}\right)\right][\epsilon_n^2] \hspace{2cm} [\epsilon_n^2] = \mathrm{diag}\left(\frac{1}{2},1,1,1,\frac{1}{2}\right) \quad \text{(D.15)}$$

$$= [C_{1e}] \hspace{3cm} \text{(biorthogonal)}$$

Make this matrix self-orthogonal.

$$\left(\frac{1}{M}[C_{1e}]\right)[C_{1e}] = [I]$$

$$\frac{1}{M}\left(2\left[\cos\left(\frac{\pi kn}{N}\right)\right][\epsilon_n^2]\right)\left(2\left[\cos\left(\frac{\pi kn}{N}\right)\right][\epsilon_n^2]\right)$$

$$= \frac{1}{M}\left(\sqrt{2N}[\epsilon_n]^{-1}[C_{IE}][\epsilon_n]\right)\left(\sqrt{2N}[\epsilon_n]^{-1}[C_{IE}][\epsilon_n]\right)$$

$$= [\epsilon_n]^{-1}[C_{IE}][C_{IE}][\epsilon_n] = [I], \hspace{1cm} [\epsilon_n] = \mathrm{diag}\left(\frac{1}{\sqrt{2}},1,1,1,\frac{1}{\sqrt{2}}\right)$$

$$\text{(D.16)}$$

$$[C_{IE}][C_{IE}] = [C_{IE}]^T[C_{IE}] = [I] \hspace{1cm} \text{(self-orthogonal)}$$

DCT-III

Let $\underline{X}$ be the type-3 DCT of an input vector $\underline{x}$ of size $(N \times 1)$.

$$\underline{X} = [G_{1/2,0}][E_{WSWA}]\underline{x} = [C_{3e}]\underline{x} \hspace{2cm} \text{(D.17)}$$

$$[G_{1/2,0}]_{kn} = \exp\left(-j\frac{2\pi\left(k+\frac{1}{2}\right)}{2N}\right) = \exp\left(-j\frac{\pi\left(k+\frac{1}{2}\right)n}{N}\right) \tag{D.18}$$

$$[G_{1/2,0}][E_{\text{WSWA}}] = \begin{bmatrix} e^{-j\frac{(1/2)\pi(0)}{N}} & e^{-j\frac{(1/2)\pi(1)}{N}} & \cdots & e^{-j\frac{(1/2)\pi(7)}{N}} \\ e^{-j\frac{(1+1/2)\pi(0)}{N}} & e^{-j\frac{(1+1/2)\pi(1)}{N}} & \cdots & e^{-j\frac{(1+1/2)\pi(7)}{N}} \\ e^{-j\frac{(2+1/2)\pi(0)}{N}} & e^{-j\frac{(2+1/2)\pi(1)}{N}} & \cdots & e^{-j\frac{(2+1/2)\pi(7)}{N}} \\ e^{-j\frac{(3+1/2)\pi(0)}{N}} & e^{-j\frac{(3+1/2)\pi(1)}{N}} & \cdots & e^{-j\frac{(3+1/2)\pi(7)}{N}} \end{bmatrix} [E_{\text{WSWA}}] \tag{D.19}$$

Note $\exp\left(-j\frac{(1/2)\pi(7)}{N}\right) = \exp\left(-j\frac{7\pi}{2N}\right) = \exp\left(j\frac{9\pi}{2N}\right) = -\exp\left(j\frac{(1/2)\pi}{N}\right)$, $N = 4$.

$$= \begin{bmatrix} e^{-j\frac{(1/2)\pi(0)}{N}} & -e^{j\frac{(1/2)\pi(1)}{N}} & \cdots & e^{-j\frac{(1/2)\pi(1)}{N}} \\ e^{-j\frac{(1+1/2)\pi(0)}{N}} & -e^{j\frac{(1+1/2)\pi(1)}{N}} & \cdots & e^{-j\frac{(1+1/2)\pi(1)}{N}} \\ e^{-j\frac{(2+1/2)\pi(0)}{N}} & -e^{j\frac{(2+1/2)\pi(1)}{N}} & \cdots & e^{-j\frac{(2+1/2)\pi(1)}{N}} \\ e^{-j\frac{(3+1/2)\pi(0)}{N}} & -e^{j\frac{(3+1/2)\pi(1)}{N}} & \cdots & e^{-j\frac{(3+1/2)\pi(1)}{N}} \end{bmatrix} \begin{bmatrix} 1 & & & \\ & 1 & & \\ & & 1 & \\ & & & 1 \\ & & & 0 \\ & & -1 & \\ & -1 & & \\ -1 & & & \end{bmatrix} \tag{D.20}$$

$$= \begin{bmatrix} \cos\left(\frac{(1/2)\pi(0)}{N}\right) & 2\cos\left(\frac{(1/2)\pi(1)}{N}\right) & 2\cos\left(\frac{(1/2)\pi(2)}{N}\right) & 2\cos\left(\frac{(1/2)\pi(3)}{N}\right) \\ \cos\left(\frac{(1+1/2)\pi(0)}{N}\right) & 2\cos\left(\frac{(1+1/2)\pi(1)}{N}\right) & 2\cos\left(\frac{(1+1/2)\pi(2)}{N}\right) & 2\cos\left(\frac{(1+1/2)\pi(3)}{N}\right) \\ \cos\left(\frac{(2+1/2)\pi(0)}{N}\right) & 2\cos\left(\frac{(2+1/2)\pi(1)}{N}\right) & 2\cos\left(\frac{(2+1/2)\pi(2)}{N}\right) & 2\cos\left(\frac{(2+1/2)\pi(3)}{N}\right) \\ \cos\left(\frac{(3+1/2)\pi(0)}{N}\right) & 2\cos\left(\frac{(3+1/2)\pi(1)}{N}\right) & 2\cos\left(\frac{(3+1/2)\pi(2)}{N}\right) & 2\cos\left(\frac{(3+1/2)\pi(3)}{N}\right) \end{bmatrix} \tag{D.21}$$

$$= 2\left[\cos\left(\frac{\pi\left(k+\frac{1}{2}\right)n}{N}\right)\right][\epsilon_n^2] \tag{D.22}$$

$$= [C_{3e}] \quad \text{(biorthogonal)}$$

Note that $[E_{\text{WSWS}}]$ in (D.18) comes from Fig. D.1c.
For the next section, DSTs are defined as

$$[S_{1e}]_{kn} = 2\left[\sin\left(\frac{\pi kn}{N}\right)\right] \qquad k, n = 1, 2, \ldots, N-1 \tag{D.23}$$

$$[S_{2e}]_{kn} = 2\left[\sin\left(\frac{\pi k\left(n+\frac{1}{2}\right)}{N}\right)\right] \qquad \begin{aligned} k &= 1, 2, \ldots, N \\ \\ n &= 0, 1, \ldots, N-1 \end{aligned} \tag{D.24}$$

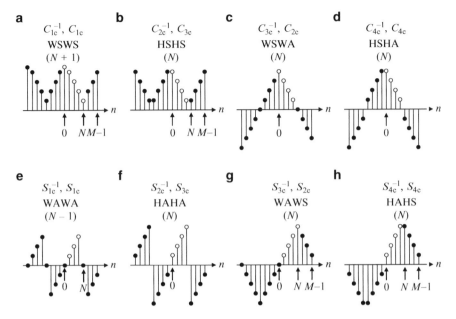

Fig. D.1 Symmetries of SPSs (symmetric periodic sequences: C_{2c}^{-1}, C_{3c}^{-1}) and DTTs (C_{2c}, C_{3c}) with $N = 4$. The representative samples for each SPS are marked by open circles. N can be even and odd. See (3.97) for an application. [I-12] © 1994 IEEE

However, derivation of a DCT from a different type of GDFT is more laborious while both ways are very useful. For example, derivation of DCT-II from the ordinary DFT (GDFT with $a = 0$ and $b = 0$) is shown in [LA5, p. 49] (see also [B19]).

Different from *symmetrically* extended input for the DCTs of types 1 and 2, *asymmetrically* extended input (minus signs in $[E_{WSWA}]$) for the DCTs of types 3 and 4 is caused by the frequency index shifted by $\frac{1}{2}$ [i.e. $\left(k + \frac{1}{2}\right)$ in $\exp\left(-j\frac{2\pi\left(k+\frac{1}{2}\right)n}{2N}\right)$ for the DCT of type 3]. The *whole-sample* symmetric extension from the representative samples ($N = 4$) of an input sequence takes a diagonal matrix $\left[\epsilon_n^2\right]$ for DCTs. Equations (D.15) and (D.22) for the DCT-I (Fig. D.1a) and DCT-III (Fig. D.1c) are examples.

Let HSHS represent a periodic sequence with half-sample symmetry at the left point of symmetry (LPOS) of the representative samples ($N = 4$) of the sequence and half-sample symmetry at the right point of symmetry (RPOS). Then an input sequence of DCT-II is implicitly HSHS as shown in Fig. D.1b. DCT-II coefficients are implicitly WSWA as shown in Fig. D.1c.

WSWA represents a periodic sequence with whole-sample symmetry at the left point of symmetry (LPOS) of the representative samples ($N = 4$) of the sequence and whole-sample asymmetry at the right point of symmetry (RPOS).

Derivation of DCT-II from DFT

Let vector $e = \left\{ \exp\left(\dfrac{j\pi 0}{2N}\right), \exp\left(\dfrac{j\pi 1}{2N}\right), \ldots, \exp\left(\dfrac{j\pi(N-1)}{2N}\right) \right\}^T$ and matrix

$$\left[\exp\left(\frac{j\pi k}{2N}\right)\right] = \operatorname{diag}\left[\exp\left(\frac{j\pi 0}{2N}\right), \exp\left(\frac{j\pi 1}{2N}\right), \ldots, \exp\left(\frac{j\pi(N-1)}{2N}\right)\right].$$

Then

$$\left[\exp\left(\frac{j\pi k}{2N}\right)\right]^{-1} = \left[\exp\left(\frac{j\pi k}{2N}\right)\right]^* = \left[\exp\left(\frac{-j\pi k}{2N}\right)\right] \tag{D.25}$$

From (D.5)

$$[G_{0,1/2}][E_{\text{HSHS}}] = \left[\exp\left(\frac{-j\pi k}{2N}\right)\right][G_{0,0}][E_{\text{HSHS}}] \tag{D.26}$$

$$= \left[\exp\left(\frac{-j\pi k}{2N}\right)\right][H][F][E_{\text{HSHS}}] = [C_{2e}] \tag{D.27}$$

where $[F]$ is the DFT matrix and $[H] = ([I_{N \times N}], [0_{N \times N}])$. The DFT of a symmetric input is

$$[H][F][E_{\text{HSHS}}] = \left[\exp\left(\frac{j\pi k}{2N}\right)\right][C_{2e}] = \sqrt{2N}\left[\exp\left(\frac{j\pi k}{2N}\right)\right][\epsilon_k]^{-1}[C_{\text{IIE}}] \tag{D.28}$$

where $[C_{2e}]$ represents unnormalized type-2 DCT and is defined in (D.9). The normalized DCT $[C_{\text{IIE}}]$ of type 2 can be expressed in terms of the DFT matrix as

$$[C_{\text{IIE}}] = \frac{1}{\sqrt{2N}}[\epsilon_k]\left[\exp\left(\frac{-j\pi k}{2N}\right)\right][H][F][E_{\text{HSHS}}] \tag{D.29}$$

See [LA5, B19] for similar derivations. This can be extended to the two dimensional case as $[X^{\text{CII}}] = [C_{\text{IIE}}][x][C_{\text{IIE}}]^T$ where $[X^{\text{CII}}]$ is the DCT coefficient matrix of an input matrix $[x]$. First make copies of an input image

$$[x_{2N \times 2N}] = [E_{\text{HSHS}}][x][E_{\text{HSHS}}]^T \tag{D.30}$$

$$[X^{\text{CII}}] = \frac{1}{2N}[\epsilon_k]\left[\exp\left(\frac{-j\pi k}{2N}\right)\right][H][F][x_{2N \times 2N}][F]^T[H]^T\left[\exp\left(\frac{-j\pi k}{2N}\right)\right][\epsilon_k] \tag{D.31}$$

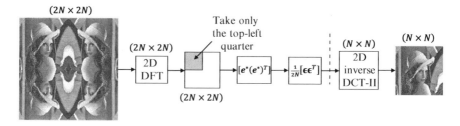

Fig. D.2 Implementation of 2-D DCT-II via 2-D DFT

By using the property in Appendix F.4, (D.31) becomes

$$[X^{\mathrm{CII}}] = \frac{1}{2N}[\boldsymbol{\epsilon}\,\boldsymbol{\epsilon}^T] \circ \left[e^*(e^*)^T \right] \circ \left([H][F][x_{2N \times 2N}][F]^T[H]^T \right) \tag{D.32}$$

where vector $\boldsymbol{\epsilon} = (1/\sqrt{2}, 1, 1, \ldots, 1)^T$ is of size $(N \times 1)$ and $\circ$ represents element-by-element multiplication of the two matrices. For an image of size $(N \times N)$ with $N = 4$

$$[\boldsymbol{\epsilon}\boldsymbol{\epsilon}^T] = \begin{bmatrix} 1/2 & 1/\sqrt{2} & 1/\sqrt{2} & 1/\sqrt{2} \\ 1/\sqrt{2} & 1 & 1 & 1 \\ 1/\sqrt{2} & 1 & 1 & 1 \\ 1/\sqrt{2} & 1 & 1 & 1 \end{bmatrix} \tag{D.33}$$

An implementation of (D.32) is shown in Fig. D.2 [LA17].

Now we are given the $(N \times N)$ DCT coefficient matrix $[X^{\mathrm{CII}}]$ of an image of size $(N \times N)$, and we want a $(2N \times 2N)$ DFT coefficient matrix $[X^F]$ with which we could compute the DCT coefficient matrix $[X^{\mathrm{CII}}]$.

$$[x_{2N \times 2N}] = [E_{\mathrm{HSHS}}][x][E_{\mathrm{HSHS}}]^T \tag{D.34}$$

$$[X^F] = [F][x_{2M \times 2N}][F]^T \tag{D.35}$$

$$[X^F] = [ee^T] \circ \left([E_{\mathrm{WSWA}}][C_{2e}][x][C_{2e}]^T[E_{\mathrm{WSWA}}]^T \right) \tag{D.36}$$

$$[X^F] = 2N[ee^T] \circ [\boldsymbol{\epsilon}\boldsymbol{\epsilon}^T] \circ \left([E_{\mathrm{WSWA}}][C_{\mathrm{IIE}}][x][C_{\mathrm{IIE}}]^T[E_{\mathrm{WSWA}}]^T \right) \tag{D.37}$$

$$\boldsymbol{\epsilon} = \left(\sqrt{2}, 1, 1, \ldots, 1 \right)^T \qquad (2N \times 1)$$

$$e = \left\{ \exp\left(\frac{j\pi 0}{2N}\right), \exp\left(\frac{j\pi 1}{2N}\right), \ldots, \exp\left(\frac{j\pi(2N-1)}{2N}\right) \right\}^T \qquad (2N \times 1)$$

where $[E_{\mathrm{WSWA}}]$ is defined in (D.19) and corresponds to Fig. D.1c.

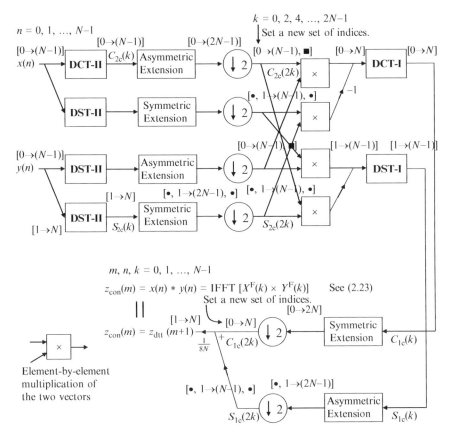

Fig. D.3 Circular convolution can be computed only using *unnormalized* DCTs and DSTs instead of the FFTs. ($\downarrow2$) represents (2:1) decimation. $*$ represents circular convolution. There is one sample delay in the resultant sequence $z_{con}(m) = z_{dtt}(m+1)$, $m = 0, 1, \ldots, N-1$. $[0 - (N-1), \bullet]$ represents a single zero is padded after N DCT/DST coefficients. *The circle* $\bullet$ represents the place always takes zero as a value. Other than this proposed method we usually overlook its existence, as it has no meaning in computation. *The square* $\blacksquare$ represents an appended zero

D.3 Circular Convolution Using DCTs and DSTs Instead of FFTs

Circular convolution of two periodic sequences in time/spatial domain is equivalent to multiplication in the DFT domain. That is also known as the convolution multiplication property of the DFT (see Section 2.4).

For discrete cosine transforms (DCTs) and discrete sine transforms (DSTs), called discrete trigonometric transforms (DTTs), such a nice property is developed by Reju *et al.* [LA6]. Since fast algorithms are available for the computation of discrete sine and cosine transforms, the proposed method is an alternative to the

DFT method for filtering applications (Fig. D.3). The MATLAB code for this method described in [LA6] is available at http://www.ntu.edu.sg/eee/home/esnkoh/.

$C_{1e}(k), C_{2e}(k), S_{1e}(k)$, and $S_{2e}(k)$ are defined in (D.15), (D.9), (D.23) and (D.24) respectively.

D.4   Circular Shifting Property of the DCT

Since the N-point DCT can be derived from the $2N$-point DFT [LA5], each basis function for which k is odd does not complete its period within N-sample length (e.g. half period for $k = 1$, one and half period for $k = 3$, ... See Fig. 5.1 in [B6]). For example, the lowest nonzero frequency element for which $k = 1$ is

$$C^{2e}(k, n) = 2 \cos\left(\frac{\pi k \left(n + \frac{1}{2}\right)}{N}\right) \tag{D. 38}$$

and its period is $2N$. The circular left shift of a sequence (advanced) by an integer sample d (d can be extended to a noninteger value with the same Eq. D.39) can be handled as if the basis functions were shifted reversely by d while the sequence itself is fixed to the original position. Under these considerations, the shifting property between the spatial-domain original data and the frequency-domain shifted data is developed by Rhee and Kang [LA17]. This can be expressed as

$$C_d^{2e}(k) = 2 \sum_{n=0}^{N-1} c(n, k) \, x(n) \cos\left[\frac{\pi k \left(n + \frac{1}{2} - d\right)}{N}\right], \quad k = 0, 1, \ldots, N-1 \tag{D.39}$$

where

$$c(n, k) = \begin{cases} (-1)^k & k = \text{odd}, d > 0 \quad \text{and} \quad 0 \le n < d, \quad \text{or} \\ & k = \text{odd}, d \le -1 \quad \text{and} \quad N + d \le n < N \\ 1 & \text{Otherwise.} \end{cases}$$

For example, $\{x(n)\} = \{1, 2, 3, 4\}$ is given, inverse DCT of $\{C_d^{2e}(k)\}$ equals to $x(n) = \{2, 3, 4, 1\}$ for $d = 1$. The property shows the relations between the original signal in the spatial domain and the frequency-domain representation of the (up to subsample/subpixel) shifted signal. For the unitary DCT, (D.39) becomes

$$C_d^{IIE}(k) = \sqrt{\frac{2}{N}} \epsilon_k \sum_{n=0}^{N-1} c(n, k) \, x(n) \cos\left[\frac{\pi k \left(n + \frac{1}{2} - d\right)}{N}\right], \quad k = 0, 1, \ldots, N-1 \tag{D.40}$$

where

$$c(n,k) = \begin{cases} (-1)^k & k = \text{odd}, \ d > 0 \quad \text{and} \quad 0 \le n < d, \quad \text{or} \\ & k = \text{odd}, \ d \le -1 \quad \text{and} \quad N + d \le n < N \\ 1 & \text{otherwise}, \end{cases}$$

$$\epsilon_k = \begin{cases} 1/\sqrt{2} & k = 0 \\ 1 & \text{otherwise}. \end{cases}$$

Problems

D.1 Prove $\left[C_N^{IV} \right] = [D] \left[S_N^{IV} \right] [J]$ shown in (D.1).

D.2 Similar to (D.3) through (D.8), derive the unitary DCT-III and DCT-IV from their corresponding forms of the GDFT. Hint: $\exp\left(-j \frac{(1/2)\pi(7+1/2)}{N} \right) = \exp\left(-j \frac{15\pi}{4N} \right) = \exp\left(j \left(1 - \frac{1}{4N} \right)\pi \right) = -\exp\left(j \frac{(1/2)\pi(1/2)}{N} \right)$, $N = 4$.

D.3 (a) See (D.5) through (D.7). Solve this problem for $N = 8$ and $N = 9$ respectively. Show how you can get the results.
Repeat part (a) for the following.
(b) DCT-I
(c) DST-I
(d) DST-II

D.4 DCT-II coefficients $C_{2e}(k)$, $k = 0, 1, \ldots, N - 1$, are given. Then how can we get $C_{2e}(2k)$, $k = 0, 1, \ldots, N - 1$ from the DCT-II coefficients $C_{2e}(k)$. Represent $C_{2e}(2k)$, $k = 0, 1, \ldots, N - 1$ in terms of DCT-II coefficients $C_{2e}(k)$. For example

$$\{C_{2e}(2k)\} = \{C_{2e}(0), C_{2e}(2), \ldots\}, \quad k = 0, 1, \ldots, N - 1$$

Projects

D.1 Run the MATLAB code in [LA3]. Show test results.

D.2 Run the MATLAB code for the circular convolution using DCTs and DSTs [LA6] that is available in http://www.ntu.edu.sg/eee/home/esnkoh/. Plot the following with $N = 8$ and $N = 9$.
(a) $C_{1e}(k), S_{1e}(k), C_{2e}(k), S_{2e}(k), C_{1e}(2k), S_{1e}(2k), C_{2e}(2k), S_{2e}(2k)$
(b) Outputs of both methods for fast circular convolution: the DFT and DTT methods.

D.3 (a) Implement 2-D DCT-II using 2-D DFT as shown in Fig. D.2 by using
          MATLAB.
     (b) Compute the $(N \times N)$ DCT coefficient matrix $[X^{CII}]$ of an image of size
          $(N \times N)$. Compute a $(2N \times 2N)$ DFT coefficient matrix $[X^F]$ with which
          we could compute the DCT coefficient matrix $[X^{CII}]$.

Appendix E
Kronecker Products and Separability

E.1   Kronecker Products

Definition E.1. Let $[A]$ be a $(p \times q)$ matrix and $[B]$ a $(k \times l)$ matrix. The *left* and *right* Kronecker products (also called *matrix product* or *direct product*) of $[A]$ and $[B]$ are the $(pk \times ql)$ matrices [B6]:

$$[A] \otimes_L [B] = \begin{bmatrix} [A]b_{00} & [A]b_{01} & \cdots & [A]b_{0,l-1} \\ [A]b_{10} & [A]b_{11} & \cdots & [A]b_{1,l-1} \\ \vdots & \vdots & \vdots & \vdots \\ [A]b_{k-1,0} & [A]b_{k-1,1} & \cdots & [A]b_{k-1,l-1} \end{bmatrix} \quad \text{(E.1a)}$$

$$[A] \otimes_R [B] = \begin{bmatrix} a_{00}[B] & a_{01}[B] & \cdots & a_{0,q-1}[B] \\ a_{10}[B] & a_{11}[B] & \cdots & a_{1,q-1}[B] \\ \vdots & \vdots & \vdots & \vdots \\ a_{p-1,0}[B] & a_{p-1,1}[B] & \cdots & a_{p-1,q-1}[B] \end{bmatrix} \quad \text{(E.1b)}$$

Example E.1   Let

$$[A] = \begin{bmatrix} 1 & 1 \\ 1 & -1 \end{bmatrix} \quad \text{and} \quad [B] = \begin{bmatrix} 1 & 2 \\ 3 & 4 \end{bmatrix}$$

then

$$[A] \otimes_L [B] = \begin{bmatrix} 1 & 1 & 2 & 2 \\ 1 & -1 & 2 & -2 \\ 3 & 3 & 4 & 4 \\ 3 & -3 & 4 & -4 \end{bmatrix} \quad \text{and} \quad [A] \otimes_R [B] = \begin{bmatrix} 1 & 2 & 1 & 2 \\ 3 & 4 & 3 & 4 \\ 1 & 2 & -1 & -2 \\ 3 & 4 & -3 & -4 \end{bmatrix}$$

Note that the left Kronecker product is denoted as $[A] \otimes_L [B]$ and the right Kronecker product is denoted as $[A] \otimes_R [B]$. As $[A] \otimes_R [B] = [B] \otimes_L [A]$, we only use the right Kronecker product in this book and MATLAB (kron).

Property E.1. If $[A]$ is $(m \times m)$ and $[B]$ is $(n \times n)$, then

$$\det([A] \otimes [B]) = (\det[A])^n (\det[B])^m \tag{E.2}$$

For example

$$[A] = \begin{bmatrix} 1 & 0 & 0 \\ 0 & 1 & 0 \\ 0 & 0 & 1 \end{bmatrix}, \quad [B] = \begin{bmatrix} 1 & 0 \\ 0 & 2 \end{bmatrix}, \quad \det([A] \otimes [B]) = 8$$

$$(\det[A])^n (\det[B])^m = 1^2 2^3 = 8 \neq (\det[A])^m (\det[B])^n = 1^3 2^2 = 4$$

Property E.2. If $[A]$ and $[B]$ are orthogonal, then $[A] \otimes [B]$ is orthogonal ($[A]^T = [A]^{-1}$ and $[B]^T = [B]^{-1}$). This is proved with the aid of following relations.

$$\begin{aligned} ([A] \otimes [B])^T ([A] \otimes [B]) &= ([A]^T \otimes [B]^T)([A] \otimes [B]) \\ &= ([A]^T [A]) \otimes ([B]^T [B]) = [I_m] \otimes [I_k] = [I_{mk}]. \end{aligned} \tag{E.3}$$

E.2   Generalized Kronecker Product

Definition E.2. Given a set of N $(m \times r)$ matrices $[A_i]$, $i = 0, 1, \ldots, N - 1$ denoted by $\{[A]\}_N$, and a $(N \times l)$ matrix $[B]$, we define the $(mN \times rl)$ matrix $(\{[A]\}_N \otimes [B])$ as [N9]

$$\{[A]\}_N \otimes [B] = \begin{bmatrix} [A_0] \otimes \underline{b}_0 \\ [A_1] \otimes \underline{b}_1 \\ \vdots \\ [A_{N-1}] \otimes \underline{b}_{N-1} \end{bmatrix} \tag{E.4}$$

where $\underline{b}_i$ denotes the ith row vector of $[B]$. If each matrix $[A_i]$ is identical, then it reduces to the usual Kronecker product.

Example E.2   Let

$$\{[A]\}_2 = \left\{ \begin{bmatrix} 1 & 1 \\ 1 & -1 \end{bmatrix} \\ \begin{bmatrix} 1 & -j \\ 1 & j \end{bmatrix} \right\}, \quad [B] = \begin{bmatrix} 1 & 1 \\ 1 & -1 \end{bmatrix}$$

then the generalized Kronecker product yields

$$\{[A]\}_2 \otimes [B] = \begin{bmatrix} \begin{bmatrix} 1 & 1 \\ 1 & -1 \end{bmatrix} \otimes [1 \quad 1] \\ \begin{bmatrix} 1 & -j \\ 1 & j \end{bmatrix} \otimes [1 \quad -1] \end{bmatrix} = \begin{bmatrix} 1 & 1 & 1 & 1 \\ 1 & -1 & 1 & -1 \\ 1 & -j & -1 & j \\ 1 & j & -1 & -j \end{bmatrix}$$

which is recognized as a (4×4) DFT matrix with the rows arranged in bit-reversed order.

E.3   Separable Transformation

Consider the 2-D transformation on an $(M \times N)$ image $[x]$

$$\begin{matrix} (M \times N) & & (M \times N) \\ [X] & = & [A] & [x] & [B]^T \\ & & (M \times M) & & (N \times N) \end{matrix}$$

or

$$X(k,l) = \sum_{m=0}^{M-1} \sum_{n=0}^{N-1} a(k,m)\, x(m,n)\, b(l,n), \quad 0 \le k \le M-1, \quad 0 \le l \le N-1 \quad (E.5)$$

This relationship is called *separable transform* for the 2-D data, where $[A]$ operates on the columns of $[x]$, and $[B]$ operates on the rows of the result. If $\underline{X}_k$ and $\underline{x}_m$ denote the kth and mth row vectors of $[X]$ and $[x]$, respectively, then the row operation becomes [B6]

$$\underline{X}_k^T = \sum_{m=0}^{M-1} a(k,m)[[B]\,\underline{x}_m^T] = \sum_{m=0}^{M-1} ([A] \otimes [B])_{k,m}\, \underline{x}_m^T \quad 0 \le k \le M-1 \quad (E.6)$$

where $[A] \otimes [B]_{k,m}$ is the (k, m)th block of $[A] \otimes [B]$, $0 \le k, m \le M-1$. Thus if $[x]$ and $[X]$ are row-ordered into vectors $\underline{x}$ and $\underline{X}$, then the separable transformation of (E.5) maps into a Kronecker product operation on a vector as

$$\begin{matrix} \underline{X} & = & ([A] \otimes [B]) & \underline{x} & (E.7) \\ (MN \times 1) & & (MN \times MN) & (MN \times 1) \end{matrix}$$

where

$$
\underline{X} = \begin{pmatrix} \underline{X}_0^T \\ \underline{X}_1^T \\ \vdots \\ \underline{X}_k^T \\ \vdots \\ \underline{X}_{M-1}^T \end{pmatrix} \quad \text{and} \quad \underline{x} = \begin{pmatrix} \underline{x}_0^T \\ \underline{x}_1^T \\ \vdots \\ \underline{x}_m^T \\ \vdots \\ \underline{x}_{M-1}^T \end{pmatrix} \tag{E.8}
$$

Theorem E.1.   Given a 2-D $(M \times N)$ image $[x]$, its 1-D transformation

$$
\underset{(MN \times 1)}{\underline{X}} \;=\; \overset{(MN \times MN)}{[A]} \quad \underset{(MN \times 1)}{\underline{x}} \tag{E.9}
$$

can be a separable transform if

$$
[A] = [A_1] \otimes [A_2] \tag{E.10}
$$

and the 2-D transformation is described by

$$
\overset{(M \times N)}{[X]} \;=\; \underset{(M \times M)}{\overset{(M \times N)}{[A_1]}} \quad [x] \quad \underset{(N \times N)}{[A_2]^T} \tag{E.11}
$$

If $[A_1]$ and $[A_2]$ are $(N \times N)$ matrices, and $[A]$ is $(N^2 \times N^2)$ matrix, then the number of operations reduces from $O(N^4)$ to $O(2N^3)$. Recall that $(N \times N)$ matrix multiplication with $(N \times N)$ data requires $O(N^3)$.

Appendix F
Mathematical Relations

F.1

From Euler's formula/identity

$$2\cos\theta = e^{j0} + e^{-j0} \tag{F.1}$$

$$j2\sin\theta = e^{j0} - e^{-j0} \tag{F.2}$$

Compound Angle formulae

$$\cos(a \pm b) = \cos a \cos b \mp \sin a \sin b \tag{F.3}$$

$$\sin(a \pm b) = \sin a \cos b \mp \cos a \sin b \tag{F.4}$$

Product identities

$$2\cos a \cos b = \cos(a - b) + \cos(a + b) \tag{F.5}$$

$$2\sin a \sin b = \cos(a - b) - \cos(a + b) \tag{F.6}$$

$$2\sin a \cos b = \sin(a - b) + \sin(a + b) \tag{F.7}$$

Hyperbolic Function

$$\cos(j\theta) = \cosh\theta \tag{F.8}$$

$$\sin(j\theta) = j\sinh\theta \tag{F.9}$$

$$2\cosh\theta = e^0 + e^{-0} \qquad \text{From (F.1) and (F.8)} \tag{F.10}$$

$$2\sinh\theta = e^0 - e^{-0} \tag{F.11}$$

$$\cosh^2\theta - \sinh^2\theta = 1 \tag{F.12}$$

F.2 If $f(x, y) \Leftrightarrow F(u, v)$, then $f(x - \sigma y, y) \Leftrightarrow F(u, v + \sigma u)$, where $\Leftrightarrow$ denotes a 2-D Fourier transform pair (see Section 5.2) [IP26].

   Proof.

$$\mathcal{F}[f(x - \sigma y, y)] = \int \int_{-\infty}^{\infty} f(x - \sigma y, y) e^{-j2\pi(ux - vy)} dx dy$$

where $\mathcal{F}$ denotes Fourier transform. Change variables as $s = x - \sigma y$ and $t = y$. Substitute $\phi(s, t) = s + \sigma t$ for x and $\psi(s, t) = t$ for y.

$$\mathcal{F}[f(x - \sigma y, y)] = \int \int_{-\infty}^{\infty} f(s, t) e^{-j2\pi(us + (u + \sigma u)t)} \begin{vmatrix} 1 & \sigma \\ 0 & 1 \end{vmatrix} ds dt$$

$$= F(u, v + \sigma u)$$

since $\begin{vmatrix} 1 & \sigma \\ 0 & 1 \end{vmatrix} = 1$.

   Corollary F.1. If $f(x, y)$ has the Fourier transform $F(u, v)$, then the inverse Fourier transform of $F(u, v + \sigma u)$ is $f(x - \sigma y, y)$.

F.3 $\dfrac{d}{du} a^u = a^u \ln a$

   Proof.

$$\frac{d}{du} a^u = \frac{d}{du} \left(e^{\ln a} \right)^u = \frac{d}{du} e^{(\ln a)u} = e^{(\ln a)u} \frac{d}{du} (\ln a)u$$

$$= \left(e^{(\ln a)u} \right) (\ln a) = a^u \ln a$$

F.4 Let a and b be vectors, and $\circ$ represent Hadamard product. Then

$$\text{diag}(a)[C]\text{diag}(b) = \left(ab^T \right) \circ [C] \tag{F.13}$$

For example let $a = (2, 1)^T$, $b = (1, 2)^T$ and $[C] = \begin{bmatrix} 1 & 2 \\ 3 & 0 \end{bmatrix}$ (see (3.74), (C.36), (C.37) and (D.32)).

$$\text{diag}(a)[C]\text{diag}(b) = \begin{bmatrix} 2 & 0 \\ 0 & 1 \end{bmatrix} \begin{bmatrix} 1 & 2 \\ 3 & 0 \end{bmatrix} \begin{bmatrix} 1 & 0 \\ 0 & 2 \end{bmatrix} = \begin{bmatrix} 2 & 8 \\ 3 & 0 \end{bmatrix}$$

$$\left(ab^T \right) \circ [C] = \begin{pmatrix} 2 \\ 1 \end{pmatrix} (1, 2) \circ \begin{bmatrix} 1 & 2 \\ 3 & 0 \end{bmatrix} = \begin{bmatrix} 2 & 4 \\ 1 & 2 \end{bmatrix} \circ \begin{bmatrix} 1 & 2 \\ 3 & 0 \end{bmatrix} = \begin{bmatrix} 2 & 8 \\ 3 & 0 \end{bmatrix}$$

Problem

F.1 Prove Corollary F.1.

Appendix G
Basics of MATLAB

This appendix describes basics of MATLAB. It also provides various websites and references related to MATLAB.

» $A = [1 \quad 2 ; 3 \quad 4]$

$$A = \begin{bmatrix} 1 & 2 \\ 3 & 4 \end{bmatrix}$$

» max(A) = [3 4]
» min(A) = [1 2]

» (A, 1, 4) = [1 3 2 4]	Returns the 1-by-4 matrix whose elements are taken columnwise from [A]
» A = [2 5 1], [B, I] = sort(A)	Sorts in ascending order. Returns an index matrix I
» A = eye(3, 4);	An identity matrix of size (3 × 4)
» A = ones(2); B = zeros(2);	(2 × 2) matrices of ones and zeros respectively

» eye (4) + eye (4) and 1 + eye(4) are different.

» sum(A) = [4 6]	For matrix A, sum(A) is a row vector with the sum over each column
» mean(A)	For matrices, mean(A) is a row vector containing the mean value of each column
» A = [1:3]	Equals to A = [1 2 3]
» diag(A)	If input is an vector, it makes a diagonal matrix. If input is a matrix, it displays the main diagonal elements of [A]
» $A * B$	Matrix multiply
» $A. * B$	Element-by-element multiplication of the two matrices
» inv	Matrix inverse
» fliplr	Flip a matrix in left/right direction
» flipud	Flip a matrix in up/down direction
» rot90	Rotate a matrix by 90° in a counter-clockwise direction

(continued)

» flipdim	Flip a matrix along specified dimension
» sin(A)	The sine of the elements of A
» asin(A)	The arcsine of the elements of A
» cosh(A)	Hyperbolic cosine. Equals to $\cos(j*A)$

o Commands for managing a session

» help topic	Gives help on the specified "topic"
» lookfor	Search all M-files for keyword
» clear	Clear variables and functions from memory
» clear all, close all	

Press the up arrow key ↑ to recall the most recent command line.

Press the keys 'h↑' sequentially to recall the most recent command line starting with 'h'.

» p ='C:\mycode' ; path(path, p)	Appends a new directory to the current path
» path(path, 'C:\mycode')	

o Operators and special characters

» A'	Complex conjugate transpose or ctranspose(A) where A is a matrix. When A is real, A' is equal to transpose(A)
» A.'	Transpose or transpose(A)
» ...	Return is ; and ... means continued line
	A command can be more than a line as follows
» $A = [1\ 2\ 3\ ...$ 4 5]	This is the same with the command $A = [1\ 2\ 3\ 4\ 5]$

o Special variables and constants

» pi	The number $\pi = 3.1415926\cdots$
» i, j	The imaginary unit. $i = j = \sqrt{-1}$
» inf	Infinity
» ans	Most recent answer

o File

» $A = [1.1\ 2.1]$	Let us save data into a text file
» save my.dat A −ascii	
» load my.dat	The data from the file "my.dat" will be load ed into the variable "my"
my $= [1.1\ 2.1]$	
» save my	For another application save workspace variables to the file "my.mat"
» load my.dat	Load workspace variables from the file "my.dat"

o Low-level file I/O function

» n = 0:.1:1; y = [n; sin(n)]
» fid = fopen('sin.dat', 'w')
» fprintf(fid, '%3.2f %3.4f\n', y)
» fclose(fid)

Creates a text file containing a short table of the sine function:

0.00	0.0000
0.10	0.0998
. . .	. . .
1.00	0.8415

» fprintf('%3.2f %3.4f\n',y);   Produces the table in the command window.

Type in '**help fprintf**' in the command window for more details. See also '**sprintf**'.

o Relational and logical operators/functions

» $A < B$, $A == B$, $A >= B$, $A <= B$
If a condition is true, **ans** is one; otherwise, **ans** is zero.

» $A = [-1, 3, 0, 5; -2, 0, 0, 9]$

» [m, n] = find(A)   Returns the row and column indices of the nonzero entries
in the matrix A.

$(m, n) = (1, 1), (2, 1), (1, 2), (1, 4), (2, 4)$

o FOR loop

» for n=1:4   n,   end
» A = [1:3:10] $A = [1 \quad 4 \quad 7 \quad 10]$

o Conditional statements

if $((attendance >= 0.90)$ & $(grad\_average >= 60))$
     pass = 1;
else
     fail = 1;
end;

o DFT/FFT functions

» dftmtx(N)	DFT matrix of size $(N \times N)$
	Its inverse is **conj(dftmtx(N))/ N**
» fft, ifft	1-D FFT and its inverse
» fft2, ifft2	2-D FFT and its inverse
» fftshift	Shift the dc coefficient to the center of the 2-D frequency
	plane. See Fig. 5.3(b). It also works for a vector

(continued)

» conv(A, B)   Aperiodic convolution of the two vectors. Use **conv2** for two
                                      matrices and **convn** for any dimensional input data

» A = [1:3], N = 2,   N determines the number of resultant rows for a row vector, A
» convmtx(A, N)

   1   2   3   0
   0   1   2   3

» xcorr(A, B)   Cross-correlation function. Use **xcorr2** for matrices A and B

» prod, prodsum, sum, cumsum, diff, gradient

» fft, fft2, fftn   Not normalized
» dct, idct   1-D normalized type-2 DCT
» dct2, idct2   2-D normalized type-2 DCT
» dctmtx(N)   The $(N \times N)$ DCT matrix
» dst, idst   1-D type-1 DST. It is not normalized
                                      Assume that **input** is a vector. Normalized DST is

» N = length(input); dst(input) $* \text{sqrt}\,(2/(N+1))$;
   In summary, dct and dct2 are normalized, whereas dst, fft and fft2 are not
normalized.

» hadamard(N)   Hadamard matrix of order N, that is, a matrix H with
                                      elements 1 or -1 such that $H' * H = N * \text{eye}(N)$
» sinc   Note sinc(0) = 1 by the definition
» chirp   Swept-frequency cosine generator

o User-defined function

---

function [MSE, PSNR] = mse(Input, Estimate)
%rComputeulnpMSEatiohPSNR (in dB) between an image and its estimate
MSE = mean (abs(Error(:).^2);
PSNR = $10 * \log_{10}(255^2/\,\text{MSE})$;

This function will be saved as the M-file "mse.m". Thus the function name becomes
its file name. The function is called as follows.

» a = mse(A, B)   To get MSE only
» [a, b] = mse(A, B)   To get both MSE and PSNR

» A = 'A text herein will be displayed.'

» disp(A)   If A is a string, the text is displayed

o Block processing

» fun = inline('sum (x(:)');
» B = blkproc(A, [2 2], fun);
» A = [1 2 3]

» T_s = toeplitz(A)   Symmetric (or Hermitian) Toeplitz matrix

$$\gg T_s = \begin{bmatrix} 1 & 2 & 3 \\ 2 & 1 & 2 \\ 3 & 2 & 1 \end{bmatrix}$$

$\gg T =$ toeplitz   Non-symmetric Toeplitz matrix. C is the first column and
   (R, C) R is the first row. See also hankel

$\gg$ kron(A, B)   Kronecker tensor product of A and B

o Image handling functions

MATLAB includes images: **cameraman.tif**, **ic.tif**, and more in the directory "imdemos".

$\gg$ I = imread('ic.tif');

$\gg$ J = imrotate(I, -3, 'bilinear', 'crop');

$\gg$ imshow(I); figure, imshow(J); title('Rotated image with \theta $= -3$')

   To have a capital Greek in a figure title or plot label, use "\Theta".

$\gg$ angle $= -3$

$\gg$ imshow (I); figure, imshow (J);

$\gg$ title (['Rotated image with \theta=', num2str (angle)]

   Argument for **title** can be a vector enclosed by '[' and ']'.

   See image, axis off, axis image.

   See also imcrop, imresize, imtransform, tformarray, zoom.

figure, by itself, creates a new figure window. If **figure** command is missed, only one figure window is created.

$\gg$ I = checkerboard(NP, M, N)   where NP is the number of pixels on the side of each square, M is half time of the number of rows of squares, and N is half time of the number of columns of squares

$\gg$ imopen, imclose   Morphological opening / closing on a grayscale or binary image

$\gg$ phantom($\ldots, N$)   Generate an $(N \times N)$ head phantom image that can be used to test the numerical accuracy of **radon** and **iradon** or other reconstruction algorithms

o Plotting graphs

$\gg$ plot (x, y)   Plots vector y versus vector x

$\gg$ bar (x, y, width)   Bar graph. Values of width > 1, produce overlapped bars

$\gg$ scatter(x, y)   Displays circles at the locations specified by the vectors x and y

See also stem, stairs, and semilogy.

$\gg$ N = hist (y, 20)   Histogram. This command bins the elements of y into 20 equally spaced containers and returns the number of elements in each container

(continued)

» hold on	Holds the current plot and all axis properties so that subsequent graphs add to the existing graph
» hold off	Returns to the default mode
» axis ([1 20 0 50])	Sets scaling for the x-axis (1, ..., 20) and y-axis (1, ..., 50) on the current plot
» C = xcorr (A, B)	Aperiodic cross-correlation function estimates
» [C, Lags] = xcorr (A, B)	Returns a vector Lags of lag indices

o Eigen vectors and eigen values

» [V, D] = eig(X) produces a diagonal matrix D of eigenvalues and a full matrix V whose columns are the corresponding eigenvectors so that
$$X * V = V * D$$

o Filter

$$H(z) = \frac{1}{1 - 0.9z^{-1}}$$ The transfer function of a lowpass filter

$$a(1)y(n) + a(2)y(n-1) = b(1)x(n)$$

$$y(n) = -a(2)y(n-1) + b(1)x(n)$$

$$= 0.9y(n-1) + x(n) = \{1, 2.9, 5.61, 9.049\}$$

» x = [1 2 3 4]	Input sequence
» A = [1 −0.9]	Lowpass filter parameters
» B = 1	
» y = filter(B, A, x)	{1, 2.9, 5.61, 9.049}: Filtered sequence

e.g. Let $H(z) = \frac{1}{2} + \frac{1}{2}z^{-1}$

» x = [1 2 3 4]	Input sequence
» A = 1	Lowpass filter parameters
» B = $\begin{bmatrix} \frac{1}{2} & \frac{1}{2} \end{bmatrix}$	
» y = filter(B, A, x)	$y = \begin{bmatrix} \frac{1}{2} & \frac{3}{2} & \frac{5}{2} & \frac{7}{2} \end{bmatrix}$ Output sequence

o Wavelet filters

» [LD, HD, LR, HR] = wfilters('w_name')
    This computes four filters associated with the orthogonal or biorthogonal wavelet named in the string 'w_name'.
    LD and HD are the decomposition low and high pass filters.
    LR and HR are the reconstruction low and high pass filters.

o Wavelet transform

» dwt	Single-level discrete 1-D wavelet transform
» idw	Inverse dwt
» dwt2	Single-level discrete 2-D wavelet transform
» idwt2	Inverse dwt2

» syms $\quad x \quad z$ — Creates the symbolic variable with name x and z

» simplify $(1/(z+1)+1/z)$ — Symbolic simplification for $\dfrac{1}{z+1}+\dfrac{1}{z}$

ans $= (2*z+1)/(z+1)/z$

» pretty(ans)

$$\frac{2z+1}{(z+1)z}$$

G.1   List of MATLAB Related Websites

G.1.1   MATLAB *Tutorial*

http://www.cyclismo.org/tutorial/matlab/
http://www.stanford.edu/~wfsharpe/mia/mat/mia_mat3.htm
http://www.contracosta.cc.ca.us/math/lmatlab.htm
http://www.cs.berkeley.edu/titan/sww/software/matlab/
http://www.cs.berkeley.edu/titan/sww/software/matlab/techdoc/ref/func_by_.html

G.1.2   MATLAB *Commands and Functions*

http://www.hkn.umn.edu/resources/files/matlab/MatlabCommands.pdf
http://www.owlnet.rice.edu/~elec241/ITMatlab.pdf Introduction to MATLAB

G.1.3   MATLAB *Summary and Tutorial*

http://www.math.ufl.edu/help/matlab-tutorial/matlab-tutorial.html

G.1.4   A MATLAB *Primer*

http://www4.ncsu.edu/unity/users/p/pfackler/www/MPRIMER.htm
http://www.glue.umd.edu/~nsw/ench250/primer.htm

G.1.5   MATLAB FAQ

http://matlabwiki.mathworks.com/MATLAB_FAQ

G.2   List of MATLAB Related Books

M1 S.D. Stearns, *Digital Signal Processing with Examples in MATLAB*® (CRC Press, Boca Raton, FL, 2003)

M2 D.G. Duffy, *Advanced Engineering Mathematics with MATLAB* (CRC Press, Boca Raton, FL, 2003)

M3 R.E. White, *Elements of Matrix Modeling and Computing with MATLAB* (CRC Press, Boca Raton, FL, 2006)

M4 A.D. Poularikis, Z.M. Ramadan, *Adaptive filtering with MATLAB* (CRC Press, Boca Raton, FL, 2006)

M5 E.W. Kamen, B.S. Heck, *Fundamentals of Signals and Systems using Web and MATLAB* (Prentice Hall, Upper Saddle River, NJ, 2007)

M6 *MATLAB and SIMULINK Student Version.* (Release 2007 – CD-ROM), Mathworks. (MATLAB 7.4 and Simulink 6.6)

M7 A.H. Register, *A Guide to MATLAB*® *Object-Oriented Programming* (CRC Press, Boca Raton, FL, 2007)

M8 M. Weeks, *Digital Signal Processing Using MATLAB and Wavelets* (Infinity Science Press LLC, Hingham, MA, 2007)

M9 A.D. Poularikis, *Signals and Systems Primer with MATALAB* (CRC Press, Boca Raton, FL, 2007)

M10 B. Hahn and D. Valentine, *Essential MATLAB for Engineers & Scientists.* III Edition, Oxford, UK: Elsevier, 2008.

M11 S.J. Chapman, *MATLAB*® *Programming for Engineers.* IV Edition, Cengage Learning, Engineering, 1120 Birchmount Rd, Toronto, ON, M1K 5G4, Canada, 2008.

M12 Amos Gilat, *MATLAB*®*: An Introduction with Applications.* III Edition, Hoboken, NJ: Wiley, 2008.

M13 Li Tan, *Digital signal processing: Fundamentals and Applications.* Burlington, MA: Academic Press (Elsevier), 2008. (This has MATLAB exercises/ programs.)

M14 M. Kalechman, *Practical MATLAB*® *Basics for Engineers* (CRC Press, Boca Raton, FL, 2008)

M15 M. Kalechman, *Practical MATLAB*® *Applications For Engineers* (CRC Press, Boca Raton, FL, 2008)

M16 J. Musto, W.E. Howard, R.R. Williams, *Engineering Computation: An Introduction Using MATLAB and Excel* (McGraw Hill, New York, NY, 2008)

M17 T.S. El Ali, M.A. Karim, *Continuous Signals and Systems with MATLAB* (CRC Press, Boca Raton, FL, 2008)

M18 A. Siciliano, *Data Analysis and Visualization* (World Scientific Publishing Co. Inc., Hackensack, NJ, 2008)

M19 B. Hahn and D. Valentine, *Essential MATLAB for Engineers and Scientists.* Elsevier, 2008

M20 W.L. Martinez, *Computational Statistics Handbook with MATLAB*, II Edition (CRC Press, Boca Raton, FL, 2008)

M21 A.D. Poularikis, *Discrete Random Signal Processing and Filtering Primer with MATLAB* (CRC Press, Boca Raton, FL, 2009)

M22 O. Demirkaya, M.H. Asyali and P.K. Sahoo, *Image Processing with MATALB* (CRC Press, Boca Raton, FL, 2009) (MATLAB codes/functions/algorithms)

M23 M.N.O. Sadiku, *Numerical Techniques in Electromagnetics with MATLAB* (CRC Press, Boca Raton, FL, 2009)

M24 S. Attaway, *MATLAB: A Practical Introduction to Programming and Problem Solving* (Elsevier, Burlington, MA, 2009)

M25 M. Corinthios, *Signals, Systems, Transforms and Digital Signal Processing with MATLAB* (CRC Press, Boca Raton, FL, 2009)

M26 A.M. Grigoryan, M.M. Grigoryan, *Brief Notes in Advanced DSP, Fourier Analysis with MATLAB* (CRC Press, Boca Raton, FL, 2009)

M27 MATLAB Educational Sites are listed on the website below. http://faculty. ksu.edu.sa/hedjar/Documents/MATLAB_Educational_Sites.htm

M28 T.A. Driscoll, *Learning MATLAB* (SIAM, Philadelphia, PA, 2009)

M29 C.F.V. Loan, K.-Y.D. Fan, *Insight through Computing: A MATLAB Introduction to Computational Science and Engineering* (SIAM, Philadelphia, PA, 2009)

Appendix H

H.1 MATLAB Code for 15-Point WFTA [A39]

```
function [ ] = fft_15( )
P_inp = zeros(15,15);
P_out = zeros(15,15);
```

% Prime factor map (PFM) indexing for input and output sequences with $N = 3 \times 5$.
% 3×5 means we first apply five-point transforms and then three-point transforms.
% This will result in 67 real additions. If we use 5×3, we get 73 real additions.
% Multiplications will be the same in either case. Hence we use the factorization
3×5.
% The frequency index map is called the *Chinese Remainder Theorem* (CRT) [B29].

```
k = 1;
for n1 = 0 : 2                        % Symbol * means matrix multiply in MATLAB.
    for n2 = 0 : 4
        inp_idx_map(k) = mod(5*n1 + 3*n2,15);
        out_idx_map(k) = mod(10*n1 + 6*n2,15);
        k = k+1;
    end
end

inp_idx_map = inp_idx_map + 1;
out_idx_map = out_idx_map + 1;
```

% Form the permutation matrices of input and output.

```
for k = 1:15
    P_inp(k, inp_idx_map(k)) = 1;
    P_out(k, out_idx_map(k)) = 1;
end
```

% Verify that the permuted transform matrix is equal to Kronecker product of prime
% factor transform matrices. P_inp $\neq$ inv(P_inp), P_out $=$ inv(P_out).

P_out $*$ fft(eye(15)) $*$ inv(P_inp) $-$ kron(fft(eye(3)), fft(eye(5)));

% Define post addition matrix for transform size 3. Refer to Winograd short-N DFT
% algorithms for the derivation of these matrices.

$$S3 \;=\; \begin{bmatrix} 1 & 0 & 0 \\ 1 & 1 & 1 \\ 1 & 1 & -1 \end{bmatrix};$$

% Multiplication matrix for length 3.

C3 $=$ diag([1 cos($-2*$pi/3)-1 i$*$sin($-2*$pi/3)]); % diag[1, cos($-2\pi/3$)-1, $j$sin($-2\pi/3$)]

% Pre additions matrix for length 3.

$$T3 \;=\; \begin{bmatrix} 1 & 1 & 1 \\ 0 & 1 & 1 \\ 0 & 1 & -1 \end{bmatrix};$$

% Post additions matrix for length 5.

$$S5 \;=\; \begin{bmatrix} 1 & 0 & 0 & 0 & 0 & 0 \\ 1 & 1 & 1 & 1 & -1 & 0 \\ 1 & 1 & -1 & 0 & 1 & 1 \\ 1 & 1 & -1 & 0 & -1 & -1 \\ 1 & 1 & 1 & -1 & 1 & 0 \end{bmatrix};$$

% Multiplication matrix for length 5.

u $= -2*$pi/5;
C5 $=$ diag([1 (cos(u)+cos(2$*$u))/2-1 (cos(u) $-$cos(2$*$u))/2 ...
                i$*$(sin(u)+sin(2$*$u)) i$*$sin(2$*$u) i$*$(sin(u) $-$sin(2$*$u))]);

% Pre additions matrix for length 5.

$$T5 \;=\; \begin{bmatrix} 1 & 1 & 1 & 1 & 1 \\ 0 & 1 & 1 & 1 & 1 \\ 0 & 1 & -1 & -1 & 1 \\ 0 & 1 & 0 & 0 & -1 \\ 0 & 1 & -1 & 1 & -1 \\ 0 & 0 & -1 & 1 & 0 \end{bmatrix};$$

% Verify (3.67) and (3.68).

kron(S3, S5)$*$kron(C3, C5)$*$kron(T3, T5) $-$ kron(fft(eye(3)), fft(eye(5)))

% Form matrix $[C_{5\times 3}]$ defined in (3.75).

```
[r_C3,temp] = size(C3);
[r_C5,temp] = size(C5);

for j = 1 : r_C5
   for q = 1 : r_C3
      C(j, q) = (C5(j, j) * C3(q, q));
   end
end

% Verify (3.73).

fft_15 = zeros(15,15);
for vec = 1 : 15                          % Test for each basis vector
   clear z15; clear x; clear y;
   x = zeros(15, 1);
   x(vec) = 1;

   % Apply the input permutation.

   x = P_inp * x;

   % Form matrix [z15] defined in (3.70).

   q = 1;
   for j = 1 : 3
      for k = 1 : 5
         z15(j, k) = x(q);
         q = q+1;
      end
   end

   z15 = S3 * (S5 * (C .* (T5 * (T3 * z15).'))).';

   % Form the output vector. Output is scrambled.

   y = zeros(15, 1);
   q = 1;
   for j = 1 : 3
      for k = 1 : 5
         y(q) = z15(j, k);
         q = q+1;
      end
   end

   % Apply inverse output permutation to get the unscrambled output.

   y = inv(P_out) * y;
   fft_15(1 : end, vec) = y;
end
fft_15 - fft(eye(15))
```

H.2   MATLAB Code for Phase Only Correlation (POC)
(see Section 8.3)

```
path(path, 'c:\ code')
f = imread('lena.jpg');              % imshow(f, [ ]);
[M, N] = size(f);                    % We use the Lena image of size 512 × 512.
imA = f(111:350, 111:350);   figure(1), imshow(imA); % 240 × 240
% imA = f(1:240, 1:240);      figure(1), imshow(imA);
S = 50;                              % Control shift amount, direction with sign + and −.
                                     % 50, 100, 140, −100, 221 (Maximum)
imB = f(111+S:350+S, 111+S:350+S);    figure(2), imshow(imB);
% imB = f( 1+S:240+S, 1+S:240+S);      figure(2), imshow(imB);
Fa = fft2(imA); Fb = fft2(imB);
Z = Fa .* conj(Fb) ./ abs( Fa .* conj(Fb) );            % Eq. (8.14)
% Z = Fa ./ Fb ;                                        % Simpler method, Eq. (8.17)
z = ifft2(Z);
max_z = max(max(z))
[m1, m2] = find(z == max_z); m1=m1−1, m2=m2−1
                                     % As MATLAB indexes start from (1, 1) instead of
                                     % (0, 0)
zz = fftshift(abs(z));               % figure(4), imshow(log(zz + 0.0001), [ ]);
figure(3), mesh(zz(1:2:240, 1:2:240)), colormap([0.6, 0.8, 0.2])
axis([1 120 1 120 0 1])
view(−37.5, 16)
xlabel('n_1'); ylabel('n_2'); zlabel('z_{poc}(n_1, n_2)');
```

Bibliography

Books

B1 H.J. Nussbaumer, *Fast Fourier Transform and Convolution Algorithms* (Springer-Verlag, Heidelberg, Germany, 1981)

B2 N.C. Geckinli, D. Yavuz, *Discrete Fourier Transformation and Its Applications to Power Spectra Estimation* (Elsevier, Amsterdam, the Netherlands, 1983)

B3 R.E. Blahut, *Fast Algorithms for Digital Signal Processing* (Addison-Wesley, Reading, MA, 1985)

B4 V. Cizek, *Discrete Fourier Transforms and Their Applications* (Adam Higler, Bristol/Boston, 1986)

B5 M.T. Heideman, *Multiplicative Complexity, Convolution and the DFT* (Springer-Verlag, Heidelberg, Germany, 1988)

B6 A.K. Jain, *Fundamentals of Digital Image Processing* (Prentice-Hall, Englewood Cliffs, NJ, 1989)

B7 R. Tolimieri, M. An, C. Lu, *Algorithms for Discrete Fourier Transform and Convolution* (Springer-Verlag, Heidelberg, Germany, 1989)

B8 C.V. Loan, *Computational Frameworks for the Fast Fourier Transform*. Uses MATLAB notation (SIAM, Philadelphia, PA, 1992)

B9 P.P. Vaidyanathan, *Multirate Systems and Filterbanks* (Prentice-Hall, Englewood Cliffs, NJ, 1993)

B10 S.K. Mitra, J.F. Kaiser, *Handbook for Digital Signal Processing*. Filter design and implementation, FFT implementation on various DSPs (Wiley, New York, 1993)

B11 O.K. Ersoy, *Multidimensional Fourier Related Transforms and Applications* (Prentice-Hall, Upper Saddle River, NJ, 1993)

B12 M.F. Cátedra et al., *The CG-FFT Method: Application of Signal Processing Techniques to Electromagnetics*. CG = Conjugate Gradient (Artech, Norwood, MA, 1994)

B13 W.W. Smith, J.M. Smith, *Handbook of Real-Time Fast Fourier Transforms: Algorithms to Product Testing* (Wiley-IEEE Press, Piscataway, NJ, 1995) (Assembly language programming, implement FFT algorithms on DSP chips, etc.)

B14 R.M. Gray, J.W. Goodman, *Fourier Transforms: An Introduction for Engineers* (Kluwer, Norwell, MA, 1995)

B15 A.D. Poularikas (ed.), *The Transforms and Applications Handbook* (CRC Press, Boca Raton, FL, 1996)

B16 O.K. Ersoy, *Fourier-Related Transforms, Fast Algorithms and Applications* (Prentice-Hall, Upper Saddle River, NJ, 1997)

B17 H.K. Garg, *Digital Signal Processing Algorithms: Number Theory, Convolutions, Fast Fourier Transforms and Applications* (CRC Press, Boca Raton, FL, 1998)

B18 T.M. Peters, J.C. Williams, *The Fourier Transform in Biomedical Engineering* (Birkhauser, Boston, MA, 1998)

B19 A.V. Oppenheim, R.W. Schafer, J.R. Buck, *Discrete-Time Signal Processing*, 2nd edn. (Prentice-Hall, Upper Saddle River, NJ, 1998)

B20 E. Chu, A. George, *Inside the FFT Black Box: Serial and Parallel Fast Fourier Transform Algorithms* (CRC Press, Boca Raton, FL, 2000)

B21 C.-T. Chen, *Digital Signal Processing* (Oxford University Press, New York, 2000)

B22 D. Sundararajan, *The Discrete Fourier Transform* (World Scientific, River Edge, NJ, 2001)

B23 K.R. Rao, P.C. Yip (eds.), *The Transform and Data Compression Handbook* (CRC Press, Boca Raton, FL, 2001)

B24 S. Wolfram, *The Mathematica Book*, 5th edn. (Wolfram Media, Champaign, IL, 2003)

B25 J.K. Beard, *The FFT in the 21st Century: Eigenspace Processing* (Kluwer, Hingham, MA, 2003)

B26 G. Bi, Y. Zeng, *Transforms and Fast Algorithms for Signal Analysis and Representations* (Birkhauser, Boston, MA, 2003)

B27 S.J. Leon, *Linear Algebra with Applications*, 6th edn. (Prentice-Hall, Upper Saddle River, NJ, 2006)

B28 V. Britanak, P. Yip, K.R. Rao, *Discrete Cosine and Sine Transforms: General Properties, Fast Algorithms and Integer Approximations* (Academic Press (Elsevier), Orlando, FL, 2007)

Books with Software

• FFT, filtering, histogram techniques, image manipulation, etc. Software on a diskette.

B29 C.S. Burrus, T.W. Parks, *DFT/FFT and Convolution Algorithms: Theory and Implementation* (Wiley, New York, 1985)

B30 H.R. Myler, A.R. Weeks, *Computer Imaging Recipes in C* (Prentice-Hall, Englewood Cliffs, NJ, 1993)

B31 C.S. Burrus et al., *Computer Based Exercises for Signal Processing Using MATLAB* (Prentice-Hall, Englewood Cliffs, NJ, 1994)

B32 O. Alkin, *Digital Signal Processing: A Laboratory Approach Using PC-DSP* (Prentice-Hall, Englewood Cliffs, NJ, 1994)

B33 F.J. Taylor, *Principles of Signals and Systems*. Includes a data disk of MATLAB and MONARCH example files (McGraw-Hill, New York, 1994)

B34 A. Ambardar, *Analog and Digital Processing*. 3.5" DOS diskette, IBM PC, PS/2 etc., compatible, requires MATLAB 3.5 or 4.0 (PWS Publishing Co., Boston, MA, 1995)

B35 H.V. Sorensen, C.S. Burrus, M.T. Heideman, *Fast Fourier Transform Database* (PWS Publishing Co., Boston, MA, 1995) (disk and hard copy)

B36 J.S. Walker, *Fast Fourier Transform*, 2nd edn. (CRC Press, Boca Raton, FL, 1996) (Software on disk)

B37 S.A. Tretter, *Communication System Design Using DSP Algorithms: With Laboratory Experiments for the TMS320C6701 and TMS320C6711* (Kluwer/Plenum, New York, 2003) (Software on disk) (800-221-9369)

B38 P.D. Cha, J.I. Molinder, *Fundamentals of Signals and Systems: A Building Block Approach* (Cambridge University Press, New York, 2006) (CD-ROM, MATLAB M-files)

B39 D.E. Dudgeon, R.M. Mersereau, *Multidimensional Digital Signal Processing* (Prentice-Hall, Englewood Cliffs, NJ, 1984)

B40 L.C. Ludeman, *Random Processes: Filtering, Estimation and Detection* (Wiley, Hoboken, NJ, 2003)
B41 W.K. Pratt, *Digital Image Processing*, 4th edn. (Wiley, New York, 2007)
B42 M. Petrou, C. Petrou, *Image Processing: The Fundamentals*, 2nd edn. (Wiley, New York, 2010)
B43 A.D. Poularikis, *Transforms and Applications Handbook*, 3rd edn. (CRC Press, Boca Raton, FL, 2010)

Interpolation Using FFT

IN1 R.W. Schafer, L.R. Rabiner, A digital signal processing approach to interpolation. Proc. IEEE **61**, 692–702 (June 1973)
IN2 M. Yeh, J.L. Melsa, D.L. Cohn, A direct FFT scheme for interpolation decimation, and amplitude modulation. in *16th IEEE Asilomar Conf. Cir. Syst. Comp.* Nov. 1982, pp. 437–441
IN3 K.P. Prasad, P. Sathyanarayana, Fast interpolation algorithm using FFT. IEE Electron. Lett. **22**, 185–187 (Jan. 1986)
IN4 J.W. Adams, A subsequence approach to interpolation using the FFT. IEEE Trans. CAS **34**, 568–570 (May 1987)
IN5 S.D. Stearns, R.A. David, *Signal Processing Algorithms* (Englewood Cliffs, NJ, Prentice-Hall, 1988). Chapter 10 – Decimation and Interpolation Routines. Chapter 9 – Convolution and correlation using FFT
IN6 D. Fraser, Interpolation by the FFT revisited – An experimental investigation. IEEE Trans. ASSP **37**, 665–675 (May 1989)
IN7 P. Sathyanarayana, P.S. Reddy, M.N.S. Swamy, Interpolation of 2-D signals. IEEE Trans. CAS **37**, 623–625 (May 1990)
IN8 T. Smith, M.R. Smith, S.T. Nichols, Efficient sinc function interpolation techniques for center padded data. IEEE Trans. ASSP **38**, 1512–1517 (Sept. 1990)
IN9 S.C. Chan, K.L. Ho, C.W. Kok, Interpolation of 2-D signal by subsequence FFT. IEEE Trans. Circ. Syst. II Analog Digital SP **40**, 115–118 (Feb. 1993)
IN10 Y. Dezhong, Fast interpolation of *n*-dimensional signal by subsequence FFT. IEEE Trans Circ. Syst II Analog Digital SP **43**, 675–676 (Sept. 1996)

Audio Coding

- *Dolby AC-2 and AC-3 audio coders*: Time Domain Aliasing Cancellation (TDAC) transform involves MDCT (modified DCT) and MDST (modified DST). Both MDCT and MDST and their inverses can be implemented via FFT. Several papers from Dolby Labs, [D13, D24, D25] http://www.dolby.com/
- Masking threshold for the psychoacoustic model is derived from an estimate of the power density spectrum obtained from a 512-point FFT. Used in MPEG Audio Coding. ISO/IEC JTC1/SC29 11172-3. See the papers [D22, D26].

D1 J.P. Princen, A.B. Bradley, Analysis/synthesis filter bank design based on time domain aliasing cancellation. IEEE Trans. ASSP **34**, 1153–1161 (Oct. 1986)
D2 J.P. Princen, A.W. Johnson, A.B. Bradley, Subband/transform coding using filter bank designs based on time domain aliasing cancellation, in *IEEE ICASSP*, Dallas, TX, Apr. 1987, pp. 2161–2164

D3 J.D. Johnston, Transform coding of audio signals using perceptual noise criteria. IEEE JSAC **6**, 314–323 (Feb. 1988)

D4 J.D. Johnston, Estimation of perceptual entropy using noise masking criteria, in *IEEE ICASSP*, vol. 5, New York, Apr. 1988, pp. 2524–2527

D5 J.H. Chung, R.W. Schafer, A 4.8 kbps homomorphic vocoder using analysis-by-synthesis excitation analysis, in *IEEE ICASSP*, vol. 1, Glasgow, Scotland, May 1989, pp. 144–147

D6 J.D. Johnston, Perceptual transform coding of wideband stereo signals, in *IEEE ICASSP*, vol. 3, Glasgow, Scotland, May 1989, pp. 1993–1996

D7 Y.-F. Dehery, A digital audio broadcasting system for mobile reception, in *ITU-COM 89*, CCETT of France, Geneva, Switzerland, Oct. 1989, pp. 35–57

D8 ISO/IEC JTC1/SC2/WG8 MPEG Document 89/129 proposal package description, 1989

D9 "ASPEC", AT&T Bell Labs, Deutsche Thomson Brandt and Fraunhofer Gesellschaft – FhG AIS, ISO/IEC JTC1/SC2/WG8 MPEG 89/205

D10 K. Brandenburg et al., Transform coding of high quality digital audio at low bit rates-algorithms and implementation, in *IEEE ICC 90*, vol. 3, Atlanta, GA, Apr. 1990, pp. 932–936

D11 G. Stoll, Y.-F. Dehery, High quality audio bit-rate reduction system family for different applications, in *IEEE ICC*, vol. 3, Atlanta, GA, Apr. 1990, pp. 937–941

D12 J.H. Chung, R.W. Schafer, Excitation modeling in a homomorphic vocoder, in *IEEE ICASSP*, vol. 1, Albuquerque, NM, Apr. 1990, pp. 25–28

D13 G.A. Davidson, L.D. Fielder, M. Artill, Low-complexity transform coder for satellite link applications, in *AES 89th Convention*, Los Angeles, CA, 21–25 Sept. 1990, http://www.aes.org/

D14 T.D. Lookabaugh, M.G. Perkins, Application of the Princen-Bradley filter bank to speech and image compression. IEEE Trans. ASSP **38**, 1914–1926 (Nov. 1990)

D15 H.G. Musmann, The ISO audio coding standard, in *IEEE GLOBECOM*, vol. 1, San Diego, CA, Dec. 1990, pp. 511–517

D16 Y. Mahieux, J.P. Petit, Transform coding of audio signals at 64 kbit/s, in *IEEE GLOBECOM*, vol. 1, San Diego, CA, Dec. 1990, pp. 518–522

D17 K. Brandenburg et al., ASPEC: Adaptive spectral perceptual entropy coding of high quality music signals, in *90th AES Convention*, Preprint 3011 (A-4), Paris, France, 19–22 Feb. 1991

D18 P. Duhamel, Y. Mahieux, J.P. Petit, A fast algorithm for the implementation of filter banks based on 'time domain aliasing cancellation', in *IEEE ICASSP*, vol. 3, Toronto, Canada, Apr. 1991, pp. 2209–2212

D19 L.D. Fielder, G.A. Davidson, AC-2: A family of low complexity transform based music coders, in *AES 10th Int'l Conference*, London, England, 7–9 Sept. 1991, pp. 57–69

D20 M. Iwadare et al., A 128 kb/s Hi-Fi audio CODEC based on adaptive transform coding with adaptive block size MDCT. IEEE JSAC **10**, 138–144 (Jan. 1992)

D21 G.A. Davidson, W. Anderson, A. Lovrich, A low-cost adaptive transform decoder implementation for high-quality audio. in *IEEE ICASSP*, vol. 2, San Francisco, CA, Mar. 1992, pp. 193–196

D22 K. Brandenburg et al., The ISO/MPEG audio codec: A generic standard for coding of high quality digital audio, in *92th AES Convention*, Preprint 3336, Vienna, Austria, Mar. 1992, http://www.aes.org/

D23 L.D. Fielder, G.A. Davidson, Low bit rate transform coder, decoder and encoder/decoder for high quality audio, U.S. Patent 5,142,656, 25 Aug. 1992

D24 A.G. Elder, S.G. Turner, A real-time PC based implementation of AC-2 digital audio compression, in *AES 95th Convention*, Preprint 3773, New York, 7–10 Oct 1993

D25 Dolby AC-3, multi-channel digital audio compression system algorithm description, Dolby Labs. Inc., Revision 1.12, 22 Feb. 1994, http://www.dolbylabs.com

D26 D. Pan, An overview of the MPEG/Audio compression algorithm, in *IS&T/SPIE Symposium on Electronic Imaging: Science and Technology*, vol. 2187, San Jose, CA, Feb. 1994, pp. 260–273

D27 C.C. Todd et al., AC-3: Flexible perceptual coding for audio transmission and storage, in *AES 96th Convention*, Preprint 3796, Amsterdam, Netherlands, Feb./Mar. 1994, http://www.aes.org/

D28 ACATS Technical Subgroup, *Grand Alliance HDTV System Specification*, Version 1.0, 14 Apr. 1994

D29 M. Lodman et al., A single chip stereo AC-3 audio decoder, in *IEEE ICCE*, Chicago, IL, June 1994, pp. 234–235 (FFT is used in time domain aliasing cancellation – TDAC)

D30 A.S. Spanias, Speech coding: A tutorial review. Proc. IEEE **82**, 1541–1582 (Oct. 1994)

D31 M. Bosi, S.E. Forshay, High quality audio coding for HDTV: An overview of AC-3, in *7th Int'l Workshop on HDTV*, Torino, Italy, Oct. 1994

D32 D. Sevic, M. Popovic, A new efficient implementation of the oddly stacked Princen-Bradley filter bank. IEEE SP Lett. **1**, 166–168 (Nov. 1994)

D33 P. Noll, Digital audio coding for visual communications. Proc. IEEE **83**, 925–943 (June 1995)

D34 Digital Audio Compression (AC-3) ATSC Standard, 20 Dec. 1995, http://www.atsc.org/

D35 M. Bosi et al., ISO/IEC MPEG-2 advanced audio coding, in *AES 101st Convention*, Los Angeles, CA, 8–11 Nov. 1996. Also appeared in J. Audio Eng. Soc., **45**, 789–814 (Oct. 1997)

D36 MPEG-2 Advanced Audio Coding, ISO/IEC JTC1/SC29/WG11, Doc. N1430, MPEG-2 DIS 13818-7, Nov. 1996

D37 K.R. Rao, J.J. Hwang, *Techniques and Standards for Image, Video and Audio Coding* (Prentice-Hall, Upper Saddle River, NJ, 1996)

D38 K. Brandenburg, M. Bosi, Overview of MPEG audio: Current and future standards for low bit-rate audio coding. J. Audio Eng. Soc. **45**, 4–21 (Jan./Feb. 1997)

D39 MPEG-2 Advanced Audio Coding (AAC), ISO/IEC JTC1/SC29/WG 11, Doc. N1650, Apr. 1997, MPEG-2 IS 13818-7. Pyschoacoustic model for NBC (Nonbackward Compatible) – AAC (Advanced Audio Coder) audio coder, IS for MPEG-2. Also adopted by MPEG-4 in T/F coder, http://www.mpeg.org/

D40 S. Shlien, The modulated lapped transform, its time-varying forms, and its applications to audio coding standards. IEEE Trans. Speech Audio Process. **5**, 359–366 (July 1997)

D41 Y. Jhung, S. Park, Architecture of dual mode audio filter for AC-3 and MPEG. IEEE Trans. CE **43**, 575–585 (Aug. 1997) (Reconstruction filter based on FFT structure)

D42 Y.T. Han, D.K. Kang, J.S. Koh, An ASIC design of the MPEG-2 audio encoder, in *ICSPAT 97*, San Diego, CA, Sept. 1997

D43 MPEG-4 WD ISO/IEC 14496-3, V4.0/10/22/1997 (FFT in 'CELP' coder)

D44 K. Brandenburg, MP3 and AAC explained, in *AES 17th Int'l Conference on High Quality Audio Coding*, Florence, Italy, Sept. 1999

D45 C.-M. Liu, W.-C. Lee, A unified fast algorithm for cosine modulated filter banks in current audio coding standards. J. Audio Eng. Soc. **47**, 1061–1075 (Dec. 1999)

D46 Method for objective measurements of perceived audio quality. Recommendation ITU-R BS.1387-1, 1998–2001

D47 R. Geiger et al., Audio coding based on integer transforms, in *AES 111th Convention*, New York, 21–24 Sept. 2001, pp. 1–9

D48 S.-W. Lee, Improved algorithm for efficient computation of the forward and backward MDCT in MPEG audio coder. IEEE Trans. Circ. Syst. II Analog Digital Signal Process. **48**, 990–994 (Oct. 2001)

D49 V. Britanak, K.R. Rao, A new fast algorithm for the unified forward and inverse MDCT/MDST computation. Signal Process. **82**, 433–459 (Mar. 2002)

D50 M. Bosi, R.E. Goldberg, *Introduction to Digital Audio Coding and Standards* (Kluwer, Norwell, MA, 2003)

D51 G.A. Davidson et al., ATSC video and audio coding. Proc. IEEE **94**, 60–76 (Jan. 2006)

D52 J.M. Boyce, The U.S. digital television broadcasting transition. IEEE SP Mag. **26**, 102–110 (May 2009)

D53 I.Y. Choi et al., Objective measurement of perceived auditory quality in multichannel audio compression coding systems. J. Audio Eng. Soc. **56**, 3–17 (Jan. 2008)

Image Processing, Fourier Phase Correlation

IP1   C.D. Kuglin, D.C. Hines, The phase correlation image alignment method, in *Proceedings on the IEEE Int'l Conference on Cybernetics and Society*, San Francisco, CA, Sept. 1975, pp. 163–165

IP2   E. De Castro, C. Morandi, Registration of translated and rotated images using finite Fourier transforms. IEEE Trans. PAMI **9**, 700–703 (Sept. 1987)

IP3   W.M. Lawton, Multidimensional chirp algorithms for computing Fourier transforms. IEEE Trans. IP **1**, 429–431 (July 1992)

IP4   M. Perry, Using 2-D FFTs for object recognition, in *ICSPAT*, DSP World Expo., Dallas, TX, Oct. 1994, pp. 1043–1048

IP5   B.S. Reddy, B.N. Chatterji, An FFT-based technique for translation, rotation, and scale-invariant image registration, IEEE Trans. IP **5**, 1266–1271 (Aug. 1996)

IP6   P.E. Pang, D. Hatzinakos, An efficient implementation of affine transformation using one-dimensional FFTs, in *IEEE ICASSP-97*, vol.4, Munich, Germany, Apr. 1997, pp. 2885–2888

IP7   S.J. Sangwine, The problem of defining the Fourier transform of a color image, in *IEEE ICIP*, vol. 1, Chicago, IL, Oct. 1998, pp. 171–175

IP8   A. Zaknich, Y. Attikiouzel, A comparison of template matching with neural network approaches in the recognition of numeric characters in hand-stamped aluminum, in *IEEE ISPACS*, Melbourne, Australia, Nov. 1998, pp. 98–102

IP9   W. Philips, On computing the FFT of digital images in quadtree format. IEEE Trans. SP **47**, 2059–2060 (July 1999)

IP10 R.W. Cox, R. Tong, Two- and three-dimensional image rotation using the FFT. IEEE Trans. IP **8**, 1297–1299 (Sept. 1999)

IP11 J.H. Lai, P.C. Yuen, G.C. Feng, Spectroface: A Fourier-based approach for human face recognition, in *Proceedings of the 2nd Int'l Conferene on Multimodal Interface (ICMI'99)*, Hong Kong, China, Jan. 1999, pp. VI115–12

IP12 A.E. Yagle, Closed-form reconstruction of images from irregular 2-D discrete Fourier samples using the Good-Thomas FFT, in *IEEE ICIP*, vol. 1, Vancouver, Canada, Sept. 2000, pp. 117–119

IP13 U. Ahlves, U. Zoetzer, S. Rechmeier, FFT-based disparity estimation for stereo image coding, in *IEEE ICIP*, vol. 1, Barcelona, Spain, Sept. 2003, pp. 761–764

IP14 A.V. Bondarenko, S.F. Svinyin, A.V. Skourikhin, Multidimensional B-spline forms and their Fourier transforms, in *IEEE ICIP*, vol. 2, Barcelona, Spain, Sept. 2003, pp. 907–908

IP15 A.H. Samra, S.T.G. Allah, R.M. Ibrahim, Face recognition using wavelet transform, fast Fourier transform and discrete cosine transform, in *46th IEEE Int'l MWSCAS*, vol. 1, Cairo, Egypt, Dec. 2003, pp. 27–30

IP16 K. Ito et al., Fingerprint matching algorithm using phase-only correlation. IEICE Trans. Fundam. **E87-A**, 682–691 (Mar. 2004)

IP17 O. Urhan, M.K. Güllü, S. Ertürk, Modified phase-correlation based robust hard-cut detection with application to archive film. IEEE Trans. CSVT **16**, 753–770 (June 2006)

IP18 R.C. Gonzalez, R.E. Woods, *Digital Image Processing*, 2nd edn. (Prentice-Hall, Upper Saddle River, NJ, 2001)

IP19 Ibid., 3rd edn. (Prentice-Hall, Upper Saddle River, NJ, 2007)

IP20 P.C. Cosman, Homework 4 and its solution on registration for digital image processing lecture, 2008, available: http://code.ucsd.edu/~pcosman/

Video/Image Coding

IP21 B.G. Haskell, Frame-to-frame coding of television pictures using two-dimensional Fourier transforms. IEEE Trans. IT **20**, 119–120 (Jan. 1974)

IP22 J.W. Woods, S.D. O'Neil, Subband coding of images. IEEE Trans. ASSP **34**, 1278–1288 (Oct. 1986) (Implementation of the FIR filters with the QMF banks by FFT)

IP23 (8×8) 2D FFT spatially on video frame/field followed by temporal DPCM for a DS-3 NTSC TV codec built by Grass Valley, P.O. Box 1114, Grass Valley, CA 95945

IP24 M. Ziegler, in *Signal Processing of HDTV*, ed. by L. Chiariglione. Hierarchical motion estimation using the phase correlation method in 140 Mbit/s HDTV coding (Elsevier, Amsterdam, 1990), pp. 131–137

IP25 H.M. Hang, Y.-M. Chou, T.-H.S. Chao, Motion estimation using frequency components, in *Proceedings of SPIE VCIP*, vol. 1818, Boston, MA, Nov. 1992, pp. 74–84

IP26 C.B. Owen, F. Makedon, High quality alias free image rotation, in *30th IEEE Asilomar Conf. on Signals, Systems and Computers*, vol. 1, Pacific Grove, CA, Nov. 1996, pp. 115–119

IP27 *JPEG Software*, The Independent JPEG Group, Mar. 1998, available: http://www.ijg.org/

IP28 M. Ghanbari, *Standard Codecs: Image Compression to Advanced Video Coding* (IEE, Hertfordshire, UK, 2003)

IP29 S. Kumar et al., Error resiliency schemes in H.264/AVC standard, J. Visual Commun. Image Represent. (JVCIR) (special issue on H.264/AVC), **17**, 425–450 (Apr. 2006)

IP30 Image Database (USC-SIPI), http://sipi.usc.edu/database/

IP31 Image Sequences, Still Images (CIPR, RPI), http://www.cipr.rpi.edu/

Bit Allocation

IP32 A. Habibi, Hybrid coding of pictorial data. IEEE Trans. Commun. **22**, 614–624 (May 1974)

IP33 L. Wang, M. Goldberg, Progressive image transmission by multistage transform coefficient quantization, in *IEEE ICC*, Toronto, Canada, June 1986, pp. 419–423 (Also IEEE Trans. Commun., vol. 36, pp. 75–87, Jan. 1988)

IP34 K. Sayood, *Introduction to Data Compression*, 3rd edn. (Morgan Kaufmann, San Francisco, CA, 2006)

Spectral Distance Based Image Quality Measure

IP35 M. Caramma, R. Lancini, M. Marconi, A perceptual PSNR based on the utilization of a linear model of HVS, motion vectors and DFT-3D, in *EUSIPCO*, Tampere, Finland, Sept. 2000, available: http://www.eurasip.org/ (EURASIP Open Library)

IP36 İ. Avcibaş, B. Sankur, K. Sayood, Statistical evaluation of image quality measures. J. Electron. Imag. **11**, 206–223 (Apr. 2002)

Fractal Image Compression

FR1 A.E. Jacquin, Image coding based on a fractal theory of iterated contractive image transformations. IEEE Trans. IP **1**, 18–30 (Jan. 1992)

FR2 D. Saupe, D. Hartenstein, Lossless acceleration of fractal image compression by fast convolution, in *IEEE ICIP*, vol. 1, Lausanne, Switzerland, Sept. 1996, pp. 185–188

FR3 M. Ramkumar, G.V. Anand, An FFT-based technique for fast fractal image compression. Signal Process. **63**, 263–268 (Dec. 1997)

FR4 M. Morhac, V. Matousek, Fast adaptive Fourier-based transform and its use in multidimensional data compression. Signal Process. **68**, 141–153 (July 1998)

FR5 H. Hartenstein et al., Region-based fractal image compression. IEEE Trans. IP **9**, 1171–1184 (July 2000)

FR6 H. Hartenstein, D. Saupe, Lossless acceleration of fractal image encoding via the fast Fourier transform. Signal Process. Image Commun. **16**, 383–394 (Nov. 2000)

Lifting Scheme

I-1 W. Sweldens, The lifting scheme: A custom-design construction of biorthogonal wavelets. J. Appl. Comput. Harmonic Anal. **3**(2), 186–200 (1996)

I-2 W. Sweldens, A construction of second generation wavelets. SIAM J. Math. Anal. **29**(2), 511–546 (1997)

I-3 I. Daubechies, W. Sweldens, Factoring wavelet transforms into lifting steps. J. Fourier Anal. Appl. **4**(3), 245–267 (1998)

Integer FFT

I-4 F.A.M.L. Bruekers, A.W.M.V.D. Enden, New networks for perfect inversion and perfect reconstruction. IEEE J. Sel. Areas Commun. **10**, 130–137 (Jan. 1992)

I-5 S.-C. Pei, J.-J. Ding, The integer transforms analogous to discrete trigonometric transforms. IEEE Trans. SP **48**, 3345–3364 (Dec. 2000)

I-6 S. Oraintara, Y. Chen, T.Q. Nguyen, Integer fast Fourier transform. IEEE Trans. SP **50**, 607–618 (Mar. 2002)

I-7 S.C. Chan, P.M. Yiu, An efficient multiplierless approximation of the fast Fourier transform using sum-of-powers-of-two (SOPOT) coefficients. IEEE SP Lett. **9**, 322–325 (Oct. 2002)

I-8 S.C. Chan, K.M. Tsui, Multiplier-less real-valued FFT-like transformation (ML-RFFT) and related real-valued transformations, in *IEEE ISCAS*, vol. 4, Bangkok, Thailand, May 2003, pp. 257–260

I-9 Y. Yokotani et al., A comparison of integer FFT for lossless coding, in *IEEE ISCIT*, vol. 2, Sapporo, Japan, Oct. 2004, pp. 1069–1073

I-10 K.M. Tsui, S.C. Chan, Error analysis and efficient realization of the multiplier-less FFT-like transformation (ML-FFT) and related sinusoidal transformations, Journal of VLSI Signal Processing Systems, **44**, 97–115 (Springer, Amsterdam, Netherlands, May 2006)

I-11 W.-H. Chang, T.Q. Nguyen, Architecture and performance analysis of lossless FFT in OFDM systems, in *IEEE ICASSP*, vol. 3, Toulouse, France, May 2006, pp. 1024–1027

DCT and DST

I-12 S.A. Martucci, Symmetric convolution and the discrete sine and cosine transforms. IEEE Trans. SP **42**, 1038–1051 (1994)

• See also [B26].

Integer DCT

I-13 W.K. Cham, Development of integer cosine transforms by the principle of dyadic symmetry. IEE Proc. I Commun. Speech Vision **136**, 276–282 (Aug. 1989). (8-point integer DCT)

I-14 W.K. Cham, Y.T. Chan, An order-16 integer cosine transform. IEEE Trans. SP **39**, 1205–1208 (May 1991). (16-point integer DCT)

I-15 Y.-J. Chen, S. Oraintara, T. Nguyen, Integer discrete cosine transform (IntDCT), in *Proceedings of the 2nd Int'l Conference on Information, Communications and Signal Processing*, Singapore, Dec. 1999

I-16 T.D. Tran, The BinDCT: fast multiplierless approximations of the DCT. IEEE SP Lett. **7**, 141–144 (June 2000)

I-17 Y.-J. Chen, S. Oraintara, Video compression using integer DCT, in *IEEE ICIP*, vol. 2, (Vancouver, Canada, Sept. 2000) (8-point integer DCT), pp. 844–847

I-18 P. Hao, Q. Shi, Matrix factorizations for reversible integer mapping. IEEE Trans. SP **49**, 2314–2324 (Oct. 2001)

I-19 J. Liang, T.D. Tran, Fast multiplierless approximations of the DCT with the lifting scheme. IEEE Trans. SP **49**, 3032–3044 (Dec. 2001)

I-20 Y.-J. Chen et al., Multiplierless approximations of transforms with adder constraint. IEEE SP Lett. **9**, 344–347 (Nov. 2002)

I-21 W. Gao et al., AVS – The Chinese Next-Generation Video Coding Standard, in *NAB*, Las Vegas, NV, Apr. 2004

I-22 G.J. Sullivan, P. Topiwala, A. Luthra, The H.264/AVC advanced video coding standard: Overview and introduction to the fidelity range extensions, in *SPIE Conference on Applications of Digital Image Processing XXVII*, vol. 5558, pp. 53–74, Aug. 2004

I-23 S. Srinivasan et al., Windows media video 9: Overview and applications. Signal Process. Image Commun. (Elsevier) **19**, 851–875 (Oct. 2004)

I-24 J. Dong et al., A universal approach to developing fast algorithm for simplified order-16 ICT, in *IEEE ISCAS*, New Orleans, LA, May 2007, pp. 281–284 (16-point integer DCT)

I-25 S.-C. Pei, J.-J. Ding, Scaled lifting scheme and generalized reversible integer transform, in *IEEE ISCAS*, New Orleans, LA, May 2007, pp. 3203–3206

I-26 L. Wang et al., Lossy to lossless image compression based on reversible integer DCT, in *IEEE ICIP*, San Diego, CA, Oct. 2008, pp. 1037–1040

• See also [I-7, I-8, I-10].

Integer MDCT

I-27 T. Krishnan, S. Oraintara, Fast and lossless implementation of the forward and inverse MDCT computation in MPEG audio coding, in *IEEE ISCAS*, vol. 2, Scottsdale, AZ, May 2002, pp. 181–184

• See also [D47].

Discrete Fourier–Hartley Transform

I-28 N.-C. Hu, H.-I. Chang, O.K. Ersoy, Generalized discrete Hartley transforms. IEEE Trans. SP **40**, 2931–2940 (Dec. 1992)

I-29 S. Oraintara, The unified discrete Fourier–Hartley transforms: Theory and structure, in *IEEE ISCAS*, vol. 3, Scottsdale, AZ, May 2002, pp. 433–436

I-30 P. Potipantong et al., The unified discrete Fourier–Hartley transforms processor, in *IEEE ISCIT*, Bangkok, Thailand, Oct. 2006, pp. 479–482

I-31 K.J. Jones, Design and parallel computation of regularised fast Hartley transform. IEE Vision Image Signal Process. **153**, 70–78 (Feb. 2006)

I-32 K.J. Jones, R. Coster, Area-efficient and scalable solution to real-data fast Fourier transform via regularised fast Hartley transform. IET Signal Process. **1**, 128–138 (Sept. 2007)

I-33 W.-H. Chang, T. Nguyen, An OFDM-specified lossless FFT architecture. IEEE Trans. Circ. Syst. I (regular papers) **53**, 1235–1243 (June 2006)

I-34 S.C. Chan, P.M. Yiu, A multiplier-less 1-D and 2-D fast Fourier transform-like transformation using sum-of-power-of-two (SOPOT) coefficients, in *IEEE ISCAS'2002*, vol. 4, Phoenix-Scottsdale, Arizona, May 2002, pp. 755–758

Image/Video Watermarking

E1 W. Bender et al., Techniques for data hiding. IBM Syst. J. **35**(3/4), 313–335 (1996)

E2 J.J.K.Ó. Ruanaidh, W.J. Dowling, F.M. Boland, Phase watermarking of digital images, in *Proceedings of the ICIP'96*, vol. 3, Lausanne, Switzerland, Sept. 1996, pp. 239–242

E3 A. Celantano, V.D. Lecce, A FFT based technique for image signature generation, in *Proc. SPIE Storage Retrieval Image Video Databases V*, vol. 3022, San Jose, CA, Feb. 1997, pp. 457–466

E4 J.J.K.Ó. Ruanaidh, T. Pun, Rotation, scale and translation invariant spread spectrum digital image watermarking. Signal Process (Elsevier) **66**, 303–317 (May 1998)

E5 H. Choi, H. Kim, T. Kim, Robust watermarks for images in the subband domain, in *IEEE ISPACS*, Melbourne, Australia, Nov. 1998, pp. 168–172

E6 F. Deguillaume et al., Robust 3D DFT video watermarking, in *SPIE Photonics West*, vol. 3657, San Jose, CA, Jan. 1999, pp. 113–124

E7 G. Voyatzis, I. Pitas, The use of watermarks in the protection of digital multimedia products. Proc. IEEE **87**, 1197–1207 (July 1999)

E8 V. Solachidis, I. Pitas, Self-similar ring shaped watermark embedding in 2D- DFT domain, in *EUSIPCO 2000*, Tampere, Finland, Sept. 2000, available: http://www.eurasip.org

E9 X. Kang et al., A DWT-DFT composite watermarking scheme robust to both affine transform and JPEG compression. IEEE Trans. CSVT **13**, 776–786 (Aug. 2003)

E10 V. Solachidis, I. Pitas, Optimal detection for multiplicative watermarks embedded in DFT domain, in *IEEE ICIP*, vol. 3, Barcelona, Spain, 2003, pp. 723–726

E11 Y.Y. Lee, H.S. Jung, S.U. Lee, 3D DFT-based video watermarking using perceptual models, in *46th IEEE Int'l MWSCAS*, vol. 3, Cairo, Egypt, Dec. 2003, pp. 1579–1582

E12 V. Solachidis, I. Pitas, Watermarking digital 3-D volumes in the discrete Fourier transform domain. IEEE Trans. Multimedia **9**, 1373–1383 (Nov. 2007)

E13 T. Bianchi, A. Piva, M. Barni, Comparison of different FFT implementations in the encrypted domain, in *EUSIPCO 2008*, Lausanne, Switzerland, Aug. 2008, available: http://www.eurasip.org/

Audio Watermarking

E14 M.D. Swanson et al., Robust audio watermarking using perceptual masking. Signal Process. **66**, 337–355 (May 1998) (Special Issue on Watermarking)

E15 B. Ji, F. Yan, D. Zhang, A robust audio watermarking scheme using wavelet modulation. IEICE Trans. Fundam. **86**, 3303–3305 (Dec. 2003)

E16 F. Yan et al., Robust quadri-phase audio watermarking, in *Acoustical Science and Technology*, Acoustical Society of Japan, vol. 25(1), 2004, available: http://www.openj-gate.org/

Speech

SP1 S. Sridharan, E. Dawson, B. Goldburg, Speech encryption using discrete orthogonal transforms, in *IEEE ICASSP-90*, Albuquerque, NM, Apr. 1990, pp. 1647–1650

SP2 S. Sridharan, E. Dawson, B. Goldburg, Fast Fourier transform based speech encryption system. IEE Proc. I Commun. Speech Vision **138**, 215–223 (June 1991)

SP3 B. Goldburg, S. Sridharan, E. Dawson, Cryptanalysis of frequency domain analog speech scramblers. IEE Proc. I Commun. Speech Vision **140**, 235–239 (Aug. 1993)

SP4 S. Nakamura, A. Sasou, A pitch extraction algorithm using combined wavelet and Fourier transforms, in *ICSPAT*, Boston, MA, Oct. 1995

SP5 C.-H. Hsieh, Grey filtering and its application to speech enhancement. IEICE Trans. Inf. Syst. **E86–D**, 522–533 (Mar. 2003)

Filtering

F1 G. Bruun, z-transform DFT filters and FFTs. IEEE Trans. ASSP **26**, 56–63 (Feb. 1978)

F2 G. Sperry, Forensic applications utilizing FFT filters, in *IS&T's 48th Annual Conference*, Washington, DC, May 1995

F3 K.O. Egiazarian et al., Nonlinear filters based on ordering by FFT structure, in *Photonics West, IS&T/SPIE Symposium on Electronic Imaging: Science and Technology*, vol. 2662, San Jose, CA, Feb. 1996, pp. 106–117

F4 A.E. Cetin, O.N. Gerek, Y. Yardimci, Equiripple FIR filter design by the FFT algorithm. IEEE SP Mag. **14**, 60–64 (Mar. 1997)

DFT Filter Banks

F5 R. Gluth, Regular FFT-related transform kernels for DCT/DST based polyphase filter banks, in *IEEE ICASSP-91*, vol. 3, Toronto, Canada, Apr. 1991, pp. 2205–2208

F6 Y.P. Lin, P.P. Vaidyanathan, Application of DFT filter banks and cosine modulated filter banks in filtering, in *IEEE APCCAS*, Taipei, Taiwan, Dec. 1994, pp. 254–259

F7 O.V. Shentov et al., Subband DFT – Part I: Definition, interpretation and extensions. Signal Process. **41**, 261–277 (Feb. 1995)

F8 A.N. Hossen et al., Subband DFT – Part II: Accuracy, complexity and applications. Signal Process. **41**, 279–294 (Feb. 1995)

F9 T.Q. Nguyen, Partial reconstruction filter banks – Theory, design and implementation, in *Technical Report 991*, Lincoln Lab, MIT, Lexington, MA, 22 June 1995 (DFT filter banks)

F10 H. Murakami, Perfect reconstruction condition on the DFT domain for the block maximally decimated filter bank, in *IEEE ICCS/ISPACS 96*, Singapore, Nov. 1996, pp. 6.3.1–6.3.3

F11 M. Boucheret et al., Fast convolution filter banks for satellite payloads with on-board processing. IEEE J. Sel. Areas Commun. **17**, 238–248 (Feb. 1999)

F12 Q.G. Liu, B. Champagne, D.K.C. Ho, Simple design of oversampled uniform DFT filter banks with applications to subband acoustic echo cancellation. Signal Process. **80**, 831–847 (May 2000)

F13 E. Galijasevic, J. Kliewer, Non-uniform near-perfect-reconstruction oversampled DFT filter banks based on all pass-transforms, in *9th IEEE DSP Workshop*, Hunt, TX, Oct. 2000, available: http://spib.ece.rice.edu/DSP2000/program.html

F14 H. Murakami, PR condition for a block filter bank in terms of DFT, in *WPMC2000*, vol. 1, Bangkok, Thailand, Nov. 2000, pp. 475–480
F15 E.V. Papaoulis, T. Stathaki, A DFT algorithm based on filter banks: The extended subband DFT, in *IEEE ICIP 2003*, vol. 1, Barcelona, Spain, Sept. 2003, pp. 1053–1056

Adaptive Filtering

F16 B. Widrow et al., Fundamental relations between the LMS algorithm and the DFT. IEEE Trans. CAS **34**, 814–820 (July 1987)
F17 J.J. Shynk, Frequency domain and multirate adaptive filtering. IEEE SP Mag. **9**, 14–37 (Jan. 1992)
F18 B. Farhang-Boroujeny, S. Gazor, Generalized sliding FFT and its applications to implementation of block LMS adaptive filters. IEEE Trans. SP **42**, 532–538 (Mar. 1994)
F19 B.A. Schnaufer, W.K. Jenkins, A fault tolerant FIR adaptive filter based on the FFT, in *IEEE ICASSP-94*, vol. 3, Adelaide, Australia, Apr. 1994, pp. 393–396
F20 D.T.M. Slock, K. Maouche, The fast subsampled updating recursive least squares (FSURLS) algorithm for adaptive filtering based on displacement structure and the FFT. Signal Process. **40**, 5–20 (Oct. 1994)
F21 H. Ochi, N. Bershad, A new frequency-domain LMS adaptive filter with reduced-sized FFTs, in *IEEE ISCAS*, vol. 3, Seattle, WA, Apr./May 1995, pp. 1608–1611
F22 P. Estermann, A. Kaelin, On the comparison of optimum least-squares and computationally efficient DFT-based adaptive block filters, in *IEEE ISCAS*, vol.3, Seattle, WA, Apr./May 1995, pp. 1612–1615
F23 K.O. Egiazarian et al., Adaptive LMS FFT-ordered L-filters, in *IS&T/SPIE's 9th Annual Symposium, Electronic Imaging*, vol. 3026, San Jose, CA, Feb. 1997, pp. 34–45. L-filters (or linear combination of order statistics)

Wavelets, Multiresolution and Filter Banks

F24 E.J. Candès, D.L. Donoho, L. Ying, Fast discrete curvelet transform, SIAM J. Multiscale Model. Simul. **5**, 861–899 (Sept. 2006). (The software CurveLab is available at http://www.curvelet.org)
F25 Y. Rakvongthai, Hidden Markov tree modeling of the uniform discrete curvelet transform for image denoising. EE5359 Project (UT–Arlington, TX, Summer 2008), http://www-ee.uta.edu/dip/ click on courses
F26 Y. Rakvongthai, S. Oraintara, Statistics and dependency analysis of the uniform discrete curvelet coefficients and hidden Markov tree modeling, in *IEEE ISCAS*, Taipei, Taiwan, May 2009, pp. 525–528
F27 R. Mersereau, T. Speake, The processing of periodically sampled multidimensional signals. IEEE Trans. ASSP **31**, 188–194 (Feb. 1983)
F28 E. Viscito, J.P. Allebach, The analysis and design of multidimensional FIR perfect reconstruction filter banks for arbitrary sampling lattices. IEEE Trans. CAS **38**, 29–41 (Jan. 1991)
F29 R.H. Bamberger, M.J.T. Smith, A filter bank for the directional decomposition of images: Theory and design. IEEE Trans. SP **40**, 882–893 (Apr. 1992)
F30 G. Strang, T. Nguyen, *Wavelets and Filter Banks*, 2nd edn. (Wellesley-Cambridge Press, Wellesley, MA, 1997)
F31 S.-I. Park, M.J.T. Smith, R.M. Mersereau, Improved structures of maximally decimated directional filter banks for spatial image analysis. IEEE Trans. IP **13**, 1424–1431 (Nov. 2004)

F32 Y. Tanaka, M. Ikehara, T.Q. Nguyen, Multiresolution image representation using combined 2-D and 1-D directional filter banks. IEEE Trans. IP **18**, 269–280 (Feb. 2009)

F33 J. Wisinger, R. Mahapatra, FPGA based image processing with the curvelet transform. Technical Report # TR-CS-2003-01-0, Department of Computer Science, Texas A&M University, College Station, TX

F34 M.N. Do, M. Vetterli, Orthonormal finite ridgelet transform for image compression, in *IEEE ICIP*, vol. 2, Vancouver, Canada, Sept. 2000, pp. 367–370

F35 J.L. Starck, E.J. Candès, D.L. Donoho, The curvelet transform for image denoising. IEEE Trans. IP **11**, 670–684 (June 2002)

F36 M.N. Do, M. Vetterli, The finite ridgelet transform for image representation. IEEE Trans. IP **12**, 16–28 (Jan. 2003)

F37 B. Eriksson, The very fast curvelet transform, ECE734 Project, University of Wisconsin (UW), Madison, WI, 2006

F38 T.T. Nguyen, H. Chauris, The uniform discrete curvelet transform. IEEE Trans. SP (Oct. 2009) (see demo code). (Under review)

Signal Processing

S1 A.K. Jain, J. Jasiulek, Fast Fourier transform algorithms for linear estimation, smoothing and Riccati equations. IEEE Trans. ASSP **31**, 1435–1446 (Dec. 1983)

S2 R.R. Holdrich, Frequency analysis of non-stationary signals using time frequency mapping of the DFT magnitudes, in *ICSPAT*, DSP World Expo, Dallas, TX, Oct. 1994

S3 V. Murino, A. Trucco, Underwater 3D imaging by FFT dynamic focusing beamforming, in *IEEE ICIP-94*, Austin, TX, Nov. 1994, pp. 890–894

S4 A. Niederlinski, J. Figwer, Using the DFT to synthesize bivariate orthogonal white noise series. IEEE Trans. SP **43**, 749–758 (Mar. 1995)

S5 D. Petrinovic, H. Babic, Window spectrum fitting for high accuracy harmonic analysis, in *ECCTD'95*, Istanbul, Turkey, Aug. 1995

S6 H. Murakami, Sampling rate conversion systems using a new generalized form of the discrete Fourier transform. IEEE Trans. SP **43**, 2095–2102 (Sept. 1995)

S7 N. Kuroyanagi, L. Guo, N. Suehiro, Proposal of a novel signal separation principle based on DFT with extended frame Fourier analysis, in *IEEE GLOBECOM*, Singapore, Nov. 1995, pp. 111–116

S8 K.C. Lo, A. Purvis, Reconstructing randomly sampled signals by the FFT, in *IEEE ISCAS-96*, vol. 2, Atlanta, GA, May 1996, pp. 124–127

S9 S. Yamasaki, A reconstruction method of damaged two-dimensional signal blocks using error correction coding based on DFT, in *IEEE APCCAS*, Seoul, Korea, Nov. 1996, pp. 215–219

S10 D. Griesinger, Beyond MLS – Occupied Hall measurement with FFT techniques, in *AES 101th Convention*, Preprint 4403, Los Angeles, CA, Nov. 1996. MLS: Maximum Length Sequence, http://www.davidgriesinger.com/

S11 G. Zhou, X.-G. Xia, Multiple frequency detection in undersampled complexed-valued waveforms with close multiple frequencies. IEE Electron. Lett. **33**(15), 1294–1295 (July 1997) (Multiple frequency detection by multiple DFTs)

S12 C. Pateros, Coarse frequency acquisition using multiple FFT windows, in *ICSPAT 97*, San Diego, CA, Sept. 1997

S13 H. Murakami, K. Nakamura, Y. Takuno, High-harmonics analysis by recursive DFT algorithm, in *ICSPAT 98*, Toronto, Canada, Sept. 1998

S14 Fundamentals of FFT-based signal analysis and measurement, Application note, National Instruments, Phone: 800-433-3488, E-mail: info@natinst.com

S15 C. Becchetti, G. Jacovitti, G. Scarano, DFT based optimal blind channel identification, in *EUSIPCO-98*, vol. 3, Island of Rhodes, Greece, Sept. 1998, pp. 1653–1656

S16 J.S. Marciano, T.B. Vu, Implementation of a broadband frequency-invariant (FI) array beamformer using the two-dimensional discrete Fourier transform (2D-DFT), in *IEEE ISPACS*, Pukhet, Thailand, Dec. 1999, pp. 153–156

S17 C. Breithaupt, R. Martin, MMSE estimation of magnitude-squared DFT coefficients with supergaussian priors, in *IEEE ICASSP*, vol. 1, Hong Kong, China, Apr. 2003, pp. 896–899

S18 G. Schmidt, "Single-channel noise suppression based on spectral weighting – An overview", Tutorial. EURASIP News Lett. **15**, 9–24 (Mar. 2004)

Communications

C1   G.M. Dillard, Recursive computation of the discrete Fourier transform with applications to a pulse radar system. Comput. Elec. Eng. **1**, 143–152 (1973)

C2   G.M. Dillard, Recursive computation of the discrete Fourier transform with applications to an FSK communication receiver, in *IEEE NTC Record*, pp. 263–265, 1974

C3   S.U. Zaman, W. Yates, Use of the DFT for sychronization in packetized data communications, in *IEEE ICASSP-94*, vol. 3, Adelaide, Australia, Apr. 1994, pp. 261–264

C4   D.I. Laurenson, G.J.R. Povey, The application of a generalized sliding FFT algorithm to prediction for a RAKE receiver system operating over mobile channels, in *IEEE ICC-95*, vol. 3, Seattle, WA, June 1995, pp. 1823–1827

C5   P.C. Sapino, J.D. Martin, Maximum likelihood PSK classification using the DFT of phase histogram, in *IEEE GLOBECOM*, vol. 2, Singapore, Nov. 1995, pp. 1029–1033

C6   A. Wannasarnmaytha, S. Hara, N. Morinaga, A novel FSK demodulation method using short-time DFT analysis for LEO satellite communication systems, in *IEEE GLOBECOM*, vol. 1, Singapore, Nov. 1995, pp. 549–553

C7   K.C. Teh, K.H. Li, A.C. Kot, Rejection of partial baud interference in FFT spread spectrum systems using FFT based self normalizing receivers, in *IEEE ICCS/ISPACS*, Singapore, Nov. 1996, pp. 1.4.1–1.4.4

C8   Y. Kim, M. Shin, H. Cho, The performance analysis and the simulation of MC-CDMA system using IFFT/FFT, in *ICSPAT 97*, San Diego, CA, Sept. 1997

C9   M. Zhao, Channel separation and combination using fast Fourier transform, in *ICSPAT 97*, San Diego, CA, Sept. 1997

C10 A.C. Kot, S. Li, K.C. Teh, FFT-based clipper receiver for fast frequency hopping spread spectrum system, in *IEEE ISCAS'98*, vol. 4, Monterey, CA, June 1998, pp. 305–308

C11 E. Del Re, R. Fantacci, L.S. Ronga, Fast phase sequences spreading codes for CDMA using FFT, in *EUSIPCO*, vol. 3, Island of Rhodes, Greece, Sept. 1998, pp. 1353–1356

C12 C.-L. Wang, C.-H. Chang, A novel DHT-based FFT/IFFT processor for ADSL transceivers, in *IEEE ISCAS*, vol. 1, Orlando, FL, May/June 1999, pp. 51–54

C13 M. Joho, H. Mathis, G.S. Moschytz, An FFT-based algorithm for multichannel blind deconvolution, in *IEEE ISCAS*, vol. 3, Orlando, FL, May/June 1999, pp. 203–206

C14 Y.P. Lin, S.M. Phoong, Asymptotical optimality of DFT based DMT transceivers, in *IEEE ISCAS*, vol. 4, Orlando, FL, May/June 1999, pp. 503–506

C15 O. Edfors et al., Analysis of DFT-based channel estimators for OFDM, in Wireless Personal Communications, **12**, 55–70 (Netherlands: Springer, Jan. 2000)

C16 M.-L. Ku, C.-C. Huang, A derivation on the equivalence between Newton's method and DF DFT-based method for channel estimation in OFDM systems. IEEE Trans. Wireless Commun. **7**, 3982–3987 (Oct. 2008)

C17 J. Proakis, M. Salehi, *Fundamentals of Communication Systems* (Prentice-Hall, Upper Saddle River, NJ, 2005)

C18 Ibid., *Instructor's Solutions Manual for Computer Problems of Fundamentals of Communication Systems* (Upper Saddle River, NJ: Prentice-Hall, 2007)

Orthogonal Frequency Division Multiplexing

O1 S.B. Weinstein, P.M. Ebert, Data transmission by frequency-division multiplexing using the discrete Fourier transform. IEEE Trans. Commun. Technol. **19**, 628–634 (Oct. 1971)

O2 W.Y. Zou, W. Yiyan, COFDM: An overview. IEEE Trans. Broadcast. **41**, 1–8 (Mar. 1995)

O3 A. Buttar et al., FFT and OFDM receiver ICs for DVB-T decoders, in *IEEE ICCE*, Chicago, IL, June 1997, pp. 102–103

O4 A. Salsano et al., 16-point high speed (I)FFT for OFDM modulation, in *IEEE ISCAS*, vol. 5, Monterey, CA, June 1998, pp. 210–212

O5 P. Combelles et al., A receiver architecture conforming to the OFDM based digital video broadcasting standard for terrestrial transmission (DVB-T), in *IEEE ICC*, vol. 2, Atlanta, GA, June 1998, pp. 780–785

O6 C. Tellambura, A reduced-complexity coding technique for limiting peak-to-average power ratio in OFDM, in *IEEE ISPACS'98*, Melbourne, Australia, Nov. 1998, pp. 447–450

O7 P.M. Shankar, *Introduction to Wireless Systems* (Wiley, New York, 2002)

O8 B.S. Son et al., A high-speed FFT processor for OFDM systems, in *IEEE ISCAS*, vol. 3, Phoenix-Scottsdale, AZ, May 2002, pp. 281–284

O9 W.-C. Yeh, C.-W. Jen, High-speed and low-power split-radix FFT. IEEE Trans. SP **51**, 864–874 (Mar. 2003)

O10 J.-C. Kuo et al., VLSI design of a variable-length FFT/IFFT processor for OFDM-based communication systems. EURASIP J. Appl. Signal Process. **2003**, 1306–1316 (Dec. 2003)

O11 M. Farshchian, S. Cho, W.A. Pearlman, Robust image transmission using a new joint source channel coding algorithm and dual adaptive OFDM, in *SPIE and IS&T, VCIP*, vol. 5308, San Jose, CA, Jan. 2004, pp. 636–646

O12 T.H. Tsa, C.C. Peng, A FFT/IFFT Soft IP generator for OFDM communication system, in *IEEE ICME*, vol. 1, Taipei, Taiwan, June 2004, pp. 241–244

O13 C.-C. Wang, J.M. Huang, H.C. Cheng, A 2K/8K mode small-area FFT processor for OFDM demodulation of DVB-T receivers. IEEE Trans. CE **51**, 28–32 (Feb. 2005)

O14 H. Jiang et al., Design of an efficient FFT processor for OFDM systems. IEEE Trans. CE **51**, 1099–1103 (Nov. 2005)

O15 A. Cortés et al., An approach to simplify the design of IFFT/FFT cores for OFDM systems. IEEE Trans. CE **52**, 26–32 (Feb. 2006)

O16 O. Atak et al., Design of application specific processors for the cached FFT algorithm, in *IEEE ICASSP*, vol. 3, Toulouse, France, May 2006, pp. 1028–1031

O17 C.-Y. Yu, S.-G. Chen, J.-C. Chih, Efficient CORDIC designs for multi-mode OFDM FFT, in *IEEE ICASSP*, vol. 3, Toulouse, France, May 2006, pp. 1036–1039

O18 R.M. Jiang, An area-efficient FFT architecture for OFDM digital video broadcasting. IEEE Trans. CE **53**, 1322–1326 (Nov. 2007)

O19 A. Ghassemi, T.A. Gulliver, A low-complexity PTS-based radix FFT method for PAPR reduction in OFDM systems. IEEE Trans. SP **56**, 1161–1166 (Mar. 2008)

O20 W. Xiang, T. Feng, L. Jingao, Efficient spectrum multiplexing using wavelet packet modulation and channel estimation based on ANNs, in *Int'l Conference on Audio, Language and Image Processing (ICALIP)*, Shanghai, China, July 2008, pp. 604–608

• See also [I-10, A-31].

General

G1 F.J. Harris, On the use of windows for harmonic analysis with the discrete Fourier transform. Proc. IEEE **66**, 51–83 (Jan. 1978)

G2 M. Borgerding, Turning overlap-save into a multiband mixing, downsampling filter bank. IEEE SP Mag. **23**, 158–161 (Mar. 2006)

Comparison of Various Discrete Transforms

G3   J. Pearl, Basis-restricted transformations and performance measures for spectral representa-
     tions. IEEE Trans. Info. Theory **17**, 751–752 (Nov. 1971)
G4   J. Pearl, H.C. Andrews, W.K. Pratt, Performance measures for transform data coding. IEEE
     Trans. Commun. **20**, 411–415 (June 1972)
G5   N. Ahmed, K.R. Rao, *Orthogonal Transforms for Digital Signal Processing* (Springer,
     New York, 1975)
G6   M. Hamidi, J. Pearl, Comparison of cosine and Fourier transforms of Markov-1 signals.
     IEEE Trans. ASSP **24**, 428–429 (Oct. 1976)
G7   P. Yip, K.R. Rao, Energy packing efficiency for the generalized discrete transforms. IEEE
     Trans. Commun. **26**, 1257–1261 (Aug. 1978)
G8   P. Yip, D. Hutchinson, Residual correlation for generalized discrete transforms. IEEE Trans.
     EMC **24**, 64–68 (Feb. 1982)
G9   Z. Wang, B.R. Hunt, The discrete cosine transform – A new version, in *IEEE ICASSP83*,
     MA, Apr. 1983, pp. 1256–1259
G10 P.-S. Yeh, Data compression properties of the Hartley transform. IEEE Trans. ASSP **37**,
     450–451 (Mar. 1989)
G11 O.K. Ersoy, A comparative review of real and complex Fourier-related transforms. Proc.
     IEEE **82**, 429–447 (Mar. 1994)

Haar Transform

G12 H.C. Andrews, K.L. Caspari, A generalized technique for spectral analysis. IEEE Trans.
     Comput. **19**, 16–25 (Jan. 1970)
G13 H.C. Andrews, *Computer Techniques in Image Processing* (Academic Press, New York,
     1970)
G14 H.C. Andrews, J. Kane, Kronecker matrices, computer implementation and generalized
     spectra. J. Assoc. Comput. Machinary (JACM) **17**, 260–268 (Apr. 1970)
G15 H.C. Andrews, Multidimensional rotations in feature selection. IEEE Trans. Comput. **20**,
     1045–1051 (Sept. 1971)
G16 R.T. Lynch, J.J. Reis, Haar transform image coding, in *Proc. Nat'l Telecommun. Conf.*,
     Dallas, TX, 1976, pp. 44.3-1 – 44.3-5
G17 S. Wendling, G. Gagneux, G.A. Stamon, Use of the Haar transform and some of its properties
     in character recognition, in *IEEE Proceedings of the 3rd Int'l Conf. on Pattern Recognition
     (ICPR)*, Coronado, CA, Nov. 1976, pp. 844–848
G18 S. Wendling, G. Gagneux, G.A. Stamon, Set of invariants within the power spectrum of
     unitary transformations. IEEE Trans. Comput. **27**, 1213–1216 (Dec. 1978)
G19 V.V. Dixit, Edge extraction through Haar transform, in *IEEE Proceedings of the 14th
     Asilomar Conference on Circuits Systems and Computations*, Pacific Grove, CA, 1980,
     pp. 141–143
G20 J.E. Shore, On the application of Haar functions. IEEE Trans. Commun. **21**, 209–216 (Mar.
     1973)
G21 D.F. Elliott, K.R. Rao, *Fast Transforms: Algorithms, Analyses, Applications* (Academic
     Press, Orlando, FL, 1982)
G22 Haar filter in "Wavelet Explorer" (Mathematica Applications Library, Wolfram Research
     Inc.), www.wolfram.com, info@wolfram.com, 1-888-882-6906.
•   See also [G5].

Fast Algorithms

A1   J.W. Cooly, J.W. Tukey, An algorithm for the machine calculation of complex Fourier series. Math. Comput. **19**, 297–301 (Apr. 1965)

A2   L.R. Rabiner, R.W. Schafer, C.M. Rader, The chirp z-transform algorithm. IEEE Trans. Audio Electroacoustics **17**, 86–92 (June 1969)

A3   S. Venkataraman et al., Discrete transforms via the Walsh–Hadamard transform. Signal Process. **14**, 371–382 (June 1988)

A4   C. Lu, J.W. Cooley, R. Tolimieri, Variants of the Winograd mutiplicative FFT algorithms and their implementation on IBM RS/6000, in *IEEE ICASSP-91*, vol. 3, Toronto, Canada, Apr. 1991, pp. 2185–2188

A5   B.G. Sherlock, D.M. Monro, Moving fast Fourier transform. IEE Proc. F **139**, 279–282 (Aug. 1992)

A6   C. Lu, J.W. Cooley, R. Tolimieri, FFT algorithms for prime transform sizes and their implementations on VAX, IBM3090VF, and IBM RS/6000. IEEE Trans. SP **41**, 638–648 (Feb. 1993)

A7   P. Kraniauskar, A plain man's guide to the FFT. IEEE SP Mag. **11**, 24–35 (Apr. 1994)

A8   J.M. Rius, R. De Porrata-Dòria, New FFT bit reversal algorithm. IEEE Trans. SP **43**, 991–994 (Apr. 1995)

A9   N. Bean, M. Stewart, A note on the use of fast Fourier transforms in Buzen's algorithm, in *Australian Telecommunications, Network and Applications Conf. (ATNAC)*, Sydney, Australia, Dec. 1995

A10 I.W. Selesnick, C.S. Burrus, Automatic generation of prime length FFT programs. IEEE Trans. SP **44**, 14–24 (Jan. 1996)

A11 J.C. Schatzman, Index mappings for the fast Fourier transform. IEEE Trans. SP **44**, 717–719 (Mar. 1996)

A12 D. Sundararajan, M.O. Ahmad, Vector split-radix algorithm for DFT computation, in *IEEE ISCAS*, vol. 2, Atlanta, GA, May 1996, pp. 532–535

A13 S. Rahardja, B.J. Falkowski, Family of fast transforms for mixed arithmetic logic, in *IEEE ISCAS*, vol. 4, Atlanta, GA, May 1996, pp. 396–399

A14 G. Angelopoulos, I. Pitas, Fast parallel DSP algorithms on barrel shifter computers. IEEE Trans. SP **44**, 2126–2129 (Aug. 1996)

A15 M. Wintermantel, E. Lueder, Reducing the complexity of discrete convolutions and DFT by a linear transformation, in *ECCTD'97*, Budapest, Hungary, Sept. 1997, pp. 1073–1078

A16 R. Stasinski, Optimization of vector-radix-3 FFTs, in *ECCTD'97*, Budapest, Hungary, Sept. 1997, pp. 1083–1086, http://www.mit.bme.hu/events/ecctd97/

A17 M. Frigo, S.G. Johnson, Fastest Fourier transform in the west. Technical Report MIT-LCS-TR728 (MIT, Cambridge, MA, Sept. 1997), http://www.fftw.org

A18 H. Guo, G.A. Sitton, C.S. Burrus, The quick Fourier transform: an FFT based on symmetries. IEEE Trans. SP **46**, 335–341 (Feb. 1998)

A19 M. Frigo, S.G. Johnson, FFTW: an adaptive software architecture for the FFT, in *IEEE ICASSP*, vol. 3, Seattle, WA, May 1998, pp. 1381–1384

A20 S.K. Stevens, B. Suter, A mathematical approach to a low power FFT architecture, in *IEEE ISCAS'98*, vol. 2, Monterey, CA, June 1998, pp. 21–24

A21 A.M. Krot, H.B. Minervina, Fast algorithms for reduction a modulo polynomial and Vandermonde transform using FFT, in *EUSIPCO-98*, vol. 1, Island of Rhodes, Greece, Sept. 1998, pp. 173–176

A22 A. Jbira, Performance of discrete Fourier transform with small overlap in transform-predictive-coding-based coders, *EUSIPCO-98*, vol. 3, Island of Rhodes, Greece, Sept. 1998, pp. 1441–1444

A23 B.G. Sherlock, Windowed discrete Fourier transform for shifting data. Signal Process. **74**, 169–177 (Apr. 1999)

A24 H. Murakami, Generalized DIT and DIF algorithms for signals of composite length, in *IEEE ISPACS'99*, Pukhet, Thailand, Dec. 1999, pp. 665–667

A25 L. Brancik, An improvement of FFT-based numerical inversion of two-dimensional Laplace transforms by means of ε-algorithm, in *IEEE ISCAS 2000*, vol. 4, Geneva, Switzerland, May 2000, pp. 581–584

A26 M. Püschel, Cooley-Tukey FFT like algorithms for the DCT, in *IEEE ICASSP*, vol. 2, Hong Kong, China, Apr. 2003, pp. 501–504

A27 M. Johnson, X. Xu, A recursive implementation of the dimensionless FFT, *IEEE ICASSP*, vol. 2, Hong Kong, China, Apr. 2003, pp. 649–652

A28 D. Takahashi, A radix-16 FFT algorithm suitable for multiply-add instruction based on Goedecker method, in *IEEE ICASSP*, vol. 2, Hong Kong, China, Apr. 2003, pp. 665–668

A29 J. Li, Reversible FFT and MDCT via matrix lifting, in *IEEE ICASSP*, vol. 4, Montreal, Canada, May 2004, pp. 173–176

A30 M. Frigo, S.G. Johnson, The design and implementation of FFTW3. Proc. IEEE **93**, 216–231 (Feb. 2005) (Free software http://www.fftw.org also many links)

A31 B.G. Jo, H. Sunwoo, New continuous-flow mixed-radix (CFMR) FFT processor using novel in-place strategy. IEEE Trans. Circ. Syst. I Reg. Papers **52**, 911–919 (May 2005)

A32 Y. Wang et al., Novel memory reference reduction methods for FFT implementations on DSP processors, IEEE Trans. SP, **55**, part 2, 2338–2349 (May 2007) (Radix-2 and radix-4 FFT)

A33 S. Mittal, Z.A. Khan, M.B. Srinivas, Area efficient high speed architecture of Bruun's FFT for software defined radio, in *IEEE GLOBECOM*, Washington, DC, Nov. 2007, pp. 3118–3122

A34 C.M. Rader, Discrete Fourier transforms when the number of data samples is prime. Proc. IEEE **56**, 1107–1108 (June 1968)

Chirp-z Algorithm

A35 X.-G. Xia, Discrete chirp-Fourier transform and its application to chirp rate estimation. IEEE Trans. SP **48**, 3122–3134 (Nov. 2000)

• See [A2, B1].

Winograd Fourier Transform Algorithm (WFTA)

A36 H.F. Silverman, An introduction to programming the Winograd Fourier transform algorithm (WFTA). IEEE Trans. ASSP **25**, 152–165 (Apr. 1977)

A37 S. Winograd, On computing the discrete Fourier transform. Math. Comput. **32**, 175–199 (Jan. 1978)

A38 B.D. Tseng, W.C. Miller, Comments on 'an introduction to programming the Winograd Fourier transform algorithm (WFTA)'. IEEE Trans. ASSP **26**, 268–269 (June 1978)

A39 R.K. Chivukula, Fast algorithms for MDCT and low delay filterbanks used in audio coding. M.S. thesis, Department of Electrical Engineering, The University of Texas at Arlington, Arlington, TX, Feb. 2008

A40 R.K. Chivukula, Y.A. Reznik, Efficient implementation of a class of MDCT/IMDCT filterbanks for speech and audio coding applications, in *IEEE ICASSP*, Las Vegas, NV, Mar./Apr. 2008, pp. 213–216

A41 R.K. Chivukula, Y.A. Reznik, V. Devarajan, Efficient algorithms for MPEG-4 AAC-ELD, AAC-LD and AAC-LC filterbanks, in *IEEE Int'l Conference Audio, Language and Image Processing (ICALIP 2008)*, Shanghai, China, July 2008, pp. 1629–1634

A42 P. Duhamel, M. Vetterli, Fast Fourier transforms: A tutorial review and a state of the art. Signal Process. (Elsevier) **19**, 259–299 (Apr. 1990)

A43 J.W. Cooley, Historical notes on the fast Fourier transform. Proc. IEEE **55**, 1675–1677 (Oct. 1967)

Split-Radix

SR1 P. Duhamel, Implementing of "split-radix" FFT algorithms for complex, real, and real-symmetric data. IEEE Trans. ASSP **34**, 285–295 (Apr. 1986)

SR2 H.R. Wu, F.J. Paoloni, Structured vector radix FFT algorithms and hardware implementation. J. Electric. Electr. Eng. (Australia) **10**, 241–253 (Sept. 1990)

SR3 S.-C. Pei, W.-Y. Chen, Split vector-radix-2/8 2-D fast Fourier transform. IEEE SP Letters **11**, 459–462 (May 2004)

• See [O9, DS5, L10]

Radix-4

R1 W. Han et al., High-performance low-power FFT cores. ETRI J. **30**, 451–460 (June 2008)

• See [E13].

Radix-8

R2 E. Bidet et al., A fast single-chip implementation of 8192 complex point FFT. IEEE J. Solid State Circ. **30**, 300–305 (Mar. 1995)

R3 T. Widhe, J. Melander, L. Wanhammar, Design of efficient radix-8 butterfly PEs for VLSI, in *IEEE ISCAS '97*, vol. 3, Hong Kong, China, June 1997, pp. 2084–2087

R4 L. Jia et al., Efficient VLSI implementation of radix-8 FFT algorithm, in *IEEE Pacific Rim Conference on Communications, Computers and Signal Processing*, Aug. 1999, pp. 468–471

R5 K. Zhong et al. A single chip, ultra high-speed FFT architecture, in *5th IEEE Int'l Conf. ASIC*, vol. 2, Beijing, China, Oct. 2003, pp. 752–756

Matrix Factoring and BIFORE Transforms

T1 I.J. Good, The interaction algorithm and practical Fourier analysis. J. Royal Stat. Soc. B **20**, 361–372 (1958)

T2 E.O. Brigham, R.E. Morrow, The fast Fourier transform. IEEE Spectr. **4**, 63–70 (Dec. 1967)

T3 W.M. Gentleman, Matrix multiplication and fast Fourier transforms. Bell Syst. Tech. J. **47**, 1099–1103 (July/Aug. 1968)

T4 J.E. Whelchel, Jr., D.R. Guinn, The fast Fourier-Hadamard transform and its signal representation and classification, in *IEEE Aerospace Electr. Conf. EASCON Rec.*, 9–11 Sept. 1968, pp. 561–573

T5 W.K. Pratt et al., Hadamard transform image coding. Proc. IEEE **57**, 58–68 (Jan. 1969)

T6 H.C. Andrews, K.L. Caspari, A generalized technique for spectral analysis. IEEE Trans. Comput. **19**, 16–25 (Jan. 1970)

T7 J.A. Glassman, A generalization of the fast Fourier transform. IEEE Trans. Comput. **19**, 105–116 (Feb. 1970)

T8 S.S. Agaian, O. Caglayan, Super fast Fourier transform, in *Proceedings of the SPIE-IS&T*, vol. 6064, San Jose, CA, Jan. 2006, pp. 60640F-1 thru 12

Miscellaneous

J1   V.K. Jain, W.L. Collins, D.C. Davis, High accuracy analog measurements via interpolated FFT. IEEE Trans. Instrum. Meas. **28**, 113–122 (June 1979)

J2   T. Grandke, Interpolation algorithms for discrete Fourier transforms of weighted signals. IEEE Trans. Instrum. Meas. **32**, 350–355 (June 1983)

J3   D.J. Mulvaney, D.E. Newland, K.F. Gill, A comparison of orthogonal transforms in their application to surface texture analysis. Proc. Inst. Mech. Engineers **200, no. C6**, 407–414 (1986)

J4   G. Davidson, L. Fielder, M. Antill, Low-complexity transform coder for satellite link applications, in *89th AES Convention*, Preprint 2966, Los Angeles, CA, Sept. 1990, http://www.aes.org/

J5   C.S. Burrus, Teaching the FFT using Matlab, in *IEEE ICASSP-92*, vol. 4, San Francisco, CA, Mar. 1992, pp. 93–96

J6   A.G. Exposito, J.A.R. Macias, J.L.R. Macias, Discrete Fourier transform computation for digital relaying. Electr. Power Energy Syst. **16**, 229–233 (1994)

J7   J.C.D. de Melo, Partial FFT evaluation, in *ICSPAT*, vol. 1, Boston, MA, Oct. 1996, pp. 137–141

J8   F. Clavean, M. Poirier, D. Gingras, FFT-based cross-covariance processing of optical signals for speed and length measurement, in *IEEE ICASSP-97*, vol. 5, Munich, Germany, Apr. 1997, pp. 4097–4100

J9   A harmonic method for active power filters using recursive DFT, in *20th EECON*, Bangkok, Thailand, Nov. 1997

J10 B. Bramer, M. Ibrahim, S. Rumsby, An FFT Implementation Using Almanet, in *ICSPAT*, Toronto, Canada, Sept. 1998

J11 S. Rumsby, M. Ibrahim, B. Bramer, Design and implementation of the Almanet environment, in *IEEE SiPS 98*, Boston, MA, Oct. 1998, pp. 509–518

J12 A. Nukuda, FFTSS: A high performance fast Fourier transform library, in *IEEE ICASSP*, vol. 3, Toulouse, France, May 2006, pp. 980–983

J13 B.R. Hunt, A matrix theory proof of the discrete convolution theorem. IEEE Audio Electroacoustics **19**, 285–288 (Dec. 1971)

FFT Pruning

J14 S. Holm, FFT pruning applied to time domain interpolation and peak localization. IEEE Trans. ASSP **35**, 1776–1778 (Dec. 1987)

J15 Detection of a few sinusoids in noise. Dual tone multi-frequency signaling (DTMF). Use pruned FFT.

• See also [A42].

CG-FFT Method for the Array Antenna Analysis

K1   T.K. Sarkar, E. Arvas, S.M. Rao, Application of FFT and the conjugate gradient method for the solution of electromagnetic radiation from electrically large and small conducting bodies. IEEE Trans. Antennas Propagat. **34**, 635–640 (May 1986)

K2   T.J. Peters, J.L. Volakis, Application of a conjugate gradient FFT method to scattering from thin planar material plates. IEEE Trans. Antennas Propagat. **36**, 518–526 (Apr. 1988)

K3 H. Zhai et al., Analysis of large-scale periodic array antennas by CG-FFT combined with equivalent sub-array preconditioner. IEICE Trans. Commun. **89**, 922–928 (Mar. 2006)

K4 H. Zhai et al., Preconditioners for CG-FMM-FFT implementation in EM analysis of large-scale periodic array antennas. IEICE Trans. Commun., **90**, 707–710, (Mar. 2007)

• See also [B12].

Nonuniform DFT

N1   J.L. Yen, On nonuniform sampling of bandwidth-limited signals. IRE Trans. Circ. Theory **3**, 251–257 (Dec. 1956)

N2   G. Goertzel, An algorithm for the evaluation of finite trigonometric series. Am. Mathem. Monthly **65**, 34–35 (Jan. 1958)

N3   J.W. Cooley, J.W. Tukey, An algorithm for the machine calculation of complex Fourier series. Math. Comput. **19**, 297–301 (Apr. 1965)

N4   F.J. Beutler, Error free recovery of signals from irregularly spaced samples. SIAM Rev. **8**, 328–335 (July 1966)

N5   A. Oppenheim, D. Johnson, K. Steiglitz, Computation of spectra with unequal resolution using the fast Fourier transform. Proc. IEEE **59**, 299–301 (Feb. 1971)

N6   A. Ben-Israel, T.N.E. Greville, *Generalized Inverses: Theory and Applications*. New York: Wiley, 1977

N7   K. Atkinson, *An Introduction to Numerical Analysis*. New York: Wiley, 1978

N8   J.W. Mark, T.D. Todd, A Nonuniform sampling approach to data compression. IEEE Trans. Commun. **29**, 24–32 (Jan. 1981)

N9   P.A. Regalia, S.K. Mitra, Kronecker products, unitary matrices and signal processing applications. SIAM Rev. **31**, 586–613 (Dec. 1989)

N10 W. Rozwood, C. Therrien, J. Lim, Design of 2-D FIR filters by nonuniform frequency sampling. IEEE Trans. ASSP **39**, 2508–2514 (Nov. 1991)

N11 M. Marcus and H. Minc, *A Survey of Matrix Theory and Matrix Inequalities*. New York: Dover, pp. 15–16, 1992

N12 A. Dutt, V. Rokhlin, Fast Fourier transforms for nonequispaced data. SIAM J. Sci. Comput. **14**, 1368–1393 (Nov. 1993)

N13 E. Angelidis, A novel method for designing FIR digital filters with nonuniform frequency samples. IEEE Trans. SP **42**, 259–267 (Feb. 1994)

N14 M. Lightstone et al., Efficient frequency-sampling design of one- and two-dimensional FIR filters using structural subband decomposition. IEEE Trans. Circuits Syst. II **41**, 189–201 (Mar. 1994)

N15 H. Feichtinger, K. Groechenig, T. Strohmer, Efficient numerical methods in non-uniform sampling theory. Numer. Math. **69**, 423–440 (Feb. 1995)

N16 S. Bagchi, S.K. Mitra, An efficient algorithm for DTMF decoding using the subband NDFT, in *IEEE ISCAS*, vol. 3, Seattle, WA, Apr./May 1995, pp. 1936–1939

N17 S. Bagchi, S.K. Mitra, The nonuniform discrete Fourier transform and its applications in filter design: Part I – 1-D. IEEE Trans. Circ. Sys. II Analog Digital SP **43**, 422–433 (June 1996)

N18 S. Bagchi, S.K. Mitra, The nonuniform discrete Fourier transform and its applications in filter design: Part II – 2-D. IEEE Trans. Circ. Sys. II Analog Digital SP **43**, 434–444 (June 1996)

N19 S. Carrato, G. Ramponi, S. Marsi, A simple edge-sensitive image interpolation filter, in *IEEE ICIP*, vol. 3, Lausanne, Switzerland, Sept. 1996, pp. 711–714

N20 G. Wolberg, Nonuniform image reconstruction using multilevel surface interpolation, in *IEEE ICIP*, Washington, DC, Oct. 1997, pp. 909–912

N21 S.K. Mitra, *Digital Signal Processing: A Computer-Based Approach*. New York: McGraw Hill, 1998, Chapters 6 and 10

N22 Q.H. Liu, N. Nguyen, Nonuniform fast Fourier transform (NUFFT) algorithm and its applications, in *IEEE Int'l Symp. Antennas Propagation Society (AP-S)*, vol. 3, Atlanta, GA, June 1998, pp. 1782–1785

N23 X.Y. Tang, Q.H. Liu, CG-FFT for nonuniform inverse fast Fourier transforms (NU-IFFTs), in *IEEE Int'l Symp. Antennas Propagation Society (AP-S)*, vol. 3, Atlanta, GA, June 1998, pp. 1786–1789

N24 G. Steidl, A note on fast Fourier transforms for nonequispaced grids. Adv. Comput. Math. **9**, 337–353 (Nov. 1998)

N25 A.F. Ware, Fast approximate Fourier transforms for irregularly spaced data. SIAM Rev. **40**, 838–856 (Dec. 1998)

N26 S. Bagchi, S.K. Mitra, *The Nonuniform Discrete Fourier Transform and Its Applications in Signal Processing* (Kluwer, Norwell, MA, 1999)

N27 A.J.W. Duijndam, M.A. Schonewille, Nonuniform fast Fourier transform. Geophysics **64**, 539–551 (Mar./Apr. 1999)

N28 N. Nguyen, Q.H. Liu, The regular Fourier matrices and nonuniform fast Fourier transforms. SIAM J. Sci. Comput. **21**, 283–293 (Sept. 1999)

N29 S. Azizi, D. Cochran, J. N. McDonald, A sampling approach to region-selective image compression, in *IEEE Conference on Signals, Systems and Computers*, Oct. 2000, pp. 1063–1067

N30 D. Potts, G. Steidl, M. Tasche, Fast Fourier transforms for nonequispaced data: A tutorial, in *Modern Sampling Theory: Mathematics and Applications*, ed. by J.J. Benedetto, P.J.S.G. Ferreira (Birkhäuser, Boston, MA, 2001), pp. 247–270

N31 G. Ramponi, S. Carrato, An adaptive irregular sampling algorithm and its application to image coding. Image Vision Comput. **19**, 451–460 (May 2001)

N32 M.R. Shankar, P. Sircar, Nonuniform sampling and polynomial transformation method, in *IEEE ICC*, vol. 3, New York, Apr. 2002, pp. 1721–1725

N33 M. Bartkowiak, High compression of colour images with nonuniform sampling, in *Proceedings of the ISCE'2002*, Erfurt, Germany, Sept. 2002

N34 K.L. Hung, C.C. Chang, New irregular sampling coding method for transmitting images progressively. IEE Proc. Vision Image Signal Process. **150**, 44–50 (Feb. 2003)

N35 J.A. Fessler, B.P. Sutton, Nonuniform fast Fourier transforms using min-max interpolation. IEEE Trans. SP **51**, 560–574 (Feb. 2003)

N36 A. Nieslony, G. Steidl, Approximate factorizations of Fourier matrices with nonequispaced knots. Linear Algebra Its Appl. **366**, 337–351 (June 2003)

N37 K. Fourmont, "Non-equispaced fast Fourier transforms with applications to tomography," J. Fourier Anal. Appl., **9**, 431–450 (Sept. 2003)

N38 K.-Y. Su, J.-T. Kuo, An efficient analysis of shielded single and multiple coupled microstrip lines with the nonuniform fast Fourier transform (NUFFT) technique. IEEE Trans. Microw. Theory Techniq. **52**, 90–96 (2004)

N39 L. Greengard, J.Y. Lee, Accelerating the nonuniform fast Fourier transform. SIAM Rev. **46**, 443–454 (July 2004)

N40 R. Venkataramani, Y. Bresler, Multiple-input multiple-output sampling: Necessary density conditions. IEEE Trans. IT **50**, 1754–1768 (Aug. 2004)

N41 C. Zhang, T. Chen, View-dependent non-uniform sampling for image-based rendering, in *IEEE ICIP*, Singapore, Oct. 2004, pp. 2471–2474

N42 Q.H. Liu, Fast Fourier transforms and NUFFT, *Encyclopedia of RF and Microwave Engineering*, 1401–1418, (Mar. 2005)

N43 K.-Y. Su, J.-T. Kuo, Application of two-dimensional nonuniform fast Fourier transform (2-D NUFFT) technique to analysis of shielded microstrip circuits. IEEE Trans. Microw. Theory Techniq. **53**, 993–999 (Mar. 2005)

N44 J.J. Hwang et al., Nonuniform DFT based on nonequispaced sampling. WSEAS Trans. Inform. Sci. Appl. **2**, 1403–1408 (Sept. 2005)

N45 Z. Deng and J. Lu, The application of nonuniform fast Fourier transform in audio coding, in *IEEE Int'l Conference on Audio, Language and Image Process. (ICALIP)*, Shanghai, China, July 2008, pp. 232–236

• See also [LA24].

Applications of DFT/FFT: Spectral Estimation

AP1 P.T. Gough, A fast spectral estimation algorithm based on FFT. IEEE Trans. SP **42**, 1317–1322 (June 1994)

Applications of DFT/FFT: Filtering

- Filtering LPF, BPF, HPF. Generalized cepstrum and homomorphic filtering. See Chapters 5 and 7 in [B6].

Applications of DFT/FFT: Multichannel Carrier Modulation

- Multichannel carrier modulation (MCM) such as orthogonal frequency division multiplexing (OFDM) for digital television terrestrial broadcasting.

AP2 Y. Wu, B. Caron, Digital television terrestrial broadcasting. IEEE Commun. Mag. **32**, 46–52 (May 1994)

Applications of DFT/FFT: Spectral Analysis, Filtering, Convolution, Correlation etc.

- Several applications in spectral analysis, filtering, convolution, correlation etc. Refer to following books among others.

AP3 See Chapter 9 – Convolution and correlation using FFT in [IN5].
AP4 S.D. Stearns, R.A. David, *Signal Processing Algorithms in Fortran and C* (Prentice-Hall, Englewood Cliffs, NJ, 1993)

Applications of DFT/FFT: Pulse Compression

- Pulse compression (radar systems – surveillance, tracking, target classification) – Matched filter with a long impulse response – convolution via FFT.

AP5 R. Cox, FFT-based filter design boosts radar system's process, *Electronic Design*, 31 Mar. 1988, pp. 81–84

Applications of DFT/FFT: Spectrum Analysis

AP6 G. Dovel, FFT analyzers make spectrum analysis a snap, *EDN*, Jan. 1989, pp. 149–155

Applications of DFT/FFT: Ghost Cancellation

AP7 M.D. Kouam, J. Palicot, Frequency domain ghost cancellation using small FFTs, in *IEEE ICCE*, Chicago, IL, June 1993, pp. 138–139

AP8 J. Edwards, Automatic bubble size detection using the zoom FFT, in *ICSPAT*, DSP World Expo., Dallas, TX, Oct. 1994, pp. 1511–1516

• Several applications of FFT in digital signal processing are illustrated in software/ hardware, books on using the software (Mathcad, MATLAB, etc.)

AP9 N. Kuroyanagi, L. Guo, N. Suehiro, Proposal of a novel signal separation principle based on DFT with extended frame buffer analysis, in *IEEE GLOBECOM*, Singapore, Nov. 1995, pp. 111–116

AP10 M. Webster, R. Roberts, Adaptive channel truncation for FFT detection in DMT systems – Error component partitioning, in *30th IEEE Asilomar Conference on Signals, Systems and Computers*, vol. 1, Pacific Grove, CA, Nov. 1996, pp. 669–673

Applications of DFT/FFT: Phase Correlation Based Motion Estimation

AP11 A. Molino et al., Low complexity video codec for mobile video conferencing, in *EUSIPCO*, Vienna, Austria, Sept. 2004, pp. 665–668

AP12 A. Molino, F. Vacca, G. Masera, Design and implementation of phase correlation based motion estimation, in *IEEE Int'l Conference on Systems-on-Chip*, Sept. 2005, pp. 291–294

FFT Software/Hardware: Commercial S/W Tools

Cs1 "Image processing toolbox" (2-D transforms) MATLAB, The MathWorks, Inc. 3 Apple Hill Drive, Natick, MA 01760, E-mail: info@mathworks.com, Fax: 508-653-6284. Signal processing toolbox (FFT, DCT, Hilbert, Filter design), http://www.mathworks.com/, FTP server ftp://ftp.mathworks.com

Cs2 "FFT tools" Software Package. Adds FFT capability to Lotus 1-2-3 & enhances FFT capability of Microsoft Excel. 1024-point FFT under a second on a 486DX/33 PC. Up to 8192-point FFT with choice of windows, Blackman, Hamming, Hanning, Parzen, tapered rectangular & triangular taper. DH Systems Inc. 1940 Cotner Ave., Los Angeles, CA 90025. Phone: 800-747-4755, Fax: 310-478-4770

Cs3 Windows DLL version of the prime factor FFT sub-routine library, Alligator Technologies, 17150 Newhope Street # 114, P.O. Box 9706, Fountain Valley, CA 92728-9706, Phone: 714-850-9984, Fax: 714-850-9987

Cs4 SIGLAB Software, FFT, correlation etc., Monarch, DSP software, The Athena Group, Inc. 3424 NW 31st Street, Gainesville, FL 32605, Phone: 904-371-2567, Fax: 904-373-5182

Cs5 Signal ++ DSP Library (C++), Several transforms including CZT, wavelet, cosine, sine, Hilbert, FFT and various DSP operations. Sigsoft, 15856 Lofty Trail Drive, San Diego, CA 92127, Phone: 619-673-0745

Cs6 DSP works-real time windows-based signal processing software. FFT, Convolution, Filtering, etc. (includes multirate digital filters, QMF bank). Complete bundled hardware and software packages, DSP operations. Momentum Data Systems Inc. 1520 Nutmeg Place #108, Costa Mesa, CA 92626, Phone: 714-557-6884, Fax: 714-557-6969, http://www.mds.com

Cs7 Version 1.1 ProtoSim, PC based software, FFT, Bode plots, convolution, filtering etc. Systems Engineering Associates Inc. Box 3417, RR#3, Montpelier, VT 05602, Phone: 802-223-6194, Fax: 802-223-6195

Cs8 Sig XTM, A general purpose signal processing package, Technisoft, P.O. Box 2525, Livermore, CA 94551, Phone: 510-443-7213, Fax: 510-743-1145

Cs9 Standard filter design software, DGS Associates, Inc. Phone: 415-325-4373, Fax: 415-325-7278

Cs10 DT VEE and VB-EZ for windows. Software for Microsoft windows. Filters, FFTs, etc., Data Translation, 100 Locke Drive, Marlboro, MA 01752-1192, Phone: 508-481-3700 or 800-525-8528

Cs11 Mathematica (includes FFT, Bessel functions), Wolfram Research, Inc., Phone: 800-441-MATH, 217-398-0700, Fax: 217-398-0747, E-mail: info@wri.com

Cs12 Mathematica 5.2, Wolfram Research, Inc. Website: http://www.wolfram.com, E-mail: info@wolfram.com, Phone: 217-398-0700, Book: S. Wolfram, *The mathematica book*. 4th ed. New York: Cambridge Univ. Press, Website: http://www.cup.org

Cs13 Mathcad 5.0, Mathsoft Inc. P.O. Box 1018, Cambridge, MA 02142-1519, Ph: 800-967-5075, Phone: 217-398-0700, Fax: 217-398-0747

Cs14 Matrix-based interactive language: Signal Processing FFTs, O-Matrix, objective numerical analysis, Harmonic Software Inc. Phone: 206-367-8742, Fax: 206-367-1067

Cs15 FFT, Hilbert transform, ACOLADE, Enhanced software for communication system, CAE, Amber Technologies, Inc. 47 Junction Square Dr., Concord, MA 01742-9879, Phone: 508-369-0515, Fax: 508-371-9642

Cs16 High-order Spectral Analysis (ISA-PC32) Software. Integral Signal Processing, Inc., P.O. Box 27661, Austin, TX 78755-2661, Phone: 512-346-1451, Fax: 512-346-8290

Cs17 Origin 7.5, 8, voice spectrum, statistics, FFT, IFFT, 2D FFT, 2D IFFT, power spectrum, phase unwrap, data windowing, Software by OriginLab Corp. One Roundhouse Plaza, Northampton, MA 01060, Phone: 413-586-2013, Fax: 413-585-0126. http://www.originlab.com

Cs18 Visilog, Image Processing & Analysis Software: FFTs and various processing operations, Noesis, 6800 Cote de Liesse, Suite 200, St. Laurent, Quebec, H4T2A7, Canada, Phone: 514-345-1400, Fax: 514-345-1575, E-mail: noesis@cam.org

Cs19 Stanford Graphics 3.0, Visual Numerics, 9990 Richmand Avenue, Suite 400, Houston, TX 77042, Phone: 713-954-6424, Fax: 713-781-9260

Cs20 V for Windows, Digital Optics Ltd., Box 35-715, Browns Bay, Optics Ltd., Auckland 10, New Zealand, Phone: (65+9) 478-5779, (65+9) 479-4750, E-mail: 100237.423@Compuserve.com

Cs21 DADiSP Worksheet (software package) (DADiSP 6.0), DSP Development Corp., 3 Bridge Street, Newton, MA 02458, Phone: 800-424-3131, Fax: 617-969-0446, student edition on the web, Website: http://www.dadisp.com

FFT Software/Hardware: Commercial Chips

Cc1 ADSP-21060 Benchmarks (@ 40 MHz) 1,024-point complex FFT (Radix 4 with digit reverse) 0.46 ms (18,221 cycles). Analog Devices, Inc. 1 Technology Way, P.O. Box 9106, Norwood, MA 02062, Phone: 617-461-3771, Fax: 617-461-4447

Cc2 TMC 2310 FFT processor, complex FFT (forward or inverse) of up to 1,024 points (514 μsec), radix-2 DIT FFT. Raytheon Semiconductor, 300 Ellis St, Mountain View, CA 94043-7016, Phone: 800-722-7074, Fax: 415-966-7742

Cc3 STV 0300 VLSI chip can be programmed to perform FFT or IFFT (up to 8,192 point complex FFT) with input f_s from 1 kHz to 30 kHz. (8,192 point FFT in 410 μsec), *SGS-Thomson Microelectronics News & Views*, no. 7, Dec. 1997, http://www.st.com

Cc4 Viper-5, FFT (IM CFFT in 21 msec), Texas Memory Systems, Inc. 11200 Westheimer, #1000, Houston, TX 77042, Ph: 713-266-3200, Fax: 713-266-0332, Website: http://www. texmemsys.com

Cc5 K. Singh, Implementing in-place FFTs on SISD and SIMD SHARC processors. Technical note, EE-267, Analog Devices, Inc. Mar. 2005 (ADSP-21065L, ADSP-21161)

Cc6 "Pipelined FFT," RF Engines Limited (RFEL), Oct. 2002 (Process data at a sample rate in 100 MSPS, the complex 4,096-point radix-2 DIF FFT core in a single 1M gate FPGA)

Cc7 G.R. Sohie, W. Chen, Implementation of fast Fourier transforms on Motorola's digital signal processors (on DSP56001/2, DSP56156, DSP96002)

FFT Software/Hardware: Commercial DSP

Cp1 ZR34161 16 bit VSP. High performance programmable 16-bit DSP. 1-D and 2-D FFTs, several DSP operations, 1,024 point radix-2 complex FFT in 2,178 µsec. ZP34325 32 bit VSP. 1-D and 2-D FFTs, several DSP operations ZR38000 and ZR38001 can execute 1,024 point radix-2 complex FFT in 0.88 msec. Zoran Corporation, 1705 Wyatt Drive, Santa Clara, CA 95054. Phone: 408-986-1314, Fax: 408-986-1240. VSP: Vector signal processor

Cp2 FT 200 series Multiprocessors 1K complex FFT < 550 µsec, 1K × 1K Real to complex FFT 782 msec. Alacron, 71 Spitbrook Road, Suite 204, Nashua, NH 03060, Phone: 603-891-2750, Fax: 603-891-2745

Cp3 IMSA 100: Programmable DSP. Implement FFT, convolution, correlation etc. SGS-Thomson Microelectronics, 1000 East Bell Road, Phoenix, AZ 85022-2699. http://www.st.com

Cp4 A41102 FFT processor, Lake DSP Pty. Ltd. Suite 4/166 Maroubra Road, Maroubra 2035, Australia, Phone: 61-2-314-2104, Fax: 61-2-314-2187

Cp5 DSP/Veclib. Vast library of DSP functions for TI's TMS 320C40 architecture, Spectro-analysis, 24 Murray Road, West Newton, MA 02165, Phone: 617-894-8296, Fax: 617-894-8297

Cp6 Toshiba IP 9506 Image Processor. High speed image processing on a single chip. Quickest FFT process. Toshiba, I.E. OEM Division, 9740 Irvine Blvd., CA 92718, Phone: 714-583-3180

Cp7 Sharp Electronics, 5700 NW Pacific Rim Blvd., Camas, WA 98607, Phone: 206-834-8908, Fax: 206-834-8903, (Real time DSP chip set: LH 9124 DSP and LH 9320 Address Generator) 1,024 point complex FFT in 80 µsec. Real and complex radix-2, radix-4 and radix-16 FFTs

Cp8 TMS 320C 6201 General-purpose programmable fixed-point-DSP chip (5 ns cycle time). Can compute 1,024-point complex FFT in 70 µsec. TI Inc., P.O. Box 172228, Denver, CO 80217. TMS 320. http://dspvillage.ti.com

Cp9 TMS 320C80 Multimedia Video Processor (MVP), 64-point and 256-point complex radix-2 FFT, TI, Market Communications Manager, MS736, P.O. Box 1443, Houston, TX 77251-1443, Phone: 1-800-477-8924

Cp10 Pacific Cyber/Metrix, Inc., 6693 Sierra Lane, Dublin, CA 94568, Ph: 510-829-8700, Fax: 510-829-9796 (VSP-91 vector processor, 1K complex FFT in 8 µsec, 64K complex FFT in 8.2 msec)

FFT Software/Hardware: Commercial H/W

H1 CRP1M40 PC/ISA-bus floating point DSP board can process DFTs, FFTs, DCTs, FCTs, adaptive filtering, etc., 1K complex FFT in 82 µsec at 40 MHz. Can upgrade up to 1 Megapoint FFT. Catalina Research, Inc. 985 Space Center Dr., Suite 105, Colorado Springs, CO 80915, Phone: 719-637-0880, FAX: 719-637-3839

H2 Ultra DSP-1 board, 1K complex FFT in 90 μsec, Valley Technologies, Inc. RD #4, Route 309, Tamaqua, PA 18252, Phone: 717-668-3737, FAX: 717-668-6360

H3 1,024 point complex FFT in 82 μsec. DSP MAX-P40 board, Butterfly DSP, Inc. 1614 S.E. 120th Ave., Vancouver, WA 98684, Phone: 206-892-5597, Fax: 206-254-2524

H4 DSP board: DSP Lab one. Various DSP software. Real-time signal capture, analysis, and generation plus high-level graphics. Standing Applications Lab, 1201 Kirkland Ave., Kirkland, WA 98033, Phone: 206-453-7855, Fax: 206-453-7870

H5 Digital Alpha AXP parallel systems and TMS320C40. Parallel DSP & Image Processing Systems. Traquair Data Systems, Inc. Tower Bldg., 112 Prospect St., Ithaca, NY 14850, Phone: 607-272-4417, Fax: 607-272-6211

H6 DSP Designer™, Design environment for DSP, Zola Technologies, Inc. 6195 Heards Creek Dr., N.W., Suite 201, Atlanta, GA 30328, Phone: 404-843-2973, Fax: 404-843-0116

H7 FFT-523. A dedicated FFT accelerator for HP's 68000-based series 200 workstations. Ariel Corp., 433 River Road, Highland Park, NJ 8904. Phone and Fax: 908-249-2900, E-mail: ariel@ariel.com

H8 MultiDSP, 4865 Linaro Dr., Cypress, CA 90630, Phone: 714-527-8086, Fax: 714-527-8287, E-mail: multidsp@aol.com. Filters, windows, etc., also DCT/IDCT, FFT/IFFT, Average FFT

H9 FFT/IFFT Floating Point Core for FPGA, SMT395Q, a TI DSP module including a Xilinx FPGA as a coprocessor for digital filtering, FFTs, etc., Sundance, Oct. 2006 (Radix-32), http://www.sundance.com

FFT Software/Hardware: Implementation on DSP

DS1 H.R. Wu, F.J. Paoloni, Implementation of 2-D vector radix FFT algorithm using the frequency domain processor A 41102, *Proceedings of the IASTED, Int'l Symposium on Signal Processing and Digital Filtering*, June 1990

DS2 D. Rodriguez, A new FFT algorithm and its implementation on the DSP96002, in *IEEE ICASSP-91*, vol. 3, Toronto, Canada, May 1991, pp. 2189–2192

DS3 W. Chen, S. King, Implementation of real-valued input FFT on Motorola DSPs, in *ICSPAT*, vol. 1, Dallas, TX, Oct. 1994, pp. 806–811

DS4 Y. Solowiejczyk, 2-D FFTs on a distributed memory multiprocessing DSP based architectures, in *ICSPAT*, Santa Clara, CA, 28 Sept. to 1 Oct. 1993

DS5 T.J. Tobias, In-line split radix FFT for the 80386 family of microprocessors, in *ICSPAT*, Santa Clara, CA, 28 Sept. to 1 Oct. 1993 (128 point FFT in 700 msec on a 386, 40 MHz PC)

DS6 C. Lu et al., Efficient multidimensional FFT module implementation on the Intel I860 processor, in *ICSPAT*, Santa Clara, CA, 28 Sept. to 1 Oct. 1993, pp. 473–477

DS7 W. Chen, S. King, Implementation of real input valued FFT on Motorola DSPs, in *ICSPAT*, Santa Clara, CA, 28 Sept. to 1 Oct. 1993

DS8 A. Hiregange, R. Subramaniyan, N. Srinivasa, 1-D FFT and 2-D DCT routines for the Motorola DSP 56100 family, *ICSPAT*, vol. 1, Dallas, TX, Oct. 1994, pp. 797–801

DS9 R.M. Piedra, Efficient FFT implementation on reduced-memory DSPs, in *ICSPAT*, Boston, MA, Oct. 1995

DS10 H. Kwan et al., Three-dimensional FFTs on a digital-signal parallel processor with no interprocessor communication, in *30th IEEE Asilomar Conference on Signals, Systems and Computers*, Pacific Grove, CA, Nov. 1996, pp. 440–444

DS11 M. Grajcar, B. Sick, The FFT butterfly operation in 4 processor cycles on a 24 bit fixed-point DSP with a pipelined multiplier, in *IEEE ICASSP*, vol. 1, Munich, Germany, Apr. 1997, pp. 611–614

DS12 M. Cavadini, A high performance memory and bus architecture for implementing 2D FFT on a SPMD machine, in *IEEE ISCAS*, vol. 3, Hong Kong, China, June 1997, pp. 2032–2036

• See also [A-32].

FFT Software/Hardware: VLSI

V1 D. Rodriguez, Tensor product algebra as a tool for VLSI implementation of the discrete Fourier transform, in *IEEE ICASSP*, vol. 2, Toronto, Canada, May 1991, pp. 1025–1028

V2 R. Bhatia, M. Furuta, J. Ponce, A quasi radix-16 FFT VLSI processor, in *IEEE ICASSP*, Toronto, Canada, May 1991, pp. 1085–1088

V3 H. Miyanaga, H. Yamaguchi, K. Matsuda, A real-time 256 × 256 point two-dimensional FFT single chip processor, in *IEEE ICASSP*, Toronto, Canada, May 1991, pp. 1193–1196

V4 F. Kocsis, A fully pipelined high speed DFT architecture, in *IEEE ICASSP*, Toronto, Canada, May 1991, pp. 1569–1572

V5 S.R. Malladi et al., A high speed pipelined FFT processor, in *IEEE ICASSP*, Toronto, Canada, May 1991, pp. 1609–1612

V6 E. Bernard et al., A pipeline architecture for modified higher radix FFT, in *IEEE ICASSP*, vol. 5, San Francisco, CA, Mar. 1992, pp. 617–620

V7 J.I. Guo et al., A memory-based approach to design and implement systolic arrays for DFT and DCT, in *IEEE ICASSP*, vol. 5, San Francisco, CA, Mar. 1992, pp. 621–624

V8 E. Bessalash, VLSI architecture for fast orthogonal transforms on-line computation, in *ICSPAT*, Santa Clara, CA, Sept./Oct. 1993, pp. 1607–1618

V9 E. Bidet, C. Joanblanq, P. Senn, (CNET, Grenoble, France), A fast single chip implementation of 8,192 complex points FFT, in *IEEE CICC*, San Diego, CA, May 1994, pp. 207–210

V10 E. Bidet, C. Joanblanq, P. Senn, A fast 8K FFT VLSI chip for large OFDM single frequency network, in *7th Int'l Workshop on HDTV*, Torino, Italy, Oct. 1994

V11 J. Melander et al., Implementation of a bit-serial FFT processor with a hierarchical control structure, in *ECCTD'95*, vol. 1, Istanbul, Turkey, Aug. 1995, pp. 423–426

V12 K. Hue, A 256 fast Fourier transform processor, in *ICSPAT*, Boston, MA, Oct. 1995

V13 S.K. Lu, S.Y. Kuo, C.W. Wu, On fault-tolerant FFT butterfly network design, in *IEEE ISCAS*, vol. 2, Atlanta, GA, May 1996, pp. 69–72

V14 C. Nagabhushan et al., Design of radix-2 and radix-4 FFT processors using a modular architecure family, in *PDPTA*, Sunnyvale, CA, Aug. 1996, pp. 589–599

V15 J.K. McWilliams, M.E. Fleming, Small, flexible, low power, DFT filter bank for channeled receivers, in *ICSPAT*, vol. 1, Boston, MA, Oct. 1996, pp. 609–614

V16 J. McCaskill, R. Hutsell, TM-66 swiFFT block transform DSP chip, in *ICSPAT*, vol. 1, Boston, MA, Oct. 1996, pp. 689–693

V17 M. Langhammer, C. Crome, Automated FFT processor design, in *ICSPAT*, vol. 1, Boston, MA, Oct. 1996, pp. 919–923

V18 S. Hsiao, C. Yen, New unified VLSI architectures for computing DFT and other transforms, in *IEEE ICASSP 97*, vol. 1, Munich, Germany, Apr. 1997, pp. 615–618

V19 E. Cetine, R. Morling, I. Kale, An integrated 256-point complex FFT processor for real-time spectrum, in *IEEE IMTC '97*, vol. 1, Ottawa, Canada, May 1997, pp. 96–101

V20 S.F. Hsiao, C.Y. Yen, Power, speed and area comparison of several DFT architectures, in *IEEE ISCAS '97*, vol. 4, Hong Kong, China, June 1997, pp. 2577–2581

V21 R. Makowitz, M. Mayr, Optimal pipelined FFT processing based on embedded static RAM, in *ICSPAT 97*, San Diego, CA, Sept. 1997

V22 C.J. Ju, "FFT-Based parallel systems for array processing with low latency: sub-40 ns 4K butterfly FFT", in *ICSPAT 97*, San Diego, CA, Sept. 1997

V23 T.J. Ding, J.V. McCanny, Y. Hu, Synthesizable FFT cores, in *IEEE SiPS*, Leicester, UK, Nov. 1997, pp. 351–363

V24 B.M. Baas, A 9.5 mw 330 µsec 1,024-point FFT processor, in *IEEE CICC*, Santa Clara, CA, May 1998, pp. 127–130

V25 S. He, M. Torkelson, Design and implementation of a 1,024 point pipeline FFT, in *IEEE CICC*, Santa Clara, CA, May 1998, pp. 131–134

V26 A.Y. Wu, T.S. Chan, Cost-effective parallel lattice VLSI architecture for the IFFT/FFT in DMT transceiver technology, in *IEEE ICASSP*, Seattle, WA, May 1998, pp. 3517–3520

V27 G. Naveh et al., Optimal FFT implementation on the Carmel DSP core, in *ICSPAT*, Toronto, Canada, Sept. 1998

V28 A. Petrovsky, M. Kachinsky, Automated parallel-pipeline structure of FFT hardware design for real-time multidimensional signal processing, in *EUSIPCO*, vol. 1, Island of Rhodes, Greece, Sept. 1998, pp. 491–494, http:// www.eurasip.org

V29 G. Chiassarini et al., Implementation in a single ASIC chip, of a Winograd FFT for a flexible demultiplexer of frequency demultiplexed signals, in *ICSPAT*, Toronto, Canada, Sept. 1998

V30 B.M. Baas, A low-power, high-performance, 1, 024-point FFT processor. IEEE J. Solid State Circ. **34**, 380–387 (Mar. 1999)

V31 T. Chen, G. Sunada, J. Jin, COBRA: A 100-MOPS single-chip programmable and expandable FFT. IEEE Trans. VLSI Syst. **7**, 174–182 (June 1999)

V32 X.X. Zhang et al., Parallel FFT architecture consisting of FFT chips. J. Circ. Syst. **5**, 38–42 (June 2000)

V33 T.S. Chang et al., Hardware-efficient DFT designs with cyclic convolution and subexpression sharing. IEEE Trans. Circ. Syst. II Analog Digital. SP **47**, 886–892 (Sept. 2000)

V34 C.-H. Chang, C.-L. Wang, Y.-T. Chang, Efficient VLSI architectures for fast computation of the discrete Fourier transform and its inverse. IEEE Trans. SP **48**, 3206–3216 (Nov. 2000) (Radix-2 DIF FFT)

V35 K. Maharatna, E. Grass, U. Jagdhold, A novel 64-point FFT/IFFT processor for IEEE 802.11 (a) standard, in *IEEE ICASSP*, vol. 2, Hong Kong, China, Apr. 2003, pp. 321–324

V36 Y. Peng, A parallel architecture for VLSI implementation of FFT processor, in *5th IEEE Int'l Conference on ASIC*, vol. 2, Beijing, China, Oct. 2003, pp. 748–751

V37 E. da Costa, S. Bampi, J.C. Monteiro, Low power architectures for FFT and FIR dedicated datapaths, in *46th IEEE Int'l MWSCAS*, vol. 3, Cairo, Egypt, Dec. 2003, pp. 1514–1518

V38 K. Maharatna, E. Grass, U. Jagdhold, A 64-point Fourier transform chip for high-speed wireless LAN application using OFDM. IEEE J. Solid State Circ. **39**, 484–493 (Mar. 2004) (Radix-2 DIT FFT)

V39 G. Zhong, F. Xu, A.N. Wilson Jr., An energy-efficient reconfigurable FFT/IFFT processor based on a multi-processor ring, in *EUSIPCO*, Vienna, Austria, Sept. 2004, pp. 2023–2026, available: http:// www.eurasip.org

V40 C. Cheng, K.K. Parhi, Hardware efficient fast computation of the discrete Fourier transform. Journal of VLSI Signal Process. Systems, **42**, 159–171 (Springer, Amsterdam, Netherlands, Feb. 2006) (WFTA)

• See also [R2].

FFT Software/Hardware: FPGA

L1 L. Mintzer, The FPGA as FFT processor, in *ICSPAT*, Boston, MA, Oct. 1995

L2 D. Ridge et al., PLD based FFTs, in *ICSPAT*, San Diego, CA, Sept. 1997

L3 T. Williams, Case study: variable size, variable bit-width FFT engine offering DSP-like performance with FPGA versatility, in *ICSPAT*, San Diego, CA, Sept 1997

L4 Altera Application Note 84, "Implementing FFT with on-chip RAM in FLEX 10K devices," Feb. 1998

L5 C. Dick, Computing multidimensional DFTs using Xilinx FPGAs, in *ICSPAT*, Toronto, Canada, Sept. 1998

L6 L. Mintzer, A 100 megasample/sec FPGA-based DFT processor, in *ICSPAT*, Toronto, Canada, Sept. 1998

L7 S. Nag, H.K. Verma, An efficient parallel design of FFTs in FPGAs, in *ICSPAT*, Toronto, Canada, Sept. 1998

L8 C. Jing, H.-M. Tai, Implementation of modulated lapped transform using programmable logic, in *IEEE ICCE*, Los Angeles, CA, June 1999, pp. 20–21

L9 S. Choi et al., Energy-efficient and parameterized designs for fast Fourier transform on FPGAs, in *IEEE ICASSP*, vol. 2, Hong Kong, China, Apr. 2003, pp. 521–524

L10 I.S. Uzun, A. Amira, A. Bouridane, FPGA implementations of fast Fourier transforms for real-time signal and image processing. IEE Vision Image Signal Process. **152**, 283–296 (June 2005) (Includes pseudocodes for radix-2 DIF, radix-4 and split-radix algorithms)

Late Additions

LA1 A.M. Grigoryan, M.M. Grigoryan, *Brief Notes in Advanced DSP: Fourier Analysis with MATLAB®* (CRC Press, Boca Raton, FL, 2009) (Includes many MATLAB codes)

LA2 H.S. Malvar et al., Low-complexity transform and quantization in H.264/AVC. IEEE Trans. CSVT **13**, 598–603 (July 2003)

LA3 M. Athineoset, The DTT and generalized DFT in MATLAB, http://www.ee.columbia.edu/~marios/symmetry/sym.html, 2005

LA4 K. Wahid et al., Efficient hardware implementation of hybrid cosine-Fourier-wavelet transforms, in *IEEE ISCAS 2009*, Taipei, Taiwan, May 2009, pp. 2325–2329

LA5 K.R. Rao, P. Yip, *Discrete Cosine Transform: Algorithms, Advantages, Applications* (Academic Press, San Diego, CA, 1990)

LA6 V.G. Reju, S.N. Koh, I.Y. Soon, Convolution using discrete sine and cosine transforms. IEEE SP Lett. **14**, 445–448 (July 2007)

LA7 H. Dutagaci, B. Sankur, Y. Yemez, 3D face recognition by projection-based methods, in *Proc. SPIE-IS&T*, vol. 6072, San Jose, CA, Jan. 2006, pp. 60720I-1 thru 11

LA8 3D Database, http://www.sic.rma.ac.be/~beumier/DB/3d_rma.html

LA9 J. Wu, W. Zhao, New precise measurement method of power harmonics based on FFT, in *IEEE ISPACS*, Hong Kong, China, Dec. 2005, pp. 365–368

LA10 P. Marti-Puig, Two families of radix-2 FFT algorithms with ordered input and output data. IEEE SP Lett. **16**, 65–68 (Feb. 2009)

LA11 P. Marti-Puig, R. Reig-Bolaño, Radix-4 FFT algorithms with ordered input and output data, in *IEEE Int'l Conference on DSP*, 5–7 July 2009, Santorini, Greece

LA12 A.M. Raičević, B.M. Popović, An effective and robust fingerprint enhancement by adaptive filtering in frequency domain, *Series: Electronics and Energetics* (Facta Universitatis, University of Niš, Serbia, Apr. 2009), pp. 91–104, available: http://factaee.elfak.ni.ac.rs/

LA13 W.K. Pratt, Generalized Wiener filtering computation techniques. IEEE Trans. Comp. **21**, 636–641 (July 1972)

LA14 J. Dong et al., 2-D order-16 integer transforms for HD video coding. IEEE Trans. CSVT **19**, 1462–1474 (Oct. 2009)

LA15 B.G. Sherlock, D.M. Monro, K. Millard, Fingerprint enhancement by directional Fourier filtering. IEE Proc. Image Signal Process. **141**, 87–94 (Apr. 1994)

LA16 M.R. Banham, A.K. Katsaggelos, Digital image restoration. IEEE SP Mag. **16**, 24–41 (Mar. 1997)

LA17 S. Rhee, M.G. Kang, Discrete cosine transform based regularized high-resolution image reconstruction algorithm. Opt. Eng. **38**, 1348–1356 (Aug. 1999)

LA18 L. Yu et al., Overview of AVS video coding standards. Signal Process. Image Commun. **24**, 263–276 (Apr. 2009)

LA19 I. Richardson, *The H.264 Advanced Video Compression Standard*, 2nd edn., Hoboken, NJ: Wiley, 2010

LA20 Y.Y. Liu, Z.W. Zeng, M.H. Lee, Fast jacket transform for DFT matrices based on prime factor algorithm. (Under review)

ة

LA21 S.-I. Cho, K.-M. Kang, A low-complexity 128-point mixed-radix FFT processor for MB-OFDM UWB systems. ETRI J **32**(1), 1–10 (Feb. 2010)

LA22 VC-1 Compressed Video Bitstream Format and Decoding Process, SMPTE 421M-2006

LA23 W.T. Cochran et al., What is the fast Fourier transform. Proc. IEEE **55**, 1664–1674 (Oct. 1967)

LA24 J.M. Davis, I.A. Gravagne, R.J. Marks II, Time scale discrete Fourier transform, in *IEEE SSST*, Tyler, TX, Mar. 2010, pp. 102–110

LA25 J. Ma, G. Plonka, The curvelet transform [A review of recent applications]. IEEE SP Mag. **27**(2), 118–133 (Mar. 2010)

FFT Software Websites

W1 Automatic generation of fast signal transforms (M. Püschel), http://www.ece.cmu.edu/~pueschel/, http://www.ece.cmu.edu/~smart/papers/autgen.html

W2 Signal processing algorithms implementation research for adaptable libraries, http://www.ece.cmu.edu/~spiral/

W3 FFTW (FFT in the west), http://www.fftw.org/index.html; http://www.fftw.org/benchfft/doc/ffts.html, http://www.fftw.org/links (List of links)

W4 FFTPACK, http://www.netlib.org/fftpack/

W5 FFT for Pentium (D.J. Bernstein), http://cr.yp.to/djbfft.html, ftp://koobera.math.uic.edu/www/djbfft.html

W6 Where can I find FFT software (comp.speech FAQ Q2.4), http://svr-www.eng.cam.ac.uk/comp.speech/Section2/Q2.4.html

W7 One-dimensional real fast Fourier transforms, http://www.hr/josip/DSP/fft.html

W8 FXT package FFT code (Arndt), http://www.jjj.de/fxt/

W9 FFT (Don Cross), http://www.intersrv.com/~dcross/fft.html

W10 Public domain FFT code, http://risc1.numis.nwu.edu/ftp/pub/transforms/, http://risc1.numis.nwu.edu/fft/

W11 DFT (Paul Bourke), http://www.swin.edu.au/astronomy/pbourke/sigproc/dft/

W12 FFT code for TMS320 processors, http://focus.ti.com/lit/an/spra291/spra291.pdf, ftp://ftp.ti.com/mirrors/tms320bbs/

W13 Fast Fourier transforms (Kifowit), http://ourworld.compuserve.com/homepages/steve_kifowit/fft.htm

W14 Nielsen's MIXFFT page, http://home.get2net.dk/jjn/fft.htm

W15 Parallel FFT homepage, http://www.arc.unm.edu/Workshop/FFT/fft/fft.html

W16 FFT public domain algorithms, http://www.arc.unm.edu/Workshop/FFT/fft/fft.html

W17 Numerical recipes, http://www.nr.com/

W18 General purpose FFT package, http://momonga.t.u-tokyo.ac.jp/~ooura/fft.html

W19 FFT links, http://momonga.t.u-tokyo.ac.jp/~ooura/fftlinks.html

W20 FFT, performance, accuracy, and code (Mayer), http://www.geocities.com/ResearchTriangle/8869/fft_{\rm s}ummary.html

W21 Prime-length FFT, http://www.dsp.rice.edu/software/RU-FFT/pfft/pfft.html

W22 Notes on the FFT (C.S. Burrus), http://faculty.prairiestate.edu/skifowit/fft/fftnote.txt, http://www.fftw.org/burrus-notes.html

W23 J.O. Smith III, *Mathematics of the Discrete Fourier Transform (DFT) with Audio Applications*, 2nd edn (W3K Publishing, 2007), available: http://ccrma.stanford.edu/~jos/mdft/mdft.html

W24 FFT, http://www.fastload.org/ff/FFT.html

W25 Bibliography for Fourier series and transform (J.H. Mathews, CSUF), http://math.fullerton.edu/mathews/c2003/FourierTransformBib/Links/FourierTransformBib_lnk_3.html

W26 Image processing learning resources, HIPR2, http://homepages.inf.ed.ac.uk/rbf/HIPR2/ hipr_top.htm, http://homepages.inf.ed.ac.uk/rbf/HIPR2/fourier.htm (DFT)

W27 Lectures on image processing (R.A. Peters II), http://www.archive.org/details/Lectures_ on_Image_Processing (DFT)

W28 C.A. Nyack, *A Visual Interactive Approach to DSP*. These pages mainly contain java applets illustrating basic introductory concepts in DSP (Includes Z-transform, sampling, DFT, FFT, IIR and FIR filters), http://dspcan.homestead.com/

W29 J.H. Mathews, CSUF, Numerical analysis: http://mathews.ecs.fullerton.edu/n2003/ Newton's method http://mathews.ecs.fullerton.edu/n2003/NewtonSearchMod.html

W30 J.P. Hornak, *The Basics of MRI* (1996–2010), http://www.cis.rit.edu/htbooks/mri/

Index

9 781402 066283